Thomas McIlwraith

The Birds of Ontario

Being a Concise Account of Every Species of Bird Known to Have Been Found in

Ontario

Thomas McIlwraith

The Birds of Ontario
Being a Concise Account of Every Species of Bird Known to Have Been Found in Ontario

ISBN/EAN: 9783337813772

Printed in Europe, USA, Canada, Australia, Japan

Cover: Foto ©berggeist007 / pixelio.de

More available books at **www.hansebooks.com**

BIRDS OF ONTARIO

BEING A CONCISE ACCOUNT OF EVERY SPECIES OF BIRD
KNOWN TO HAVE BEEN FOUND IN ONTARIO

WITH A

DESCRIPTION OF THEIR NESTS AND EGGS

AND INSTRUCTIONS FOR COLLECTING BIRDS AND PREPARING
AND PRESERVING SKINS, ALSO DIRECTIONS HOW
TO FORM A COLLECTION OF EGGS

By THOMAS McILWRAITH

MEMBER OF THE AMERICAN ORNITHOLOGISTS' UNION

SECOND EDITION—ENLARGED AND REVISED TO DATE

WITH ILLUSTRATIONS

Toronto
WILLIAM BRIGGS, WESLEY BUILDINGS

MONTREAL: C. W. COATES   HALIFAX: S. F. HUESTIS

MDCCCXCIV

TO

HER EXCELLENCY

The Countess of Aberdeen

IN VIEW OF THE

INTELLIGENT ATTENTION SHE HAS GIVEN TO

SCIENCE AND LITERATURE IN CANADA

THIS WORK IS RESPECTFULLY

DEDICATED

CONTENTS.

PREFACE.

———

THE first edition of "THE BIRDS OF ONTARIO" had its origin in the Hamilton Association, a local scientific society of which I am one of the oldest members.

In the spring of 1885, I read the introductory part of this book as a paper at one of the ordinary meetings. So few people devote any time to the study of Ornithology, that the subject was quite new to the Association, and, at the request of those present, I afterwards supplemented the paper with a record of the names of all the birds observed during my excursions near the city. By giving a technical description of each, I hoped to enable anyone desirous of pursuing the study to identify those birds likely to be found in the same district. The whole was subsequently published in book form by the Association, under the name of "The Birds of Ontario." Each member received a copy, and the balance of the issue was placed in the hands of the booksellers, but the number published was limited, and the book is now entirely out of the market.

The kind reception of the first edition by the public, and the numerous inquiries which have recently been made for copies of the book, have induced me to prepare this second edition, which I hope may be equally fortunate in meeting with public favor.

In the first edition the accounts of the birds were, to a great extent, the result of my own observations made in the vicinity of Hamilton, where I have resided for the past forty years. Nearly all of our native birds being migratory, the record given of each species was necessarily incomplete. Some were mentioned as winter visitors, others as summer residents, and a numerous class as spring and autumn migrants, visiting southern Ontario on their annual journey to and from their breeding places farther north.

In the present edition, it has been my object to place on record, as far as possible, the name of every bird that has been observed in Ontario; to show how the different species are distributed throughout the Province; and, especially, to tell where they spend the breeding season. To do this, I have had to refer to the notes of those who have visited the remote homes of the birds, at points often far apart and not easy of access, and to use their observations, published or otherwise, when they tend to throw light on the history of the birds observed in Ontario.

Of the works I have found most useful in this connection, I have pleasure in mentioning Mr. Ernest E. Thompson's "Birds of Manitoba," published by the Smithsonian Institute at Washington. In it the author records his ornithological notes made during a three years' residence in Manitoba, as well as the numerous observations by others of similar tastes in different parts of the Province with whom he was in correspondence.

Mr. E. W. Nelson, an officer of the United States Signal Service, has furnished the material for a work on the birds of an entirely new field, and has greatly extended our knowledge of many species which are common at different points as migrants. The work is a history of the collection of birds made by the author in Alaska during the years 1877 to 1881. It is edited by Mr. W. H. Henshaw, and handsomely illustrated. Issued by the Signal Service at Washington, it has been liberally distributed among the lovers of birds.

The ornithological work which has attracted the greatest amount of attention lately is entitled, "The Hawks and Owls of the United States in their Relation to Agriculture," prepared under the direction of Dr. C. Hart Merriam, Ornithologist, by Dr. A. K. Fisher, Assistant. This is a book of two hundred pages, enriched with colored illustrations of most of the birds it describes.

I have also found much that is interesting in the "Life Histories of North American Birds, with special reference to their Breeding Habits and Eggs," by Captain Charles Bendire, United States Army (retired). This work, which has now reached four hundred pages, is still in progress, and promises to be the most useful work we have on the subjects of which it treats.

To Mr. Oliver Davie I am greatly indebted for the privilege of using the information contained in his "Nests and Eggs of North American Birds," without which my record in that department would have been incomplete.

With regard to the distribution of the birds, I have found a great deal of information in a "Catalogue of Canadian Birds," with notes on the distribution of species by Montague Chamberlain.

In the Annual Reports of the Ornithological Sub-section of the Canadian Institute are many interesting notices of rare birds found near Toronto and elsewhere throughout the country.

I have also had frequent occasion to refer to the writings of Dr. Coues, Robert Ridgway, J. A. Allen, Dr. Merriam, Dr. Fisher, Wm. Brewster; Dr. Bell, Prof. Macoun, and Geo. R. White, of Ottawa; Dr. Macallum, of Dunnville; J. M. Lemoine, Quebec; Amos W. Butler, Indiana; Manly Hardy, of Brewer, Maine; W. P. Peacock, Marysville, California; Dr. P. L. Hatch, Minnesota; A. J. Cook, Michigan, to all of whom I now return my best thanks for the privilege of using their writings, as well as to many others who have in various ways contributed toward the completion of this work. I hope it may be found useful to those beginning the study of Ornithology. I wish for their sakes, as well as for the interest of the subject, that it were better than it is, but it has been written at spare hours amid the frequently recurring calls of business, and this may account for some of its imperfections.

The classification and nomenclature used correspond with the "Check List of North American Birds," published by the American Ornithologists' Union.

THOMAS McILWRAITH.

CAIRNBRAE, HAMILTON, CANADA,
February 1st, 1894.

INTRODUCTION.

To EVERY lover of outdoor life the birds are familiar objects, and their society is a continual source of enjoyment. Not only are their colors pleasing to the eye and their voices agreeable to the ear, but their various habits, when observed, excite our admiration of the unerring instinct which directs them in all their movements.

Upon my younger readers whose tastes may lead them in this direction, I would strongly urge the necessity for keeping a diary in which to make a record of everything connected with bird-life seen during each outing. Various shapes and sizes of books ruled in various ways have been suggested for this purpose, but for a beginner I would recommend a plain page on which to enter notes of the birds seen at such a place on such a date, with any facts relating to their numbers, occupation or manners which may have been observed.

When this book has been kept for two or three seasons, even the writer of it will be astonished at the amount and variety of the information accumulated, and the reading of it in after years will recall many of the pleasant experiences of the time when the entries were made. The book will also be very valuable as a work of reference while the history of each species is being investigated. Even the mention of the names of the birds seen, showing that certain species were noticed at a given place on a given date, becomes important, taken in connection with their migration, which is perhaps the most interesting part of their history.

We have become accustomed to speak of certain birds as being migratory, and we can tell, within a day or two, the time at which they will arrive from the south in spring to take possession of their former box or other nesting place near our houses. By referring to the observations of others made at different points on the continent, we can tell where certain species spend the winter, and likewise how far north they go in summer, and the date of their appearance at the various places where they pause by the way. By following the observations recently made by Professor Cook, we can also judge pretty accurately at what rate of speed the birds travel when coming up from the south in spring. All this information is the result of much time and care having been devoted to the subject by a host of observers, and it is very valuable to those interested in the subject, but the general questions relating to bird migration still remain unanswered, and the student may as well begin at the beginning and try to discover : " Why do birds migrate ? How do they know where to go ?"

With regard to those hatched in the north, we can readily understand that the failure of food supply at the time when insect life is wiped out, and the fields and marshes are buried under deep snow, makes a change of climate an absolute necessity. Then comes the second question : " How do they know where to go ?" The answer to this is, that they are guided by the observations of the older members of the flock, and to a great extent by the exercise of inherited memory, that is, the memory inherited of what has been done by other birds of the same species for countless generations.

Of inherited memory we cannot speak from experience, for it is not inherent in our nature, but that birds do possess the faculty is evident from many of their common habits. Take, for instance, that of nest-building. Different species of birds build very different kinds of nests, and often in very different situations, but birds of any given species usually build the same kind of nest, year after year, in the same kind of situation. That this regularity is not the result of personal observation is also quite apparent. For example, a pair of Grassfinches build their nest on the ground under a bunch of weeds or a little bush. The eggs are therein deposited, and in due time the birds are hatched. They grow quickly, and soon begin to feel crowded in the limited space. Presently, along comes a sportsman's dog with glaring eyes, and his tongue lolling out of his open mouth. He stares and snuffs at the tiny objects, and they hurry off in terror to hide among the brambles, where their wants are attended to by their parents for a few days until they are able to shift for themselves, which they soon do without having once seen the nest in which they were hatched. Yet in spring, when the different members of this little family start housekeeping on their own account, each one builds a nest precisely like that from which they were scared by the dog.

We can see at once the great value of this gift to the birds while pursuing their long and hazardous journey, much of which is performed during the hours of darkness, the daylight being necessary to enable them to secure the food which they require by the way.

Regarding the services rendered by the more experienced members in guiding the flock, the value of these will be seen when we remember that the bulk of our birds are born in the north, and are called upon to provide for the coming winter at an age when they are without personal experience of any kind. That all birds migrate in flocks is generally admitted, although it is only the larger species of water-fowl whose movements on such occasions come within the range of our observation.

The migrations of the smaller birds are noticed chiefly by their absence or presence in their usual haunts. For instance, when in the marsh on an afternoon toward the end of September, we noticed that the Sora Rails, birds of apparently weak and uncertain flight, were very abundant. Every few steps we made one would get up, fly a few yards, and again drop, apparently exhausted, among the reeds. During the night a sharp frost set in, the first of the season, and on visiting the same part of the marsh next day, not a single Rail could be found. All had gone during the night. The migrations of this species are always performed at night, when the birds cannot be seen, but we have occasionally heard the weak, whimpering note the birds utter when travelling, to prevent the weaklings from getting lost.

The geese are the most conspicuous of our migratory birds, the Λ-shaped flocks and the hoarse, honking cry being familiar to all Canadians. In the Hudson's Bay regions, where these birds are raised, we are told that at the approach of winter there are great gatherings of old and young along the shores, and great gabbling and apparent discussions relating to the journey in which they are about to start. All finally soar aloft and assume the usual Λ-shape, at the apex of which is the leader, always an old male, more or less familiar with the route. It is also said that the senior members of the flock take this position by turns, and relieve each other of the responsibility which for a time they assume. So they press on toward the south, lakes and rivers, which at night all show clearer and brighter than the land, being never-failing guides throughout the journey.

Some of the water-fowl seem so reluctant to leave their northern home that they remain until they are actually frozen out. On the other hand some of the waders leave their summer haunts long before we can see any necessity for their doing so. In the latter part of August, while lying awake with the windows open during the warm summer night, we can hear the skirling of the Sandpipers as they pass overhead on their way southward, leaving a land of plenty behind them.

In the movements of some of the warblers, too, are seeming irregularities which we are at a loss to explain. These, like other migrants which raise their young in the north, retire before the approach of winter, and we should expect to hear of their resting when a temperate region was reached : but many of them follow up the southern route till they reach the equator, and pass on two or three hundred miles beyond it.

We have so far taken a cursory view of the *southern* movement of the birds at the approach of winter. We see the necessity for it and admire the means they possess to enable them to carry it out. Eventually, all the species find suitable winter-quarters, where they quietly remain for a time and soon get to be in excellent condition, both as regards flesh and plumage, having nothing to engage their attention but dress and diet.

Soon, however, the time arrives when another change of habitat must be made, and the migratory feeling again stirs within the different species. There is a flapping of wings, a stretching of necks, a reiteration of their peculiar calls, and an occasional flight with no apparent object save exercise. " To the north, to the north," is now the general cry, and to the north they go, often fighting their way through storms and fogs, but still bent on making the journey. We should respect their courage and admire their intelligence more highly if we knew that the change was necessary. but that is the point we are unable to decide. The birds were living comparatively undisturbed with abundant fare and pleasant surroundings, why should they change? Why do they expose themselves to the vicissitudes and fatigue of this tedious journey of thousands of miles, to reach a land where they know that they cannot remain? We have heard "love of the nesting ground," and "strong home affection," mentioned among the leading causes of the desire to return, and there may be instances where such feelings, fully developed, have great influence in this connection. These are usually strongest in the female sex, but we find that the males always lead the northern movement, and are often in their former haunts a week or more before the females arrive. The home affection must, in certain cases, be of short duration, for we find in the history of some of the ducks, that no sooner are the eggs deposited than the female assumes the duties of incubation, and the males, getting together in flocks, keep outside in the open water, and ignore all further family responsibilities.

There may be something in the increased temperature in the south which the birds have found to be unsuitable while raising their young, and a change has become a matter of necessity, though the cause may not be to us apparent.

How are we to account for the habit in such birds as the Little Bittern, very many of which are resident in the south, and raise their young in tropical America, while those we see in Ontario are regular migrants, generally distributed, some straggling as far north as Manitoba and Hudson's Bay, but all leaving the country before the first touch of frost?

With these facts in view, it is not surprising to find some difference of opinion among ornithologists regarding the causes of migration. It seems as if the habit were, to some extent, coincident with the origin of the species, had extended very gradually through a long succession of ages, to meet the various climatic and other changes which have taken place in the surroundings of this part of the animal kingdom since "the beginning." Even within our short lifetime we have seen changes taking place in the distribution of the birds, some of which we can account for, and for others we have no explanation to offer; but the whole subject is one about which we have yet much to learn.

I commend it to the special consideration of my youthful readers, who, I feel sure, will find it most interesting, and I hope that they may yet be able to explain many of the difficulties which at present surround the subject of bird migration.

COLLECTING AND PREPARING SPECIMENS.

Since it is possible that the perusal of these pages may create in some of my younger readers the desire to collect and preserve specimens of the birds whose history they have been considering, I would advise them, by all means, to cultivate the taste, for I know of no pastime so conducive to health, nor one that will afford so much rational enjoyment. An outing in our bracing Canadian air is enjoyable at any season of the year, and the capture of a rare bird is an event productive of feelings which only the enthusiastic collector can understand.

In spring, to watch the daily arrival of migrants from the south, clad in their gayest attire and uttering their sweetest notes, is a constant source of delight. In summer the nests and eggs engage the attention for a time. In autumn the return of the birds seen passing north in the spring is again looked for with interest, and the changes in the dress of many are carefully recorded in the note-book which should be always at hand. But winter, after all, is the season in which we expect to find the rarest of our birds. We say *our* birds, for those we look for at that season are Canadian in the fullest sense of the term, having been born and brought up in the Dominion, but only on very rare occasions do they come so far south as our southern boundary. The collector in Southern Ontario who is fortunate enough to secure such birds as the Gyrfalcon, Ptarmigan, Three-toed Woodpecker, or Greater Red Poll, will not soon forget his agreeable sensations on the occasion, but he will gaze on the interesting strangers with regret if he does not know how to preserve their skins. It was probably some such experience that first suggested the attainment of this accomplishment, and in order to place it within the reach of all, I shall here give a brief account of how anyone may, with a little practice, become proficient in the art.

Since bird collecting can be successfully practised only by the use of the gun, let me here, for the guidance of beginners, repeat the directions so often given to guard against accidents in its use. The excuse for three-fourths of the mishaps which occur is, " Didn't know it was loaded," but the safe way to avoid this is at all times to handle the gun as if it were known to be loaded, for in the Irishman's way of putting it, " It may go off, whether it's loaded or not." When in company with others, never under any circumstances allow the gun for an instant to be pointed toward anything you do not wish to shoot. Never for any purpose blow into the muzzle, and do not have it "on cock " till the moment you expect to use it.

With regard to the choice of a gun, I am supposed to be speaking to a reader who has made up his mind to make a collection of the skins of those birds he finds near his home in Ontario. Water-fowl shooting, I may here remark, is a special department by itself. A 12-bore double breech-loader, and cartridges charged with No. 5 shot, with a few of No. 1 or BB, would be a suitable equipment for ducks, with the possibility of a chance shot at geese or swans.

The collector going into the country may unexpectedly meet with some very desirable bird, and should be prepared to take it, whatever be its size or shape, and to do so with the least possible injury to its plumage.

The birds met with on such excursions range in size from a horned owl to a humming-bird, the majority being intermediate between the two. Supposing that only one gun is desirable, the most suitable weapon is a No. 14 double breech-loader of good make. This will be just right for the majority, and with a little care in loading the cartridges, it can be made to suit the two extremes. I have found three sizes of shot to be sufficient for ordinary collecting trips,— Nos. 5, 8 and 12,—but the size of the charge must be varied to suit circumstances. For instance, a charge of No. 12 will bring down a snipe at 30 yards, but to shoot a kinglet, or a warbler, for preserving, with the same size shot, the charge would have to be very much lighter. Just how much lighter is a point to be learned by experience. It depends to some extent on the individual peculiarities of the gun, equal measure of powder and shot being in all charges the usual rule. I have often used dust shot for very small birds, but to be sure of getting them with that, one has to be pretty close to the birds, and then their feathers are a good deal cut up and broken. I find that a warbler killed by a single pellet of No. 12, is in better condition to make into a specimen than one that is killed with a dozen pellets of dust. No. 5 is big enough for hawks, owls, etc., and No. 8 is right for plovers, sandpipers, rails, etc., but the nature of the locality and the size of the birds most likely to be met, are the best guides in such matters.

To approach birds without alarming them, a mild form of deception is sometimes practised with advantage. They have keen sight, sharp hearing, and are at all times on the alert to escape danger, so that a direct approach is almost sure to make them take wing : but by walking as if intending to pass and yet gradually slanting nearer, a better chance may be obtained. Birds are used to the sight of horses and cows, and do not usually object to their presence near their haunts. I once knew an aged gunner who was aware of this fact, and for a time turned it to good account. He lived on the Bay shore not far from Hamilton, at a part of the beach which was a favorite resort of curlews, plovers and sandpipers. His old nag used to graze on the sward close by, and the gunner got into the way of steering him by the tail till he was within shot of the birds, when he would step out from behind and blaze away without alarming "Jerry" in the least. The same flock would rarely be deceived twice, but fresh arrivals were sure to be taken unawares.

A game bag, such as used by sportsmen, is not suitable for a collector, for the specimens are injured by the pressure to which they

2

are often subjected.   I have found a fishing basket very suitable for
carrying small birds.   It is not heavy to handle, and the birds, when
once placed therein, are beyond the reach of injury.   In this basket,
when leaving home, should be placed some sheets of brown paper
about the size of letter paper, and a little cotton wadding.   When a
bird is killed, the shot holes should be plugged with cotton to stop
the bleeding, and a pellet of the same material put into the mouth to
prevent the juices of the stomach oozing out and soiling the feathers.
If the bird is wing-broken or otherwise wounded, it should be killed
at once, and the simplest way of doing this is to catch it firmly across
the small of the back and press hard with the fingers and thumb
under the wings, which will suffocate the bird in a few seconds.   The
throat and shot holes can then be filled as described.   A paper cone
of suitable size is made next, the bird dropped into it headforemost,
and the outer edges of the paper turned inwards to prevent it slipping
out, and so it is placed in the basket.

Having reached home, the collector divests himself of his muddy
boots, gets a pair of slippers and a change of coat, and sets himself to
work to prepare his specimens.   In his tool box should be the things
he needs and nothing more, for the surplus only causes confusion :
a very fine penknife suitable for the smallest birds, and a larger
one for larger specimens : two or three knitting needles of different
sizes, a pair of cutting pliers, a few needles and some thread, a paper
of pins, a bottle with the preserve, and a bag with some cotton
and a small lot of tow.

The birds being removed from their cones are laid out in order,
and the one most desired for a specimen is selected to be first
operated upon.   The cotton is taken from the mouth and a fresh
pellet put in.   The wing bones are then broken close to the body
with the pliers, and the bird laid on its back on the table with the
bill towards the operator.   The middle finger, slightly moistened,
will separate the feathers from the breast-bone downward, leaving a
bare space exposed.   About the end of the breast-bone the point of
the knife is inserted, back downward, under the skin, and a clean
cut of the skin made from this point down to the vent.   The skin is
then loosened from the body till the thighs are exposed, which are
here cut through at the joints.   The tail is next separated from the
body and the skin turned carefully down, the turn-over being greatly
facilitated by the wings being loose, and they can now be separated
from the body, and the skin turned back to the base of the bill.
This should be done with the finger nails, assisted here and there

with the knife, but it must not be pulled or drawn or it will be injured by being stretched. The neck is then cut off where it joins with the skull, and the body laid aside for the present. The brains and eyes are next scooped out, and all the flesh and muscles removed from the bones of the legs and wings. The skin is now ready for the preserve, which is simply *arsenic* procured from the drug store. If the skin is thick and greasy it may be dusted on dry, giving the skin as much as will adhere to it : but if it is a thin, dry skin it is better to mix the arsenic with water to about the thickness of cream and put it on with a brush. When the preserve is thus applied to all the parts, a little cotton is wrapped around the bones of the legs and wings, the eye-sockets are filled with the same, and a piece about the length and thickness of the neck pressed firmly into the skull. The wings and legs are then pulled outward till they take their proper place, and the skin turned backward till it assumes its natural position, and it is filled with wadding to its former size. The mandibles are then fastened together by a thread passed through the nostril and tied under the bill. The legs are crossed and tied together with a thread just above the feet ; the feathers of the body are drawn together so as to cover the opening. Just in proportion to the care and pains bestowed on the specimen at this point, will it be a good skin or the reverse, for whatever position the feathers now get, that position they will retain. The body should next be examined to ascertain the sex of the specimen. This is done by cutting a hole in the side opposite the small of the back. Attached to the backbone at this point may be found the testicles of the male, two round bodies of dirty white, varying in size according to the size of the bird, but always largest in spring. Those of such birds as the warblers are no larger than pin heads. If the specimen is a female, the ovaries will be found in the same position, a mass of flattened spheres, similar in color, ranging in size according to the size of the bird, and also with the season. There should be attached to the legs of the specimen by a thread, a "tag" giving its scientific and common name, the sex, date of capture, locality where found, and name of collector. This is the course followed with birds up to the size of a Robin, the time required to complete the operation being twenty minutes. Three in an hour is about my rate of progress, but I have seen statements made of a much larger number being done in that time.

The specimen is now put away to dry, and the position in which it is left will decide its future appearance. I have a setting-board which I find very useful at this stage. It may be described as a series of zinc tubes, varying in size from one inch to six inches in diameter and two feet long, cut in two lengthwise, and placed side by side in a pine frame made to receive them. In these grooves the skins are placed according to size. The semicircular shape of the bed keeps the back and wings in the natural position, and the feathers smooth and regular. The front part of the specimen being uppermost, it is always visible, and if need be the skin can be taken up while drying and any irregularities adjusted.

Another way of disposing of the skin when made is to form a cosy bed for it in a soft sheet of cotton. Arrange the feathers and shape of the skin as desired, lay the specimen gently into the bed prepared for it, and do not touch it again till it is perfectly dry.

In larger specimens a different mode of treatment has to be adopted. With ducks, for instance, where the head is large and the neck small, the skin cannot be turned over as described. The plan then is to skin as far down the neck as possible, and to cut it off, leaving the head still inside of the skin. An opening is then made in the skin from the centre of the crown, a little way over the back of the head, through which the head is turned out, cleaned and then preserved. The eye-sockets and other apertures are filled with cotton, the head put back into its place, the opening carefully sewed up, and the skin filled out as already described. In birds of this size it is necessary to sew up the original opening, bringing the feathers together so that the disturbance may not be observed.

In skinning oily water-fowl, it is sometimes necessary to sew a strip of cloth around the opening as soon as it is made, to prevent the feathers being soiled by contact with the grease of the body. With birds of large size, a hook is sometimes passed through the bony part of the body just where the tail has been separated. To this a string is attached, by which the body is suspended while the skin is being removed. This plan gives more freedom in handling large birds, the only objection to it being that the skin is apt to be stretched.

Some birds are so fat that it is always necessary to have an absorbent of some kind at hand to take up the grease which would otherwise soil the feathers. Cornmeal is the best thing to use, but plaster of Paris does very well, except for birds of black plumage which it leaves slightly shaded with gray. The latter material may also be used with advantage for removing blood stains and other impurities which the birds may have contracted from their surroundings. The soiled places are moistened with hot water before the absorbent is applied, and they are patted with the hand while the plaster is drying off. This must not be done while the birds are in the flesh, for the application of hot water will increase the flow of blood from the openings.

In putting away birds which are too large for the setting-board, a good way is to make a paper cylinder which will slip easily over the bird while in the flesh. When skinned and filled out in the usual way, the slip is again drawn over the specimen and allowed to remain till it is dry. In making skins of birds with long necks and heavy heads, it is necessary to put a long wire, or a tough sapling, through the whole length of the body and neck, because the skin becomes very brittle when it is dry, and runs the risk of being broken in handling.

Proficiency in the art of mounting birds can only be acquired by long practice, and a lot of poorly mounted specimens sitting about a house are neither useful nor ornamental; but anyone, by following these instructions, should be able, with a little practice, to make up skins fit to appear in any public museum or private collection.

NESTS AND EGGS.

Next in interest to a good collection of mounted birds, or skins, is a collection of nests and eggs. By this I do not mean simply an accumulation of nests and eggs, the number of which constitutes the value of the collection, but a carefully handled nest, and a correctly identified set of eggs of every bird known to breed within the district over which the collector extends his observations.

The suggestions following are for the guidance of those who may wish to include them in their collections:

Nest-building I regard as most interesting and important in the study of a bird's history, for it tends to bring out all the intelligence and taste of a species better than anything else in which we see it engaged. There are some actions in a bird's life over which it has no control, such as the depositing of eggs of a certain size and color, and the hatching of them in a given time. These things are fixed already, but the selection of the site for the nest, the material of which it is to be composed, and the careful finish bestowed on

it, are all matters which vary with the different species. Between
individuals of the same species there is also, frequently, a great
difference in the skill exhibited in nest-building.

Among the lowest types of bird-life, we find species which make
no nest of any kind, but deposit their eggs on the rocks, or on the
sand by the seashore. Others make the merest apology for one, and
it consists only of a few straws ; while still another species admits
the desirability of a nest, but dislikes the trouble of building it, and
therefore appropriates that of another species. But as we advance
upward in the scale, we find, especially among the nests of the
smaller birds, some beautiful specimens of bird architecture, one of
the finest being the work of our tiniest, the Humming-bird. The
Summer Yellow Bird builds a beautiful nest in the fork of a lilac in
the shrubbery ; and while observing a handsome elm tree budding
out in spring, I never think it complete unless it has the pretty,
pensile nest of the Baltimore Oriole swaying at the point of one of its
long, pendulous branches.

The principal objection to a collection of nests is the amount of
room that they require, but the finest nests are those of the smallest
birds, and a great number of these can be kept in a tray subdivided
as required, and they are never-failing objects of interest. Some
nests are found saddled on a limb, and are loosely built of twigs.
This kind should be removed very carefully, and afterwards sewn
together with inconspicuous thread, to keep the fabric as near as
possible in its original shape. Others are in the forks of thick bushes,
and the branches to which they are attached should be cut and the
nests and their connections lifted out. Those of the ground-building
birds require very careful handling, and often the safest way is to
cut out a piece of the sod in which the nest is placed and bring it
along, to keep the domicile from falling to pieces. Occasionally, a
rare nest is found which it is impossible to remove. Of this the
collector should make a careful record in his note-book, giving the
general surroundings, date of occurrence, situation, size of nest,
materials of which it is composed, number of eggs or young, action of
the parents during the examination, and any other item of interest
observed at the time, which will soon be forgotten if not recorded.

The impossibility of making a full collection of nests has been
shown, but there is nothing to prevent anyone making a full collection of eggs. The most important point in this work is the correct
identification of the eggs, and unless the collector makes up his mind

at the start to be sure of this in every case, he had better leave the matter alone, for without this, his collection, whatever the extent of it may be, will be absolutely worthless.

For anyone observant of the habits of the birds in the woods, the best guide to their nest is found by watching the actions of the birds, especially the female, during the breeding season, for she is sure eventually to betray the whereabouts of her treasures. But birds' nests are often close together, and their eggs so nearly alike that the only way to be absolutely certain of the identity of the rarer species is to secure the parent, either while she is on the nest or when she is seen fluttering away from it. This seems a cruel act, and the collector should not repeat it unless he is actually in need of the species.

On taking a set of eggs, the first thing to do is to remove any stains which they may have acquired from their surroundings : but this must be done with care, for in fresh eggs the colors sometimes yield to the process and disappear. Eggs for a collection are now emptied through one hole about the middle of the side. It should first be pricked with a needle, and then the drill applied and worked with the finger and thumb till a smooth round hole is made, large enough to allow the contents to escape. The blowpipe should then be used, either close to the hole or a little way inside. In the latter case the blowpipe must be loose in the hole to allow the contents to escape around it. If the material inside is thick, and does not run freely, it should be cut up with a piece of fine wire with a sharp bent point, and removed with the aid of a small syringe.

When thoroughly clean, the egg should be laid aside to dry on some substance which will absorb the moisture, cornmeal being recommended because it does not adhere to the shell. Where eggs are in an advanced stage of incubation, those of small size can rarely be saved, but the embryos in larger ones may be cut to pieces by widening the hole a little and introducing a sharp hook. The contents can then be removed piecemeal, but great care and much time are necessary to accomplish the task. In some cases it is considered safer to allow the embryo to remain in the shell till it is decomposed. To assist the process of decay, a strong solution of caustic potash is introduced, and this is allowed to remain till the embryo becomes soft and pulpy, when it will yield to the ordinary treatment.

The safest mode of carrying eggs from the field, or sending them away by mail when prepared, is to have each egg wrapped in cotton-

batting and placed in a box of suitable size subdivided for the purpose. In some collections, the species to which they belong, and
other particulars, are given by marking them on the shell with a
soft pencil, but I dislike this mode, for it destroys the look of the
egg. A better way, I think, is to have a number attached to the
box or nest containing the eggs, and a corresponding number in a
catalogue in which can be given all the necessary particulars in
detail.

THE BIRDS OF ONTARIO.

Order PYGOPODES. Diving Birds.

Suborder PODICIPEDES. Grebes and Loons.

Family PODICIPIDÆ. (Grebes.

Genus COLYMBUS Linn.

Subgenus COLYMBUS.

COLYMBUS HOLBŒLLII (Reinh.).

*1. Holbœll's Grebe. (2)†

Tarsus about four-fifths the middle toe and claw; bill little shorter than tarsus; crests and ruff moderately developed. Length, about 18; wings, 7-8; bill, 1⅞ to nearly 2; tarsus, 3; middle toe and claw, 2⅜. *Adult:*—Front and sides of neck rich brownish-red; throat and sides of head ashy, whitening where it joins the dark color of the crown, the feathers *slightly* ruffed; top of head with its *slight* occipital crest, upper parts generally, and wings dark brown, the feathers of the back paler edged; primaries brown; part of inner quills white; lower parts pale silvery-ash, the sides watered or obscurely mottled, sometimes obviously speckled with dusky; bill black, more or less yellow at base. The young may be recognized by these last characteristics, joined with the peculiar dimensions and proportions.

Hab.—North America at large, including Greenland. Also Eastern Siberia, and southward to Japan. Breeds in high latitudes, migrating south in winter.

Nest, a floating clump of vegetable material fastened to the reeds in shallow water.

Eggs, two to five, rough-dull white, shaded with greenish.

In Ontario, the Red-necked Grebe is only a transient visitor, its summer home being far to the north of this province, and its winter-quarters to the south.

It breeds abundantly along the borders of the Yukon River in Alaska, and has also been found by Macoun so engaged on the Waterhen River and south end of Waterhen Lake.

' Current number | Number in American Ornithologists' Union "Check List."

Holbœll's has the habit peculiar to other members of the Grebe family of covering its eggs with vegetable matter when it has occasion to be away from the nest, so that incubation goes on during the absence of the mother-bird.

The migrations of the species extend across the Province, for it is reported as a visitor in spring and fall at Ottawa, Hamilton, London and the Detroit River. The visits are always short, and the manners of the birds while here are shy and retiring.

For many years the young of this species was described as the *Crested Grebe*, owing to the close resemblance it bears to the British bird of that name. Dr. Brewer was the first to point out the error, which is now corrected in all modern works on American Ornithology.

SUBGENUS DYTES KAUP.

COLYMBUS AURITUS LINN.

2. **Horned Grebe.** (3)

Tarsus about equal to the middle toe without its claw; bill much shorter than the head, little more than half the tarsus, *compressed*, higher than wide at the nostrils, rather obtuse; crests and ruffs highly developed. Small, length about 14; extent, 24; wing, 6 or less; bill, about ¾; tarsus, 1¼. *Adult:*— Above, dark brown, the feathers paler edged; below, silvery-white, the sides mixed dusky and reddish; most of the secondaries white; fore neck and upper breast brownish-red; head glossy black, including the ruff; a broad band over the eye, to and including occipital crests brownish-yellow; bill black, yellow-tipped; the eye fiery-red. The young differ as in other species, but always recognizable by the above measurements and proportions.

HAB.— Northern hemisphere. Breeds from the northern United States northward.

Nest, a floating mass of vegetable material fastened to the reeds or rushes in shallow water.

Eggs, two to seven, whitish, shaded with greenish-blue.

Generally distributed, breeding in all suitable places throughout Ontario, notably at St. Clair Flats. The nest is so completely isolated, that the young when hatched may be said to tumble out of the shell into the water. The birds arrive in spring, as soon as the ice begins to break up, and remain quite late in the fall, individuals being occasionally seen on Lake Ontario during the winter.

The Grebes upon land are the most awkward of birds, but in the water they are surpassed by none in the swiftness and grace of their movements.

One of the accomplishments possessed by this species is the ability, when alarmed, to sink under water without leaving so much as a ripple to mark where it has gone down, the point of the bill being last to disappear.

It has a wide breeding range, having been found by Dr. Bell at Fort George, on the east side of James' Bay, and also at Fort Severn and York Factory, on the west coast of Hudson's Bay. Nelson says of it, in the "Birds of Alaska": "Like the preceding, this handsome species occurs along the eastern shore of the Behring Sea in very small numbers in the breeding season, but is not rare in autumn. It is also a common summer resident along the Yukon, and occurs rarely on the Commander Islands."

COLYMBUS NIGRICOLLIS CALIFORNICUS (HEERM.).

3. American Eared Grebe. (4)

Adult male:—Long ear tufts of rich, yellowish brown; head and neck all round, black; upper parts, grayish-black; sides, chestnut; lower parts, silvery gray; primaries, dark chestnut; secondaries, white, dusky at the base; length, 13 inches. Young similar, the ear tufts wanting, and the colors generally duller.

Hab.—Northern and western North America, from the Mississippi Valley westward.

The nest and eggs cannot be distinguished from those of the preceding species.

I mention this as an Ontario species on the authority of Dr. Garnier, of Lucknow, Bruce Co., who informs me that a specimen was sent to him in the flesh from Colpoy's Bay, as being something different from those usually seen at that point. It was too far gone for preservation when received, but the Doctor, who has long been an ardent collector, assures me that he is quite satisfied of the correctness of his identification.

This species is a comparatively new acquaintance to American ornithologists, for although described by Audubon, it was not found by him. It is now known to breed in Texas, Kansas, Illinois, Dakota, and Colorado; and Macoun reports it breeding abundantly in the prairie pools of the North-West, so that we need not be surprised if a straggler is now and then wafted a little way out of its ordinary course.

GENUS PODILYMBUS Lesson.

PODILYMBUS PODICEPS (Linn.).

4. Pied-billed Grebe. (6)

Length, 12 to 14; wing, about 5; bill, 1 or less; tarsus, 1½. *Adult:* Bill bluish, dusky on the ridge, encircled with a black bar; throat with a long black patch; upper parts blackish-brown; primaries ashy-brown; secondaries ashy and white; lower parts silky white, more or less mottled or obscured with dusky; the lower neck in front, fore breast and sides, washed with rusty. Young lacking the throat-patch and peculiar marks of the bill, otherwise not particularly different; in a very early plumage with the head curiously striped.

HAB. –British Provinces southward to Brazil, Buenos Ayres and Chili, including West Indies and the Bermudas, breeding nearly throughout its range.

Nest, a little floating island of withered reeds and rushes mixed with mud, fastened to the aquatic plants, raised two or three inches above water.

Eggs, five to seven, whitish, clouded with green.

The Dab Chick is not quite so numerous as the Horned Grebe, neither is it so hardy, being a little later in arriving in spring, and disappearing in the fall at the first touch of frost. It is generally distributed, and is the only one of the family which breeds in Hamilton Bay, where it may often be seen in the inlets in summer accompanied by its young with their curiously striped necks. From its small size and confiding manners it is not much disturbed, but if alarmed it has a convenient habit of sinking quietly under water, not to reappear till danger is past.

In summer these Grebes breed commonly throughout the North-West.   Here is what is said about them in Mr. Thompson's "Birds of Manitoba": "Very abundant summer residents on every lake, slough or pond large enough to give them sufficient water privilege, arriving as soon as the ice is gone and departing when their haunts freeze over."

FAMILY URINATORIDÆ.   LOONS.

GENUS URINATOR CUVIER.

URINATOR IMBER (GUNN.).

5. Loon. (7)

Black; below from the breast white, with dark touches on the sides and vent; back with numerous square white spots; head and neck iridescent with violet and green, having a patch of sharp white streaks on each side of the neck and another on the throat; bill black.   *Young:* - Dark gray above, the feathers with paler edges; below, white from the bill, the sides dusky; bill yellowish-green and dusky.   Length, 2½-3 feet; extent, about 4; wing, about 14 inches; tarsus, 3 or more; longest toe and claw, 4 or more; bill, 3 or less, at base 1 deep and ¾ wide, the culmen, commissure and gonys all gently curved.

HAB. --Northern part of northern hemisphere.   In North America breeds from the northern tier of States northward; ranges in winter south to the Gulf of Mexico.

Nest, a hollow in the sand near the water's edge, sometimes lined with grass, weeds, etc.

Eggs, two or three, olive-brown, spotted and blotched with very dark brown.

The Loon, on account of his large size, is conspicuous wherever he appears, and his loud and melancholy cry is often heard at night during rough weather, when the bird himself is invisible. Many pairs raise their young by the remote lakes and ponds throughout the country, but they all retire farther south to spend the winter. As soon as the ice disappears they return, mostly in pairs, and by the end of May have chosen their summer residence. The Loon, in common with some other water-fowl, has a curious habit, when its curiosity is excited by anything it does not understand, of pointing its bill straight upwards, and turning its head rapidly round in every direction, as if trying to solve the mystery under consideration. Once when in my shooting skiff, behind the rushes, drifting down the bay before a light wind, I came upon a pair of these birds feeding about twenty yards apart. They did not take much notice of what must have seemed to them a clump of floating rushes, and being close enough to one of them I thought to secure it, but the cap snapped. The birds hearing the noise, and still seeing nothing living, rushed together, and got their bills up, as described, for a consultation. So close did they keep to each other that I shot them both dead at forty yards with the second barrel.

In his notes on the "Birds of Hudson's Bay," Dr. Bell says: "The Loon, or Great Northern Diver, is at home in and around Hudson's Bay. In the spring, as soon as the water opens at the mouths of the rivers, these birds appear in incredible numbers, as if by a previous understanding, about a common meeting place. At such a time they may be much more easily approached than usual. These birds are said to spear the fish with the *bill closed*, and to bring them to the surface so that they may turn them endways for the purpose of swallowing The gulls, hovering overhead, and seeing what is going on down in the clear water, watch for the moment the fish is raised to the surface, when they swoop down and carry it off. When many hungry gulls are present, this process is repeated till the patience of the Loon is quite exhausted."

In Alaska, Nelson says: "Throughout the interior of the territory this bird is a common summer resident, and was found breeding abundantly at the western extremity of the Aleutian Islands by Dall. The skins of the birds are used by the natives in their bird-skin clothing, and are specially prized for tool-bags."

URINATOR ARCTICUS (Linn.).

6. **Black-throated Loon.** (9)

Back and under parts much as in the last species; upper part of head and hind neck, *bluish-ash* or hoary-gray ; fore neck purplish-black, with a patch of white streaks ; bill, black. The young resemble those of that species, but will be known by their inferior size. Length, under 2½ feet ; extent, about 3 ; wing, 13 inches or less ; tarsus, 3 ; bill, about 2½.

Hab. Northern part of the northern hemisphere. In North America migrating south in winter to the northern United States.

Nest, a hollow in the ground not far from the water's edge.

Eggs, two, dark olive, marked with black spots towards the larger end.

This is a much more northern bird than the preceding, for it is seldom met with in the United States, and then mostly in winter in immature plumage. In its migratory course it no doubt visits the waters of Ontario, and should be looked for by those who have opportunity to do so. A pair of these birds, found in the neighborhood of Toronto, was included in a collection that was sent to the Paris Exposition in 1866, and I once saw an individual in Hamilton Bay under circumstances which prevented me from shooting it, though I was quite close enough, and satisfied of its identity. It was on a still, dull day in the early part of April, and the ice on the bay was broken up and floating about in loose flakes. Water-fowl of different kinds were coming rapidly in and pitching down into the open water. I was out in my shooting skiff in search of specimens, when the wind suddenly blew up from the east, and I was caught among the drifting ice. Everything in the skiff got soaking wet. I broke both paddles trying to force a passage, and for a time was at the mercy of the elements. While drifting along in this condition I came close to a Black-throated Diver in similar trouble, for it was caught among the ice, unable to rise, and evidently afraid to dive, not knowing where it might come up. We looked sympathizingly at each other, it uttered a low whining cry, and we drifted apart. I got safe to land, and it is to be hoped the rare bird reached the open water and got off in safety. We did not meet again. Not having seen the species recently, nor heard of its capture by others, I consider it a very rare visitor to these inland-waters. In Dr. Wheaton's exhaustive report on the "Birds of Ohio," mention is made of an individual having been shot in Sandusky Bay in the fall of 1880, but the line of its migratory course is probably along the sea coast.

Dr. Coues, when speaking in his "Birds of the North-West" of the familiarity of the Pacific Black-throated Diver in the harbor of San Pedro, in southern California, says : "They even came up to the wharves, and played about as unconcernedly as domestic ducks. They constantly swam around the vessels lying at anchor in the harbor, and all their motions both *on* and *under* the clear water could be studied to as much advantage as if the birds had been placed in artificial tanks for the purpose. Now, two or three would ride lightly over the surface, with the neck gracefully curved, propelled with idle strokes of their broad paddles to this side or to that, one leg after the other stretched at ease almost horizontally backwards, while their flashing eyes first directed upwards with curious sidelong glances, then peering into the depths below, sought for some attractive morsel. In an instant, with the peculiar motion impossible to describe, they would disappear beneath the surface, leaving a little foam and bubbles to mark where they had gone down, and I could follow their course under water; see them shoot with marvellous swiftness through the limpid element, as, urged by powerful strokes of the webbed feet and beats of the half open wings, they *flew* rather than *swam* ; see them dart out the arrow-like bill, transfix an unlucky fish and lightly rise to the surface again. While under water the bubbles of air carried down with them cling to the feathers, and they seem bespangled with glittering jewels, borrowed for the time from their native element, and lightly parted with when they leave it. They arrange their feathers with a shiver, shaking off the last sparkling drop, and the feathers look as dry as if the bird had never been under the water. The fish is swallowed headforemost with a peculiar jerking motion, and the bird again swims at ease with the same graceful curve of the neck."

It is said to be common in Norway and Sweden, and more rarely has been found breeding near some of the lonely lochs in the north of Scotland.

In the "Birds of Alaska," Mr. Nelson says of this species : "This Loon is very common all along the American shore of the sea, and about Kotzebue Sound ; they are also numerous on the large streams and marshes of the interior. The skins of these birds, as of other heavily plumaged water-fowl, are much used by the natives from St. Michaels south for clothing. The natives snare and spear them in the shallow ponds and lagoons where they breed, and Dall mentions having seen one dress containing the skins from over one hundred loons' throats."

3

URINATOR LUMME (Gunn.).

7. Red-throated Loon. (11)

Blackish ; below, white; dark along the sides and on the vent and crissum ; most of the head and fore neck, bluish-gray ; the throat with a large *chestnut* patch ; hind neck, sharply streaked with white on a blackish ground ; bill, black. Young have not these marks on the head and neck, but a profusion of small, sharp, circular or oval white spots on the back. Size of the last, or rather less.

Hab.—Northern part of northern hemisphere, migrating southward in winter nearly across the United States.

Breeds in high latitude. Eggs, two in number, pale green spotted with brown, deposited in a hollow in the ground close to the water's edge.

Audubon found this species breeding at Labrador, and in the *Fauna Boreali Americana* it is spoken of as "frequenting the shores of Hudson's Bay up to the extremity of Melville Peninsula."

Large numbers of these birds visit the waters of Southern Ontario in March and April, about the time of the breaking up of the ice, but an adult with the red-throat patch is scarcely ever seen. The one in my collection was procured out on Lake Ontario at midsummer, having for some reason failed to follow the flocks to the far north. In the fall very few are seen, their route to the south being in some other direction.

All the birds of this class have a most ungainly gait on land, and when surprised away from the water are often taken by the hand before they can get up to fly. On the water or under its surface their motions are exceedingly graceful.

Dr. Bell found this species on both sides of Hudson's Bay, but only in spring and autumn.

Mr. Nelson says regarding it : "Throughout Alaska the present bird is by far the most abundant species of Loon. At St. Michael's and the Yukon Delta they arrive with the first open water, from May 12th to 20th, and by the end of the month are present in large numbers. Their arrival is at once announced by the hoarse, grating cries which the birds utter as they fly from place to place or float upon the water. When the ponds are open in the marshes the Red-throated Loons take possession, and are extremely noisy all through the first part of summer. The harsh cry arising everywhere from the marshes during the entire twenty-four hours, renders this note one of the most characteristic which greets the ear in spring in those northern wilds. The Russian name, *Gagara*, derived from the birds' notes, is a very appropriate one.

" The Red-throated Loon is one of the few birds which raises its voice in the quiet of the short Arctic night.

" In spring, with the Cranes, they foretell an approaching storm by the increased repetition and vehemence of their cries."

FAMILY ALCIDÆ. AUKS, MURRES, AND PUFFINS.

SUBFAMILY FRATERCULINÆ. PUFFINS.

GENUS FRATERCULA BRISSON.

FRATERCULA ARCTICA (LINN.).

8. Common Puffin. (13)

Adult male :—Entire upper parts, and a collar passing round the fore neck, black ; sides of the head and throat, grayish-white ; lower parts, white ; a horny protuberance on the upper eyelid. In the young the white of the plumage is shaded with dusky, and the curiously shaped bill is less fully developed. Length, 13 inches.

HAB.—Coasts and islands of the North Atlantic, breeding from the Bay of Fundy northward. South in winter to Long Island and occasionally farther.

Nest, in a burrow underground, or in a hole among the rocks, one egg, dull white, sometimes veined or spotted with brown.

The Puffin is essentially a bird of the sea coast, which it seldom leaves except under stress of weather. They breed in immense numbers in Labrador, Newfoundland, and sparingly in the Bay of Fundy. In winter they scatter along the sea coast and are found as far south as Long Island. In the report of The Ottawa Field Naturalists' Club for 1882 and 1883, it is stated that "a young bird of this species was shot on the Ottawa, towards the end of October, 1881. It had probably been blown inland by a severe storm which took place some days previous." This is the only Ontario record we have of its occurrence so far from the sea, or so far west, for this species does not occur on the Pacific coast.

My first observations of this bird were made during my school days, but may be worth recording, for they show the habits of the bird, which is identical with our American species. One of the principal breeding places, which I frequently visited, was Ailsa Craig, on the west coast of Scotland. The Craig is an isolated, circular rock two or three miles off the coast, about as high as it is wide, and inhabited only by a keeper and many different sorts of sea-fowl.

Around the base of the rock at the water's edge is a belt of loose
rocks which, during ages past, have fallen from above. Higher up
there are patches of soil on which the keeper raises grass enough
to feed two or three goats. Divers, gulls, auks, petrels, loons,
etc., are found there in great abundance, but the most numerous are
the Puffins, which, in local parlance, are called the "Paties" or
"Coulternebs," from the fancied resemblance the bill of the bird
bears to the cutting part of a plough. The birds are to some extent
migratory, going south during winter, but in February they return
in crowds and at once select their breeding places, no nest being
required. Each pair chooses an opening among the loose rocks at
the base of the cliff, within which is duly deposited their one white
egg. Others dig holes for themselves, like rabbit holes, in the ground
higher up, in which to lodge their treasure, and these they vigorously
defend against all intruders. Pleasure parties often visit the Craig
during the summer, some of whom make a point of taking back
specimens from the island, but woe betide the hand that is thought-
lessly pushed into one of these holes if "Patie" is at home, for he has
a most powerful beak, and once taking a hold he can hardly be made
to let go.

One of the keepers, who made a business of selling the birds, had
a dog trained to the work of catching them. He was a rough Scotch
terrier, and it was no uncommon sight to see him come out from
among the loose rocks with several of the birds hanging on to his
hair. He did not need to catch *them*, as they caught *him* and held
on till taken off by the keeper. The Craig is of limited extent, but
the number of birds which frequented it was almost incredible.

Many of the birds spent the day in fishing out at sea, but all
returned about the same time in the evening, and that was the time
to see the multitudes gathered together. Looking at the rock it
seemed as if there was scarcely room for another bird, so completely
was every available spot covered. It was a custom with the keeper,
when making up a lot to send to the market, to take his place on
some prominent point in the evening, and with a club knock down as
many birds as he needed while they flew past. It is on record that
one of the keepers, a strong, active man, used to the work, undertook
for a wager to kill eighty dozen paties in one day, and he actually
managed to do it, in the manner described.

The young birds are fed on fish brought by the old ones, often
from a great distance. They seldom brought in fewer than five or
six at once, and all were killed by a squeeze on the head, but how

"Patie" disposed of No. 1 while he was killing No. 2, and so on, we could never understand. Guillemots and auks were also common on the Craig, and cormorants were often there as visitors fishing. At night they roosted on the shelves of the rocks along the shore at Mochrum, where their gaunt, grim figures were seen in rows in the evening, giving a chance to some local Scotch "wut" to christen them the "Mochrum Elders," a name which, in that district, has clung to them ever since.

SUBFAMILY PHALERINÆ.

GENUS CEPPHUS PALLAS.

CEPPHUS GRYLLE (LINN.).

9. Black Guillemot. (27)

Adult male :--In full plumage, black, shaded with dull green; a white patch on the wings. In all other stages, a marbled mixture of black and white. Length, 13 inches.

HAB.--Coasts of northern Europe, south to Denmark and British Islands. Coast of Maine, south in winter to Philadelphia; Newfoundland (?)

Eggs laid on the rocks near the sea, two in number, sea-green blotched with brown.

There is an old record of an individual of this and one of the succeeding species being found in Hamilton Bay in a state of extreme exhaustion about twenty-five years ago. I did not see the birds, but inquired into the circumstances at the time and considered the report correct. As none of this family has been observed since that time, these two can only be regarded as waifs carried away against their wishes by the force of the wind.

This species is very common along the west coast of Scotland, where I have seen the females with the bare spot on the under parts, the feathers having, according to custom, been plucked off to allow the heat of the body to be conveyed more directly to the eggs.

SUBFAMILY ALCINÆ.

GENUS URIA BRISSON.

URIA LOMVIA (LINN.).

10. Brünnich's Murre. (31)

Adult male: Head and neck, brown; upper parts, grayish-brown; secondaries tipped with white, lower parts white from the throat downwards. Length. 17 inches.

HAB.—Coasts and islands of the North Atlantic and Arctic Oceans; south on the Atlantic coast of North America to New Jersey, breeding from the Gulf of St. Lawrence northward.

Eggs on the cliff near the sea, pale green.

Found on Hamilton Bay, under circumstances similar to the preceding.

The two species of Guillemot which frequent the North Atlantic resemble each other closely in appearance, but one is much more abundant than the other. For many years the one which is comparatively rare was supposed to be the one which is abundant, a mistake which has only recently been corrected. In the "Birds of New England," Dr. Coues says regarding *Uria troile* (common Guillemot): "Contrary to the general impression, the "common" Guillemot appears to be a rare bird in New England, most of the Murres occurring in winter along our coasts being of the following species: *Uria lomvia* (Brünnich's Murre)." And again, in referring to the present species, Dr. Coues says: "This is the common winter Guillemot of the New England coast, and probably most of the references made to *Uria troile* really apply to the present species."

Mr. Brewster, following in the same strain, says: "At different times during the past ten years I have examined specimens from different points along the shores of Maine, New Hampshire, Massachusetts, and all of the numerous birds which have come under my notice have proved to be Brünnich's Guillemot; indeed, the example of *Uria troile* mentioned by Mr. Merrill, is the only New England one of which I have any knowledge."

While the foregoing was in the hands of the printer, we had quite a number of this species as visitors to Lake Ontario, very few of which, I fear, were able to return to their favorite seaboard.

The first I saw was in the hands of a local gunner, who killed it on Hamilton Bay on the 20th November, and a few days later I heard from Mr. White that five had been taken at Ottawa. The next report came from Toronto, where about thirty were killed, and finally, early in December, I obtained three which were found on Hamilton Bay so much reduced and exhausted for lack of proper food, that they were taken alive by the hand. I believe that about fifty were captured altogether. This is the only occasion on which I have heard of these birds appearing in Ontario, except the one mentioned at the beginning of this notice.

Genus ALCA Linnæus.

ALCA TORDA (Linn.).

11. Razor-billed Auk. (32)

Adult, in summer:—Upper parts, black, glossed with green; head and neck, brownish-black, without gloss; tips of the secondaries and all the lower parts, white; a white line from the eye to the base of the culmen; feet, black; mouth, yellow; eye, bluish. Length, about 18 inches; wing, about 7.75.

In winter, the white covers the throat and encroaches on the sides of the head.

Hab.- Coasts and islands of north Atlantic. South in winter, along the coast to New England.

Nest, none.

Eggs, one or two, deposited in caverns or deep fissures of the rocks; creamy-white, spotted and blotched with black toward the larger end.

The first notice we have of the occurrence of this species in Ontario is in the published proceedings of the Canadian Institute, where Mr. Wm. Cross reports the capture of a specimen in Toronto Bay, on the 10th December, 1889.

A second specimen was shot off the beach at the west end of Lake Ontario, in November, 1891. This was afterwards mounted, and is now in possession of Captain Armstrong, who resides near the Hamilton reservoir.

This species keeps close to the sea coast. It has not been observed at Ottawa, and is not named among the birds found by Dr. Bell at Hudson's Bay.

We can only account for the presence of these isolated individuals in Lake Ontario by supposing that they have been driven from their usual habitat by an easterly blow.

Order LONGIPENNES. Long-winged Swimmers.

Family STERCORARIIDÆ. Skuas and Jaegers.

Genus STERCORARIUS Brisson.

STERCORARIUS POMARINUS (Temm.).

12. Pomarine Jaeger. (36)

Middle tail feathers finally projecting about four inches, *broad to the tip.* Length, about 20 inches; wing, 14; bill, 1½-1¾; tarsus, about 2. *Adult:*— Back, wings, tail, crissum and lower belly, brownish-black; below from bill to belly, and neck all round, pure white, excepting acuminate feathers of sides of neck, which are pale yellow; quills, whitish basally, their shafts largely white; tarsi—above, blue; below, with the toes and webs, black. *Not quite adult:*—As before, but breast with dark spots, sides of the body with dark bars, blackish of lower belly interrupted; feet, black. *Younger:*—Whole under parts, with upper wings and tail coverts, variously marked with white and dark; feet, blotched with yellow. *Young:*—Whole plumage transversely barred with dark brown and rufous; feet, mostly yellow. *Dusky stage* (coming next after the barred plumage just given?); fuliginous, unicolor; blackish-brown all over, quite black on the head, rather sooty-brown on the belly; sides of the neck slightly shaded with yellow.

Hab.—Seas and inland waters of northern portion of the northern hemisphere; chiefly maritime. South in North America to the Great Lakes and New Jersey.

Nest composed of grass and moss placed on an elevated spot in a marsh.

Eggs, two or three, grayish-olive, with brown spots.

The Pomarine Skua is occasionally seen in company with the large gulls, which spend a short time during the fall around the west end of Lake Ontario, following the fishing boats and picking up the loose fish that are shaken out of the nets. It is spoken of by the fishermen as a bird of a most overbearing, tyrannical disposition, one which they would gladly punish, but on these trying trips all hands are occupied with matters of too much importance to think of shooting gulls.

The home of this species is in the far north. Mr. Nelson says, in the " Birds of Alaska ": " They are abundant in spring off the mouth of the Yukon. Along both shores of the Arctic to the north they are very numerous, and to a great extent replace the other two species.

" They are especially common along the border of the ice-pack and about the whaling fleet, where they fare abundantly.

" The peculiar twirl in the long tail feathers of this species renders it conspicuous and easily identified as far away as it can be seen."

STERCORARIUS PARASITICUS.

13. Parasitic Jaeger. (37)

Middle tail feathers finally projecting about four inches, tapering, acuminate smaller; wing, 12-13; tarsus, 1.75 to 1.87; bill, 1.33 to 1.5; tail, 5-6, the long feathers up to 9. *Adult:*—Upper parts, including top of head, slight occipital crest, and crissum blackish-brown, deeper on wings and tail; chin, throat, sides of head, neck all round, and under parts to the vent, white; sides of the neck, pale yellow; quills and tail feathers with whitish shafts; feet, blue and black. *Younger:*—Clouded with dusky in variable pattern and amount. *Young:*—Barred crosswise with rufous and dusky; feet, mostly yellow. There is a fuliginous stage, same as described in last species.

Hab.—Northern part of northern hemisphere, southward in winter, to South Africa and South America. Breeds in high northern districts, and winters from the Middle States and California, southward to Brazil and Chili.

Nest, on the ground on the margin of lakes or on islands, a depression in the ground, lined with grasses, a few withered leaves and feathers.

Eggs, two or three, varying much in markings; olive-drab to green, gray and brown, marked with several shades of chocolate, brown, and an obscure shade of stone-gray distributed over the entire egg.

The breeding place of the Arctic Skua is in the far north, but many go a long way south to spend the winter, and a few call at the lakes in Ontario by the way. At such places they are occasionally seen singly, mixing with the gulls and terns which pass to the south in the fall, but they are very watchful and seldom obtained. On the 16th October, 1886, Dr. Macallum secured a young female, which was shot on the Grand River, near Dunnville, just after a severe storm.

In the report of the proceedings of the Ornithological Sub-section of the Biological Section of the Canadian Institute, Ernest E. Thompson mentions a specimen shot at Ashbridge's Bay, near Toronto, in September, 1885, which is now mounted and in possession of Mr. Loane of that city.

In the same report mention is made of a fine adult specimen which is in the museum of the Toronto University, marked "Toronto."

Dr. Bell also reports taking a specimen at Fort George, James' Bay.

Turning to our usual authority for northern birds, Mr. Nelson says : "This tyrannical bird occurs along the entire coast-line of the Behring Sea, but is most numerous along the low marshy coast of Norton Sound, and thence south to Kuskoquim River.

"Its breeding range covers the entire region from the Aleutian Islands north to the extreme northern part of the mainland.

"At all times jaegers are given to wandering, and one is likely to find them anywhere along the coast. They are frequently seen harrying terns or gulls, to make them disgorge fish just caught. If successful, they dart down and, rising under the falling morsel, catch it in their capacious mouths. This robbery is often performed by two birds acting in unison, but whether the birds alternate in disposing of the spoil has not been observed. They are very greedy, and often swallow so much that they cannot rise to fly till a portion is disgorged."

FAMILY LARIDÆ.   GULLS AND TERNS.

SUBFAMILY LARINÆ.   GULLS.

GENUS GAVIA BOIE.

GAVIA ALBA (GUNN.).

11. Ivory Gull. (39)

Adult male :—Pure white all over; quills of the primaries, yellow; feet and legs, black; bill, dull greenish, yellow at the tip. *Young:*--Plumage clouded with dusky. Primaries and tail feathers, spotted with dusky. Length, 20 inches.

HAB.—Arctic seas, south in winter on the Atlantic coast of North America to Labrador and Newfoundland. Not yet found on the coast of the Pacific.

Nest, on some inaccessible cliff. It is made of dry grass lined with moss and a few feathers.

Eggs, yellowish olive, with small blotches of dark brown clouded with lilac.

Having received interesting accounts from fishermen of pure white gulls following their boats out on the lake, I tried in vain for two seasons to persuade them to take my large single gun, and bring me a specimen. Finally I got them to attach a long line to the stern of one of the boats, with a hook at the end, baited with a ciscoe, and in this way they succeeded in getting me a fine adult male of the Ivory Gull, the only one I ever obtained.

This species of gull is said to breed farther north than any other. On the islands and along the coasts of Spitzbergen it occurs sparingly. In like places on the coast of northern Siberia it is abundant, and in Greenland it is resident. It is seldom found as far south as Ontario, but the movements of all such birds are liable to be affected by wind and weather.

GENUS RISSA STEPHENS.

RISSA TRIDACTYLA (LINN.).

15. **Kittiwake.** (40)

Hind toe appearing only as a minute knob, its claw abortive. Mantle, rather dark grayish-blue ; first primary, with the whole outer web, and the entire end for about two inches, black ; next one, with the end black about as far, but outer web elsewhere light, and a white speck at extreme tip; on the rest of the primaries that have black, this color decreases in extent proportionally to the shortening of the quills, so that the base of the black on all is in the same line when the wings are closed (a pattern peculiar to the species of *Rissa*) ; and these all have white apex. Bill, yellow, usually clouded with olivaceous ; feet, dusky olivaceous. Rather small; 16-18; wing, 12; bill, $1\frac{1}{8}$-$1\frac{1}{2}$; tarsus, about the same; middle toe and claw, longer; tail, usually slightly emarginate. In winter, nape and hind neck shaded with the color of the mantle. *Young:*— Bill, black; a black bar on the tail, another across the neck behind; wings and back variously patched with black; dark spots before and behind the eyes; quills mostly black.

HAB.—Arctic regions, south on the Atlantic coast in winter to the Great Lakes and the Middle States.

Nest of grass and seaweed, usually placed on cliffs or ledges of rock overhanging the water.

Eggs, two or three, greenish-gray, marked irregularly with varying shades of brown.

The Kittiwake is a species of wide distribution, being common along the coast of New England, while in the north it is found in Greenland, and has been reported from nearly all parts of the Arctic regions and many intermediate points. It breeds throughout its range, but always returns from the far north at the approach of winter. At that season it is very common around the west end of Lake Ontario, and in the summer it is one of those which assemble on the bird rocks of the St. Lawrence in such numbers as to give the rocks, when seen from a distance, the appearance of being covered with snow.

LARUS GLAUCUS (BRUNN.).

16. **Glaucous Gull.** (42)

Adult:—Plumage, pure white except the mantle, which is grayish-blue. Bill, gamboge yellow with a carmine patch toward the end of the lower mandible; feet, flesh color. In the young the upper parts are yellowish-white,

mottled with pale brown; breast and lower parts. gray; tail, white, mottled
with brown.   Length, 27 inches.

HAB.—Arctic regions, south in winter in North America to the Great Lakes
and Long Island.   North Pacific.

Nest of seaweed and moss placed on the ground, or rocks, or even on blocks
of ice.

Eggs, two or three, varying in color from grayish-brown to white.

During the winter months the "Burgomaster," as this species is
usually named, may be seen roaming around the shores of Lake
Ontario, seeking what it may devour, and it is not very scrupulous
either as regards quantity or quality.   In the *Fauna Boreali Americana*, it is described as being "notoriously greedy and voracious,
preying not only on fish and birds but on carrion of every kind.
One which was killed in Capt. Ross' expedition disgorged an auk
when it was struck, and on dissection was found to have another in
its stomach."

In March, when the days begin to lengthen and the ice begins to
soften, these large gulls rise from Lake Ontario, and soaring around
in wide circles at a great height, pass away towards the north.

In the spring of 1884, a specimen was shot near Toronto by Mr.
George Guest of that city.

On the 25th March, 1889, a fine female specimen which had been
shot on the Island at Toronto, was brought to Mr. Cross, and on the
8th December, of the same year, he received another which was shot
off the Queen's Wharf.

This is one of the largest and most powerful birds of the family.
It is of circumpolar distribution, and is said to be one of the noisiest
of those which gather together during the breeding season in the far
north.

Regarding it, Mr. Nelson says : "The solitary islands of Behring
Sea and all its dreary coast line are familiar to this great gull.   In
summer it occurs from the Aleutian Islands north to the farthest
points reached by the hardy navigators in the Arctic Ocean adjoining.
According to Murdoch, it is numerous at Point Barrow.   At St.
Michael's they appear each year from the 12th to the 30th April,
following the leads in the ice as they open from the south."

"They are the first of the spring birds to occur in the north, and
their hoarse cries are welcome sounds to the seal hunter as he wanders
over the ice-fields far out to sea in early spring.

"They become more and more numerous until they are very common.   They wander restlessly along the coast until the ponds open
on the marshes near the sea, and then, about the last half of May,

they are found straying singly or in pairs about the marshy ponds where they seek their future summer homes."

" During the cruise of the *Corwin*, in the summer of 1881, the writer found this fine bird at every point which he visited along the coast of Behring Sea and the Arctic Ocean."

They are among the last birds to leave the marshes, which they do only when the ice shuts them out in October.

Recently, Mr. Ridgway has sought to separate the eastern from the western forms of this species, owing to a slight difference in size and some other points of distinction, but it is doubtful if the change will be generally accepted.

GENUS LARUS LINNÆUS.

LARUS MARINUS (LINN.).

17. Great Black-backed Gull. (47)

Feet, flesh-colored; bill, yellow with red spot. Mantle, blackish slate-color; first primary, with the end white for 2-3 inches; second primary, with a white sub-apical spot, and like the remaining ones that are crossed with black, having the tip white (when not quite mature, the first with small white tip and sub-apical spot, the second with white tip alone). In winter, head and neck streaked with dusky. *Young:*— Whitish, variously washed, mottled and patched with brown or dusky; quills and tail, black, with or without white tips; bill, black. Very large; length, 30 inches; wing, $18\frac{1}{2}$; bill, above $2\frac{1}{4}$.

HAB.—Coast of the north Atlantic; south in winter to Long Island.

Nest on the ground, built of grasses, bulky and deeply hollowed.

Eggs, three, bluish-white or olive-gray, irregularly spotted, and blotched with reddish-brown and lilac.

This species is common to both continents, breeding abundantly in Norway, as far as North Cape. It is also very common in Scotland, especially on the indented coasts and islands of the north and west, and also on the outer Hebrides. There they often have their nests on the heathery hillsides, and are seen coursing along in search of wounded grouse, sickly or injured ewes, weakly lambs, eggs or carrion, nothing in the way of animal food coming amiss.

In Labrador, Mr. Abbot M. Frazer found them breeding on small islands, usually placing their nests on some elevated spot. During the breeding season there, they feed largely on the eggs of other birds, especially on those of the murre, and on young eider ducks.

They are seen roaming round the western end of Lake Ontario till the end of March, when they all disappear for the season.

LARUS ARGENTATUS SMITHSONIANUS Coues.

18. American Herring Gull. (51)

Feet, flesh color; bill, yellow with red spot; mantle, pale dull blue (darker than in *leucopterus*, but nothing like the deep slate of *marinus*, much the same as in all the rest of the species); primaries marked as in *marinus* (but the great majority of specimens will be found to have the not quite mature or final condition); length, 22-27; wings, 15-18; tarsus, 2¼-2¾; bill, about 2¼ long, about ⅜-¾ deep at base, and about the same at the protuberance. In winter, head and hind neck streaked with dusky. *Young:*—At first almost entirely fuscous or sooty-brown, the feathers of the back white tipped or not; size, at the minimum above given. As it grows old it gradually lightens; the head, neck and under parts are usually quite whitish, before the markings of the quills are apparent, and before the blue begins to show, as it does in patches mixed with brown; the black on the tail narrows to a bar, at the time the primaries are assuming their characters, but this bar disappears before the primaries gain their perfect pattern. At one time the bill is flesh color or yellowish, black-tipped.

HAB.—North America generally, breeding on the Atlantic coast from Maine northward; in winter, south to Cuba and Lower California.

The original nest was on the ground, not far from water, simply a hollow lined with grass. In some places, where the birds have been persecuted by the robbery of their eggs, they have taken to building in trees, fifty or sixty feet from the ground. There the nests are firmly put together and warmly lined.

The eggs are three, bluish-white, irregularly spotted with brown of different shades.

This is the most abundant bird of its class on the inland lakes, and it may be seen at nearly all seasons of the year, either soaring in wide circles overhead, or passing along in front of the wharves, always on the alert to examine any offal which may be thrown overboard from the vessels. It breeds abundantly along the sea coast and also in suitable paces inland, as shown by the following, which occurs in the transactions of the Ottawa Field Naturalists' Club for 1881: "On this excursion, which was held about the 21st of May, we succeeded in discovering on one of the many small lakes near the Cave, a nest of the common Gull (*Larus argentatus*), but we were unfortunately too late, as not only were the eggs hatched, but the young had already left the nest; from this fact, it is probable that, with this species, the period of incubation is very early in the season. The nest, which was very shallow, was built almost altogether of dried moss, and was placed on the top of a small rock, which stood about a foot and a half out of the water towards one end of the lake."

It breeds commonly in Lake Manitoba, and in suitable places intermediate as far as Alaska.

In Southern Ontario it is seen only during the winter. When Hamilton Bay becomes frozen over, there is always a patch of open water where the city sewer empties into the bay, and there for several years past two or three Herring Gulls have daily resorted, to feed on the rich morsels which come down from the city. They become comparatively tame, and being regarded by the "boys" as belonging to the place, are not much disturbed. The birds in possession of the spot agree about the division of the spoil, but should a stranger seek to share it, he is at once attacked and driven off ere he has time to taste the dainties.

They arrive here about the end of October, and leave early in April.

LARUS DELAWARENSIS Ord.

19. Ring-billed Gull. (54)

Adult plumage precisely like that of the Herring Gull, and its changes substantially the same; bill, *greenish*-yellow, encircled with a *black band* near the end, usually complete, sometimes defective; the tip and most of the cutting edges of the bill, yellow; in high condition, the angle of the mouth and a small spot beside the black, red; *feet, olivaceous,* obscured with dusky or bluish, and partly yellow; the webs, bright chrome. Notably smaller than *argentatus;* length, usually 18-20 inches; extent, about 48; wing, about 15; bill, *under* 2, and only about ½ deep at the protuberance; tarsus, about 2, obviously longer than the middle toe.

Hab.—North America at large; south in winter to Cuba and Mexico.

Nest, on the ground, a hollow lined with grass; sometimes on cliffs of rocks.

Eggs, three, dark cream color, blotched with purple, umber and black.

This is one of the common gulls which frequent Lake Ontario during the winter, whose numbers help to make up the vast crowd which is seen assembled on the edge of the ice at the western extremity of the lake, or in Hamilton Bay, near the canal.

In all stages of plumage it bears a strong resemblance to the Herring Gull, but the ring round the bill and its smaller size serve as distinguishing marks.

This is, perhaps, the most abundant of the gulls. It has been taken at Toronto and other points in Southern Ontario, but only in the winter, its breeding ground being rather farther north. It breeds in the interior and on both shores, and its nesting places are spoken of as nurseries of very great extent. Mr. Stebbins, who visited an island of about an acre in extent, in Devil's Lake, Dakota, in the first week in June, found the entire island covered with eggs of gulls and terns. He says: "I don't suppose you could lay down a two-feet rule without each end of it touching a nest. The terns and gulls were here breeding side by side. Most of the gulls' nests were in the grass, those of the tern in the sand. I did not find a gull's nest with more than three eggs, and very few with two; whereas several hollows had as many as eighteen terns' eggs in them, which had rolled together."

Mr. Frazer also found the Ring-billed Gulls breeding in Labrador, and he remarked that the number of eggs did not exceed four.

Macoun reports it breeding in all the lakes of any size in the North-West.

LARUS ATRICILLA Linn.

20. Laughing Gull. (58)

Adult, in summer: –Bill and edges of eyelids, deep carmine; legs and feet, dusky red; iris, blackish. Hood, deep plumbeous, grayish-black, extending farther on the throat than on the nape. Eyelids, white, posteriorly. Neck all round, rump, tail, broad tips of secondaries and tertials, and whole under parts white, the latter with a rosy tinge which fades after death. Mantle, grayish plumbeous; outer six primaries, black, their extreme tips white; their bases for a short distance on the first, and only on the inner web, and for a successively increasing distance on both webs of the others, of the color of the back.

HAB. –Tropical and warm temperate America, chiefly along the sea coast, from Maine to Brazil.

Nest, in a tussock of grass, the cavity nicely lined with fine dry grasses.

Eggs, three to five, bluish white, spotted and blotched with brown, umber and lilac of various shades.

In the report of the proceedings of the Ornithological Sub-section of the Canadian Institute for 1890-91, occurs the following:

"On May 23rd, 1890, a gull was brought to my store. It had been shot on Toronto Island, and, being unlike any of our native species, I had it thoroughly examined, and it proved to be a male Laughing Gull (Larus atricilla). This is, I believe, the first record of this bird for Ontario."--WILLIAM CROSS.

The Laughing Gull is a southern bird, whose centre of abundance is along the shores of the Gulf of Mexico. It is also common in the South Atlantic and Gulf States, and is found breeding as far north as the coast of New England, but this, so far as I know, is the first record of its occurrence in Ontario. Speaking of this species, Mr.

Maynard says : " The notes of gulls are loud and startling, but those of the Laughing Gull are the most singular of them all, for their cries, especially when the bird is excited, sound like peals of prolonged and derisive laughter."

LARUS FRANKLINII Sw. & Rich.

21. Franklin's Gull. (59)

Adult male:—Eyelids, neck, rump, tail and lower parts white, the latter with the under part of the wings deeply tinged with rich rosy red; hood, black, descending downwards on the nape and throat; mantle and wings, bluish-gray; a band of black crosses the five outer primaries near the end: all the quill feathers are tipped with white. *Young:*—Changing with age as in other birds of this class. Length, 15 inches.

Hab.—Interior of North America, breeding chiefly north of the United States; south in winter to South America.

Nest in a marsh, or wooded swamp, built of sedges and grass a little above the water level.

Eggs, four, greenish-gray with numerous brown markings, heaviest at the larger end.

When questioning that indefatigable sportsman, John Dynes, about the rare birds he had seen on his many excursions round Hamilton, he told me of a gull with a pink breast, which he had sometimes seen in the fall, and finally in October, 1865, he brought me one of the birds thus referred to, which proved to be of this species. Subsequently I shot another in the month of April, about the time the ice was breaking up. The latter was in the more advanced stage of plumage, but neither was mature.

This is not a *sea* gull in the ordinary use of the term, for it prefers the interior to the coast, breeding in the inland swamps far from the sea, and making its annual journey north and south entirely inland. A short time ago I had a beautiful pair sent to me from Minnesota, where they breed. I understand that they also breed abundantly in the marshes of the Red River valley in western Manitoba.

The species has not been observed in the Atlantic States, its line of route north and south being chiefly west of the Mississippi. The few seen in Ontario can thus be regarded only as stragglers making their migratory journey a little farther to the east than usual.

4

LARUS PHILADELPHIA (Ord.).

22. Bonaparte's Gull. (60)

Tarsus about equal to middle toe and claw. Small; 12-14; wing, 9½-10½;
tarsus, 1⅛; bill, 1¼-1⅓; very slender, like a Tern's. *Adult in summer:*—Bill,
black; mantle, pearly blue, much paler than in *atricilla*; hood, slaty-plumbeous
with white touches on the eyelids; many wing coverts white; feet, chrome-
yellow, tinged with coral red; webs, vermilion. Primaries finally:—The first
5-6 with the shafts white except at tip; first white, with outer web and
extreme tip black; second white, more broadly crossed with black; 3rd to 6th-
8th with the black successively decreasing. In winter no hood, but a dark
auricular spot. *Young:*—Mottled and patched above with brown or gray, and
usually a dusky bar on the wing; the tail with a black bar, the primaries with
more black, the bill dusky,. much of the lower mandible flesh-colored or
yellowish, as are the feet.

Hab.—Whole of North America, breeding mostly north of the United
States; south in winter to Mexico and Central America.

The nest is usually placed on an elevation, in a tree, bush. or on a high
stump; it is composed of sticks and grass with a lining of soft vegetable
material.

Eggs, three or four, greenish-gray spotted, and blotched with brown and
lilac of various shades.

About the middle of May this dainty little gull arrives in small
flocks, and for a week or two enlivens the shores of Hamilton Bay
with its airy gambols, but soon passes on farther north to its breeding
grounds. In the fall it returns, subdued in dress and manners,
remains till the weather begins to get cold, and then retires to the
south to spend the winter.

It has a wide distribution, being found at some period of the year
at almost every point on the continent. Speaking of this species in
the "Birds of the North-West," Dr. Coues says: "This little gull
holds its own, from the Labrador crags, against which the waves of
an angered ocean ceaselessly beat, to the low, sandy shores of the
Gulf, caressed by the soothing billows of a tropical sea."

Macoun mentions it as breeding on all the lakes of any size
throughout the North-West, and Dr. Bell has found it along the
Nelson River and at York Factory on Hudson's Bay.

In Lake Erie, a little way out from the mouth of the Grand River,
is Mohawk Island, where Dr. Macallum says this gull used in former
years to breed regularly along with Forster's and the Common Tern.
It is still seen there in small numbers during the summer, but of late
the "boys" have got into the habit of visiting the island on Sundays
during the nesting time, bringing away large quantities of eggs, so

that now the place is comparatively deserted. In the spring its plumage is so perfect, and its flight so light and graceful, that it attracts notice wherever it appears. In Southern Ontario it is seen only in small numbers, the migratory route being mostly along the sea coast.

SUBFAMILY STERNINÆ TERNS.

GENUS STERNA LINNÆUS.

SUBGENUS THALASSEUS BOIE.

STERNA TSCHEGRAVA LEPECH.

23. Caspian Tern. (64)

Adult male : - Crown, sides of the head, and hind head, black. glossed with green; back and wings. light bluish-gray ; the outer primaries, dark bluish-gray on the inner webs; upper tail coverts and tail grayish-white; neck and lower parts, pure white; bill, rich vermilion; legs and feet, black; tail, slightly forked. *Young:*—Mottled and barred with dull brown. Length, 20 inches.

HAB. - Nearly cosmopolitan; in North America breeding southward to Virginia, Lake Michigan, Nevada and California.

Eggs. two or three, laid in a hollow in the sand; pale olive buff. marked with spots of dark brown, and lilac of various shades.

The harsh cry, long pointed wings, and coral red bill of this
species at once attract the attention of anyone who may happen to
be close enough for observation. In spring, when the departure of
the ice gives them the privilege of roving about over the inland
waters, they visit Hamilton Bay in small numbers, and are seen
fishing about the mouths of the inlets, or more frequently basking in
the sun on a sandy point which runs out into the bay opposite Dynes"
place. In the fall they pay a similar visit, but at that season they
are less attractive in appearance, the bill having lost much of its
brilliancy, and the plumage being comparatively dull.

This is the largest of the terns, and it has a very wide distribution, being found breeding at different points, from the Gulf of
Mexico to Alaska, and along the entire Atlantic coast. It is also
said to occur in various portions of the eastern hemisphere, including
Australia. It does not breed in communities like many of the other
terns, being mostly found in retired places in single pairs.

STERNA SANDVICENSIS ACUFLAVIDA (Cabot.).

24. Cabot's Tern. (67)

Bill, rather longer than the head, slender, black, with the tip yellow; mouth
inside, deep blue; feet, black; wings longer than tail, which is deeply forked;
upper part of the head and hind neck, bluish-black; sides of the head, neck all
round, and rest of the lower parts, white; the sides and breast tinged with
pink; fore part of the back, scapulars and upper surface of the wings, pale
bluish-gray; the tips and greater part of the inner web of the scapulars and
quills, white, as are the rump and tail; the four outer quills blackish, but
covered with light gray down on the outer webs, and over a considerable portion of the inner, their shafts white. Length, 15-16; wing, 12-50.

Hab.—Tropical America, northward along the Atlantic coast irregularly to
southern New England.

Eggs, two or three, dropped on the dry sand, rather pointed, yellowish-drab,
spotted with dark and reddish-brown.

In the spring of 1882, Dr. Garnier noticed three terns of this
species coursing around a mill-pond not far from his residence at
Lucknow. The Doctor attended to them at once, and the result was
that one went clear off toward Lake Huron, another wriggled with
difficulty after it, and the third fell dead on the pond. I afterward
saw this specimen mounted, and satisfied myself of its identity. It is
difficult to account for birds wandering away at times beyond their

usual limit, but we might with as much truth say that it is difficult to account for birds so regularly keeping within certain limits. When those of this class find themselves farther from home than they intended, it does not cost them much labor to correct the mistake.

This is the only record I have of the species in Ontario, and the visit can only be considered accidental, as the summer home of the species is far to the south.

They breed in colonies on the coast of Central America and on the larger West India Islands.

SUBGENUS STERNA.

STERNA FORSTERI Nutt.

25. Forster's Tern. (69)

Like the Common Tern; larger, tail longer than wings. Wing of adult, 9½-10½; tail, 6½-8, thus often beyond the extreme of *hirundo*, and nearly as in *paradisæa*: bill, 1⅜ (1½-1⅜), and about 2-5 deep at base (in *hirundo* rarely if ever so deep); tarsus seldom down to ⅞; whole foot, about 2. Little or no plumbeous wash below; inner web of the outer tail feathers darker than outer web of the same. Young and winter birds may be distinguished from *hirundo* at gunshot range; the black cap is almost entirely wanting, and in its place is a broad black band on each side of the head through the eye; several lateral tail feathers are largely dusky on the inner web; their outer webs are white.

HAB.—North America generally, breeding from Manitoba southward, in the United States to Virginia, Illinois, Texas and California; in winter, southward to Brazil.

Eggs, two or three, from greenish-white to drab, blotched and spotted with brown and lilac of different shades.

This is another of the sea swallows, a name which is indiscriminately applied to all the terns. The species is abundant and widely distributed. It is one of those found by Dr. Macallum breeding on Mohawk Island, in Lake Erie. Mr. Saunders reports it breeding on the St. Clair Flats, and Macoun speaks of it breeding abundantly in Lake Manitoba, Waterhen River and Lake Winnipegosis. Farther south, it is known to breed in Virginia, Texas, Illinois and California.

Throughout Ontario generally, it is only a migrant in spring and fall. It is usually in company with the Common Tern, which it closely resembles, the points of difference being readily observed when the birds are placed side by side.

STERNA HIRUNDO Linn.

26. Common Tern. (70)

Bill, red, blackening on the terminal third, the very point usually light; feet, coral red; mantle, pearly grayish-blue; primary shafts white, except at the end; below white, washed with pale pearly plumbeous blanching on throat and lower belly; tail mostly white, the *outer* web of the outer feather darker than inner web of the same. Length of male, $14\frac{1}{2}$ (13-16); extent, 31 (29-32); wing, $10\frac{1}{2}$ ($9\frac{3}{4}$-$11\frac{3}{4}$); tail, 6 (5-7); tarsus, $\frac{3}{4}$ ($\frac{5}{8}$-$\frac{7}{8}$); bill, $1\frac{1}{4}$-$1\frac{1}{2}$; whole foot, averaging $1\frac{3}{4}$; female rather less, averaging toward these minima; young birds may show a little smaller, in length of tail particularly, and so of total length: length, 12 or more; wing, 9 or more; tail, 4 or more; bill, $1\frac{1}{2}$ or more. In winter this species does not appear to lose the black-cap, contrary to a nearly universal rule. *Young:*—Bill mostly dusky, but much of the under mandible yellowish; feet simply yellowish; cap more or less defective; back and wings patched and barred with gray and light brown, the bluish showing imperfectly if at all, but this color shading much of the tail; usually a blackish bar along the lesser coverts, and several tail feathers dusky on the *outer* web: below, pure white, or with very little plumbeous shade.

HAB.—Greater part of northern hemisphere and Africa. In North America chiefly confined to the Eastern Province, breeding from the Arctic coast, somewhat irregularly, to Florida and Texas, and wintering farther south.

Eggs, two or three, deposited in a hollow in the sand, light brown, tinged with green and blotched with dark brown and lilac of various shades.

This species is common to both continents, and has been found breeding as far north as Greenland and Spitzbergen. It migrates south in the cold weather, and its return to its summer haunts is hailed as a sure indication that winter is really gone. For a time many a quiet bay and inlet is enlivened by its presence.

> " Swift by the window skims the Tern,
> On light and glancing wing,
> And every sound which rises up
> Gives token of the Spring."

On Hamilton Bay the terns make their appearance about the 10th of May, and in company with the black-headed gulls go careering around the shores in merry groups, or settle on the sand bars to rest and plume their feathers in the sun. By the end of the month they have all gone to the north and west to raise their young, but they pay us a short visit in the fall on their way south.

The range of this species is very similar to that of Forster's Tern. Gull Island in Lake Erie, St. Clair Flats, and especially Waterhen River and adjoining waters, are named among its breeding places.

D. Gunn, writing in the "Birds of Manitoba," on the habits of these birds in Lake Winnipeg, says :

" When at Lake Winnipeg, in 1862, I noticed that the terns which occupied sandy or gravelly islands made their nests as those do on the gravelly islands of Shoal Lake, while those found on the rocky island on the east side of the lake chose for their nesting places depressions and clefts in the surface of the rocks. These they carefully lined with moss, thus giving a remarkable example of the instinct which teaches them that their eggs when laid on the sand or gravel are safe, but when placed on these hard and often cold materials, a warm lining for the protection of the eggs and young birds is indispensable."

STERNA PARADISÆ Brünn.

27. Arctic Tern. (71)

Bill, carmine ; feet, vermilion ; plumage, like that of *hirundo*, but much darker below, the plumbeous wash so heavy that these parts are scarcely paler than the mantle ; crissum, pure white ; throat and sides of the neck, white or tinged with gray. In winter, cap defective ; in young the same, upper parts patched with gray, brown or rufous ; under parts paler or white ; a dark bar on the wing; outer webs of several tail feathers, dusky; bill, blackish or dusky red, with yellow on the under mandible; feet, dull orange, smaller than *hirundo*, but tail much longer. Length, 14-17; wing, 10-12; tail, 5-8; bill, 1.20-1.40.

HAB.—Northern hemisphere; in North America breeding from Massachusetts to the Arctic regions, and wintering southward to Virginia and California.

Eggs, two or three; laid on the bare rock or sand; drab, spotted and dashed with brown of different shades, indistinguishable from those of the common tern.

For several reasons the terns which visit Ontario are less known than birds belonging to other classes. They are not sought after by sportsmen, and at present the number of collectors is so few that the sea swallows (as they are here called) are little molested. There are several species, such as the Common Tern, Forster's Tern, and the one we are now considering, which resemble each other so closely that the difference can only be made out on careful examination by one who is familiar with their appearance. Compared with the Common Tern, the present species is a bird of more slender make, the tail feathers being usually much longer, and the under parts of a much darker shade.

In the spring and fall flocks of terns resembling each other in general appearance are seen frequenting Hamilton Bay and the inlets along the shores of Lake Ontario. Considering the range of this species, it is likely that it is here with the others, but among the few which I have killed I have not found an Arctic.

In the collection of birds brought together under the direction of the late Prof. Hincks, and sent to the Paris Exposition in 1867, a pair of Arctic Terns was included which were said to have been procured near Toronto.

The species is of circumpolar distribution. Dr. Bell found it on Hudson's Bay, and it occurs on the coast of California, but is not named among the birds of Manitoba, being perhaps strictly maritime in its haunts.

Around the shores of Great Britain it is the most abundant of its class, and here, too, it is remarked that it does not occur inland.

Mr. Gray, in his "Birds of the West of Scotland," says regarding it : "On the western shores of Ross, Inverness and Argyleshire, there are numerous breeding places for this bird, especially on the rocky islands in the sea-lochs stretching inland, such as Loch Sunart, Loch Alsh and Loch Etive. These nurseries are equally numerous off the coast of Mull, and others of the larger islands forming the Inner Hebrides."

"In the first week in August, 1870, when travelling from North Uist to Benbecula, and crossing the ford which separates the two islands, I witnessed a very interesting habit of this tern. I had been previously told by a friend to look out for the birds which he said I should find waiting for me on the sands. Upon coming within sight of the first ford, I observed between twenty and thirty terns sitting quietly on the banks of the salt water stream, but the moment they saw us approaching they rose on the wing to meet us, and kept hovering gracefully over our heads till the pony stepped into the water. As soon as the wheels of the conveyance were fairly into the stream, the terns poised their wings for a moment, then precipitated themselves with a splash exactly above the wheel tracks and at once arose, each with a sand eel wriggling in its bill. Some had been caught by the head and were unceremoniously swallowed, but others which had been seized by the middle were allowed to drop, and were again caught properly by the head before they reached the water.

"I was told by the residents that it is a habit of the birds to be continually on the watch for passing vehicles, the wheels of which bring the sand eels momentarily to the surface, and the quick eyes of the terns enable the birds to transfix them on the spot."

SUBGENUS STERNULA BOIE.

STERNA ANTILLARUM (LESS.).

28. Least Tern. (74)

Bill, yellow, usually tipped with black; mantle, pale pearly grayish-blue, unchanged on the rump and tail: *a white frontal crescent*, separating the cap from the bill, bounded below by a black loral stripe reaching the bill; shafts of two or more outer primaries, *black* on the upper surface, white underneath; feet, orange. *Young:*—Cap, too defective to show the crescent; bill, dark, much of the under mandible pale; feet, obscured. Very small, only 8-9; wing, 6-6½; tail, 2-3½; bill, 1-1¼; tarsus, ⅔.

HAB.—Northern South America, northward to California and New England, and casually to Labrador, breeding nearly throughout its range.

Eggs, two or three, variable in color, usually drab, speckled with lilac and brown; left in a slight depression in the dry beach sand beyond the reach of water.

This is a refined miniature of the Common Tern, and a very handsome, active little bird. It is common along the sea coast to the south of us, but probably does not often come so far north as

Lake Ontario. Dr. Wheaton mentions its irregular occurrence on Lake Erie, and Dr. Brodie reports it being found near Toronto. In the month of October, several years ago, I shot an immature specimen as it rose from a piece of driftwood in Hamilton Bay, during a southerly blow of several days' duration, and that is the only time I have ever seen the species here.

In the report of the proceedings of the Canadian Institute for 1889, the following passage occurs: "A Least Tern (*sterna antillarum*) was shot here by Mr. Wm. Loane on September 5th. This, with Dr. Brodie's former record for Toronto, and Mr. McIlwraith's for Hamilton, makes the third for Lake Ontario."

The Little Striker, as it is called along the sea coast, we may not expect to see often in Ontario, for its principal breeding ground is on the Gulf coast, and on the islands of the Atlantic coast of Florida. The eggs are placed in a slight hollow in the sand and broken shells of the beach, which they resemble so much that they are often passed without being observed, and thus escape trouble which might otherwise befall them.

GENUS HYDROCHELIDON BOIE.

HYDROCHELIDON NIGRA SURINAMENSIS (GMEL.).

29. Black Tern. (77)

Adult in breeding plumage:—Head, neck and under parts, uniform jet black; back, wings and tail, plumbeous; primaries, unstriped; crissum, pure white; bill, black. In winter and young birds, the black is mostly replaced by white on the forehead, sides of head and under parts, the crown, occiput and neck behind, with the sides under the wings, being dusky-gray; a dark auricular patch and another before the eye; in a very early stage, the upper parts are varied with dull brown. Small; wing, 8-9, little less than the whole length of the bird; tail, 3½, simply forked; bill, 1-1½; tarsus, ¾; middle toe and claw, 1¼.

HAB. Temperate and tropical America. From Alaska and the fur countries to Chili, breeding from the middle United States northward.

No nest. Eggs, on the bog, two or three; brownish-olive, splashed and spotted with brown.

Common to both continents, and extending its migrations far north. The Black Tern has been found in Iceland, and, according to Richardson, is known to breed in the fur countries. It enters Southern Ontario early in May, visiting the various feeding resorts along the route, in company with the smaller gulls, and retires to the marshes to raise its young. At the St. Clair Flats it breeds abundantly, its eggs being often apparently neglected, but they are said to be covered by the female at night and in rough weather.

Dr. Macallum reports it as breeding also on Mohawk Island, though from being frequently disturbed and robbed of its eggs, it is not so numerous as formerly. In the North-West, according to Macoun, it "breeds in all the marshes from Portage la Prairie westward, in less numbers in the wooded region, but generally distributed."

At Ottawa, Toronto and Hamilton it occurs as a regular migrant in spring and fall.

ORDER STEGANOPODES. TOTIPALMATE SWIMMERS.

FAMILY SULIDÆ. GANNETS.

GENUS SULA BRISSON.

SUBGENUS DYSPORUS ILLIGER.

SULA BASSANA (LINN.).

30. Gannet. (117)

Adult male:– White, the head and hind neck tinged with yellowish-brown; primaries, black. *Young:*—Dark brown, spotted with white; lower parts, grayish-white. Length, 30 inches.

HAB.—Coasts of the North Atlantic, south in winter to the Gulf of Mexico and Africa; breeds from Maine and the British Islands northward.

Breeds in communities on rocks near the sea. One egg, pale greenish-blue.

We have very few records of the Gannet, or Solan Goose, in Ontario, because it is a bird of the North Atlantic, where it is found on the shores on both sides, on the east, perhaps, in greater abundance than on the west.

Many years ago, an individual of this species was found in Hamilton Bay in a state of extreme exhaustion, after a severe "northeaster." In the proceedings of the Canadian Institute for 1890, it is stated, "one specimen in immature plumage was shot at Oshawa, Ontario, in 1862, by Mr. A. Dulmage," which, so far as I know, completes the list for the Province.

These birds are in the habit of breeding in colonies on large isolated rocks, which are called Gannet Rocks, wherever they occur. One of these is situated off the coast of Maine, southward of Grand

Manan Island, at the mouth of the Bay of Fundy. Another lies to the south-west of Yarmouth, Nova Scotia. The great breeding resorts of the species, however, are the Bird Rocks, in the Gulf of St. Lawrence, and Bonaventure Island, near Gaspé.

This bird takes its name (*Sula bassana*) from one of its first and best known breeding places, the Bass Rock, in the Firth of Forth, where I have seen them in thousands engaged in completing their domestic arrangements in the early summer. They used to have many breeding places round the rocky coasts of Scotland, but Mr. Gray tells us that the number has now been reduced to five. These are Ailsa Craig, St. Kilda, North Barra, Stack of Suleskerry, and the Bass Rock, Firth of Forth. The number of birds which frequent these places is still very great. A moderate calculation of those seen on the Bass Rock Mr. Gray placed at 20,000, and I feel sure that the number frequenting Ailsa Craig cannot be much less. Mr. Gray further states, that on the more remote breeding places very little change has taken place, but that on the Bass Rock, where in former years a large colony had possession of the grassy slopes on which they built their nests, the number of birds is greatly reduced. The intrusion of visitors has driven the birds entirely to the rocky ledges on the west side of the island, where the nests cannot be examined as they used to be. Before this change took place in their breeding grounds, the birds were greatly changed in their habits, the old birds having become dreadfully vociferous, and in some cases showing fight. Professor Macgillivray well describes their cry in comparing the torrent of sounds to the words, "Kirra kirra, cree cree, grog, grog, grog." Surrounded by a multitude of open bills, and noticing the guide apparently absorbed in thought, he inquired, "Is there any risk of them biting?" "Oh, no, sir," he rejoined, "I was only thinking how like they are to *oursel's*." He stated also that these birds sometimes lay two eggs, fourteen nests on the grassy slope already referred to having been found to contain that number. This statement may be easily believed, when we are told that during the breeding season the rock is visited daily by excursionists accompanied by guides, and that the number of eggs in a nest is made to suit the wishes of the visitor.

Notwithstanding the protection now afforded to this bird, in common with other water-fowl, the numbers have greatly diminished, and it is very doubtful if the Bass ever again has as many tenants as it once had.

Long ago the breeding places appear to have been more numerous than at present, for we find frequent reference to them in the works of writers long departed. For instance, Dean Munroe, who visited nearly the whole of the British islands between 1540 and 1549, has left a quaint account of what he saw. In describing Eigg Island, he says: "North from Ellan about foure myles lyes ane iyle called iyle of egga, four myle lange and twa myle braid, guid maine land with a Paroch kirk in it and maney Solane geese." One cannot exactly see how the two things should in this curt way be associated.

The Solan Geese live chiefly on herring and other fish which are very irregular in their movements, and the birds have, therefore, often to travel a long distance to obtain food for their young. As soon as the young birds are able to fly, they all leave the rocks, and follow the shoals of fish, wherever they are to be found, returning again to their familiar rocks in March or April.

FAMILY PHALACROCORACIDÆ. CORMORANTS.

GENUS PHALACROCORAX BRISSON.

PHALACROCORAX CARBO (LINN.).

31. Cormorant. (119)

General plumage, black, glossed with blue, a white patch on the throat and another on the sides of the body; in summer the head is crested with long narrow feathers, which fall off when the breeding season is over; the white patches on the throat and sides also disappear about the same time. Length, 36 inches.

HAB.—Coasts of the North Atlantic, south in winter on the coast of the United States, casually to the Carolinas; breeding from the Bay of Fundy to Greenland.

Nest, on precipitous rocks, built of sticks and sea-weed, kept in a filthy condition from the refuse of the larder, etc.

Eggs, four or five, pale bluish-green.

Although the Cormorants are generally birds of the sea coast, when not specially engaged at home, they make periodical excursions to the lakes, where no doubt they find the change of food and scenery very agreeable. In spring and fall they are occasionally seen in Hamilton Bay, and at other points in Southern Ontario, following their usual avocation of fishing. Not long since, with the aid of a powerful glass, I watched one sitting on a buoy out off the wharves, and could not but admire the graceful motions of his long, lithe neck, as he preened his plumage in conscious safety. The inspection

at that distance was more pleasant than it might have been closer
by, for these birds, though apparently cleanly, carry with them a
most unsavory odor.

This is another North Atlantic species which is found breeding on
the rocky ledges along the shores of both continents. It is gregarious, living in thickly-settled communities, the sanitary condition of
which prevents their being popular with excursionists, so that the
birds when at home are seldom disturbed.

On the rocky shore of Newfoundland it is especially abundant,
and was also found by Mr. Frazer in Labrador. It has been observed
singly or in pairs, at Ottawa, Hamilton, London and other points,
but the species seldom leaves the sea coast. Those observed inland
are apparently stragglers which by chance or choice have wandered
for a time away from their usual habitat.

PHALACROCORAX DILOPHUS (Sw. & Rich.).

32. Double-crested Cormorant. (120)

Tail of twelve feathers, gular sac convex or nearly straight-edged behind;
glossy greenish-black; feathers of the back and wings, coppery-gray, black-
shafted, black-edged; adult with curly black *lateral* crests, and in the breeding
season other filamentous white ones over the eyes and along the sides of the
neck; white flank-patch, not observed in the specimens examined, but probably
occurring; gular sac and lores, orange; eyes, green. Length, 30-33 inches;
wing, 12 or more; tail, 6 or more; bill along gape, 3½; tarsus, a little over 2.
Young:—Plain dark brown, paler or grayish (even white on the breast) below,
without head plumes.

HAB. -Eastern coast of North America, breeding from the Bay of Fundy
northward; southward in the interior to the Great Lakes and Wisconsin.

Eggs, two or three, bluish-green.

This, like the common species, occasionally visits the inland lakes,
and is distinguished by its smaller size and richer plumage. The
specimen in my collection I shot off Huckleberry Point, when it rose
from a partially submerged stump that it had been using for a short
time as a fishing station. All the Cormorants have the reputation of
being voracious feeders, and they certainly have a nimble way of
catching and swallowing their prey, but it is not likely that they consume more than other birds of similar size.

Though this species breeds along the sea coast on both sides of
the Atlantic, it has also been found breeding in colonies in the

nterior. It is the one we see most frequently in Southern Ontario, and Macoun mentions having found it breeding abundantly in Lake Winnipegosis.

Dr. Hatch, in the " Birds of Minnesota," describes it as a common summer resident which "local observers from all parts of the State report from 'occasional' to 'innumerable,' according to the nearness of the observer to the breeding places of the birds." The preparations for incubation are made about the 10th of May, in large communities on islands in the lakes and ponds and almost impenetrable marshes, where there are some large old trees, in which they most frequently build their coarse but substantial nests. These are usually bulky from having been added to every year, and consist of weeds, vines and sticks, piled together carelessly around a deep depression, in which is deposited the three pale greenish or bluish eggs. It is not an uncommon sight to see one or more of these nests on the same tree in which are a number of heron nests, and the owners seem to live in harmony.

When the young are sufficiently grown, they gather into immense flocks in unfrequented sections, and remain until the ice-lid has closed over their supply of food, when they go away, not to return till the cover is lifted up in the spring.

FAMILY PELECANIDÆ.

SUBGENUS CYRTOPELICANUS REICHENBACH.

PELECANUS ERYTHRORHYNCHOS GMEL.

33. American White Pelican. (125)

White; occiput and breast, yellow; primaries, their coverts, bastard quills and many secondaries, black; bill, sac, lores and feet, yellow. Length, about 4 feet; expanse, 7-9; wing, 2; bill, 1 or more; tail, ½, normally 24-feathered.

HAB.—Temperate North America, north in the interior to about Lat. 61, south to Central America; now rare or accidental in the north-eastern States; abundant in the Middle Province and along the Gulf coast; common on the coast of California and western Mexico.

Nest, on the ground or in a low bush near the water.

Eggs, one to three, dull white.

Early in the month of May, 1864, five of these large, odd-looking birds were observed on Hamilton Bay, and were accorded the attention that is usually bestowed upon visitors of this description. John Dynes was the first to give them a salute, and he captured two of their number, one of which came into my possession. The other three remained for a day or two, but were much disturbed, and finally got away. On the 13th of March, 1884, a similar visit was made by a like number, about the time the ice was breaking up. Mr. Smith, who was in charge of the Ocean House at the time, saw them flying heavily up the lake. They seemed much exhausted, and on alighting on the ice near the edge of the water, at once squatted flat, with their heads resting between their shoulders. When two or three rifle bullets were landed uncomfortably near them, they got up reluctantly, and went off eastward down the lake, hugging the shore for shelter from the wind, which was blowing fresh at the time.

Dr. Macallum writes that "on the 28th of September, 1889, a very fine female White Pelican was shot at the mouth of the Grand River, near Dunnville, which came into my hands. There had not been one shot here for twenty-two years. It was in a very emaciated condition, but in good plumage, and now adorns one of my cases."

So the stragglers are picked up, but the bulk of the species is found to the north and west of Ontario. Macoun found them breeding in Old Wives, Gull and Long Lakes in the North-West. It is also said that several thousands of these birds are permanent residents on Great Salt Lake, Utah, where they breed on the islands twenty miles out in the lake.

ORDER ANSERES. LAMELLIROSTRAL SWIMMERS.

FAMILY ANATIDÆ. DUCKS, GEESE AND SWANS.

SUBFAMILY MERGINÆ. MERGANSERS.

GENUS MERGANSER BRISSON.

MERGANSER AMERICANUS (CASS.).

34. **American Merganser.** (129)

Nostrils, nearly median; frontal feathers reaching beyond those on sides of bill; male with the head scarcely crested, glossy green; back and wings, black and white, latter crossed by one black bar; under parts, salmon-colored; length, about 24; wing, 11, female smaller, occipital crest better developed, but still flimsy; head and neck, reddish-brown; black parts of the male, ashy gray; less white on the wing; under parts less tinted with salmon.

HAB. -North America generally, breeding south to the northern United States.

Nest, in a hollow tree. It is composed of weeds and moss, and lined with down from the breast of the bird.

Eggs, six to eight, buff or dark cream.

This is the largest, and by many considered the handsomest, of the three saw-bills which visit us. It is never plentiful, being a bird of the sea coast, but it is usually seen singly, or in pairs, among the flocks of water-fowl which crowd up from the south as soon as the ice begins to move in the lakes and rivers in spring.

In the fall they are again observed in company with their young, which at this stage all resemble the female in plumage. The flesh of the saw-bills being fishy, the gunners often allow them to pass when a blue-bill or a red-head would not get off so easily.

They are reported from Ottawa, Toronto, Hamilton and other points in Southern Ontario. In the "Birds of Manitoba," Macoun says of them: "Breeds here abundantly on the rivers emptying into Lake Winnipegosis, and on all the rivers visited by me in Manitoba. I never observed this bird on still water during the breeding season. They feed only on fish, and are found only on clear running streams where fry are abundant."

They are generally but not equally distributed, being more common at some points than others. They are among the first to arrive when the ice breaks up, and indeed Dr. Macallum mentions that so long as there is open water in the Grand River they will remain all winter.

MERGANSER SERRATOR (LINN.).

35. **Red-breasted Merganser.** (130)

Nostrils, sub-basal; frontal feathers not reaching beyond those on sides of bill; a long, thin, pointed crest in both sexes. Smaller than the last; wing, 8-9; general coloration, sexual difference the same, but the male with the jugulum rich reddish-brown, black-streaked, the sides conspicuously finely waved with black, a white, black-bordered mark in front of the wing, and the wing crossed by two black bars.

HAB.—Northern portions of northern hemisphere; south in winter, throughout the United States.

Nest, among the weeds, built of grass, and warmly lined with down.

Eggs, nine or ten, creamy buff.

5

Rather more numerous than the preceding, being often seen in spring and fall in flocks of six or eight, fishing about the mouths of the inlets in Hamilton Bay.

This species is common to both continents, and breeds on the rocky islets on many of the inland lochs in the north of Scotland. All the young birds appear for the first season in the plumage of the female, but the male can readily be distinguished by a peculiar bony enlargement in the windpipe, which does not occur in the opposite sex.

It is said that in this, and in the preceding species, as soon as the female has completed her set of eggs, the male has the ungallant habit of ignoring all family responsibilities, and leaving the entire care of the youngsters to their mother, who leads them carefully to the water, and gives them their first lesson at a very early age.

In Manitoba, Macoun says that the species "breeds on all the northern streams and ponds, feeds largely on vegetable matter, and is quite edible."

In the fall they occur in small flocks along the southern border of Ontario, but none are observed to remain over the winter.

Mr. Nelson says, regarding this species : " During the summer of 1881, I found them breeding upon St. Lawrence Island and along the Siberian coast from Plover Bay to Cape North through Behring Strait. On the Alaskan coast they breed everywhere in suitable places, from Sitka to Icy Cape, and perhaps to Point Barrow."

Genus LOPHODYTES Reichenbach.

LOPHODYTES CUCULLATUS (Linn.).

36. Hooded Merganser. (131)

Nostrils, sub-basal; frontal feathers, reaching beyond those on sides of bill; a compact, erect, semicircular, laterally compressed crest in the male, smaller and less rounded in the female. *Male:*—Black, including two crescents in front of wing, and bar across speculum; under parts, centre of crest, speculum and stripes on tertials, white; sides, chestnut, black-barred. Length, 18-19; wing, 8. *Female:*—Smaller; head and neck, brown; chin, whitish; back and sides, dark brown, the feathers with paler edges; white on the wing less; bill, reddish at base below.

Hab.—North America generally, south to Mexico and Cuba, breeding nearly throughout its range.

Nest, in a hole in a tree or stump, warmly lined with soft grass, feathers and down.

Eggs, six to eight, buff or dark cream color.

This beautiful little Saw-bill is a regular visitor to Hamilton Bay, where it spends a short time in the beginning of April, before retiring to its more remote breeding grounds.

The habit of raising its young in a hole in a tree seems rather a singular one for a bird of this class, but in this retired position the female spends the anxious hours of incubation, beyond the reach of danger to which she might elsewhere be exposed. As soon as the young are old enough to bear transportation, she takes them one after another by the nape of the neck and drops them gently into the water. Like the other saw-bills, this species feeds on fish, on account of which its flesh is not considered a delicacy.

Such is the record for Southern Ontario, but a change of residence and surroundings may bring about other changes. In the "Birds of Manitoba," Macoun says regarding this bird: "Found in all the smaller ponds and lakes, very common in streams around Porcupine Mountain, feeding on vegetable substances, and quite edible, in this respect unlike *M. americanus*."

SUBFAMILY ANATINÆ.   RIVER DUCKS.

GENUS ANAS LINNÆUS.

ANAS BOSCHAS LINN.

37. Mallard. (132)

Male: With the head and upper neck, glossy green, succeeded by a white ring; breast, purplish-chestnut; tail feathers, mostly whitish; greater wing coverts tipped with black and white, the speculum violet; feet, orange red. *Female:*—With the wing as in the male; head, neck and under parts, pale ochrey, speckled and streaked with dusky.   Length, about 24; wing, 10-12.

HAB.—Northern parts of northern hemisphere: in America, south to Panama and Cuba, breeding southward to the northern border of the United States.

Nest, on the ground, built of dry grass, lined with feathers.

Eggs, eight to ten, dull drab or olivaceous green.

This, the parent of the domestic duck, is an abundant species and widely distributed, but it is found in greatest numbers at certain points, where its food abounds. In Hamilton Bay it occurs sparingly during the migratory season, but at Rond Eau, at Long Point on Lake Erie, and on the flats along the River St. Clair it assembles in vast flocks in the fall to feed on the wild rice. At the flats a few pairs remain during summer to rear their young, but the greater number pass farther north.

A few years ago Mr. John Bates, whose farm is on the shore of Hamilton Bay, near the waterworks, noticed a female of the species late in the fall associating with his tame ducks. It was shy, and kept away from the house for a time, but as the season advanced and the water froze over, it came into the sheds and remained permanently with the others. In the spring it built a nest in an out-of-the-way place, and in due time came forth followed by a brood of young ones, which in time grew up and bred with the domestic species.

Mr. Bates pointed out to me some of the stock, which he could always recognize by their sitting deeper in the water, by their comparatively long, slim necks, and by a certain wild look of suspicion and mistrust which clung to them through several generations. Mr. Bates thought the individual referred to had been wounded in the wing, and thus incapacitated for performing the usual journey south.

The Mallard is reported breeding abundantly throughout Michigan and Minnesota, while in the North-West Macoun says regarding it : "The most abundant duck of the North-West, breeding in nearly all the marshes north of the boundary."

It has also been observed at Hudson's Bay, and rarely in Alaska and on the Fur Seal Islands.

ANAS OBSCURA Gmel.

38. Black Duck. (133)

Size of the Mallard, and resembling the female of that species, but darker and without decided white anywhere, except under the wings. Tail, with 16 to 18 feathers.

Hab.—Eastern North America, west to Utah and Texas, north to Labrador, breeding southward to the northern United States.

Nest, on the ground, built of grass, weeds and feathers.

Eggs, eight to ten, yellowish drab or buff, shaded with green.

Although there are several other ducks darker in color than this species, it is still the "Black Duck" of the gunners all over the continent, and is excelled by no other in the excellence of its flesh. It is not so plentiful throughout Ontario as the Mallard, being rather a bird of the sea coast, frequenting the salt marshes along the coast of Maine, where it breeds abundantly. A few pairs have also been found mating in the marsh along the River St. Clair, but such an occurrence is by no means common.

We are told that long ago the Black Duck was a regular visitor to the marshy inlets around Hamilton Bay, but now there is so much to disturb, and so little to attract them, that their visits are few and far between.

In the "Birds of Manitoba," they are spoken of as being very rare, only two specimens having been obtained in ten years. At Hudson's Bay only rare stragglers have been noticed.

SUBGENUS CHAULELASMUS Bonaparte.

ANAS STREPERA Linn.

39. Gadwall. (135)

Male:—With most of the plumage barred or half-ringed with black and white or whitish; middle coverts, *chestnut;* greater coverts, *black: speculum, white.* *Female:*—Known by these wing marks. Length, 19-22; wing, 10-11.

HAB.—Nearly cosmopolitan. In North America breeds chiefly within the United States.

Nest, usually on the ground, sometimes in trees.

Eggs, buff or dull cream color.

The Gadwall is rare throughout Ontario. When a large mixed lot of ducks is sent down in the fall from any of the shooting stations in the west, a pair or two of this species may sometimes be picked out, but that is all.

The pair in my collection were shot in Hamilton Bay many years ago, but since that time I have not heard of any having been obtained there. It is common to both continents, but it is nowhere abundant.

The only place I have seen that adjective applied to it is in Macoun's Annual Report of the Department of the Interior for the year ending December, 1880, page 28, where he says: " *Chaulelasmus streperus,* gray, Gadwall, gray duck, abundant throughout the interior." In the " Birds of Manitoba," the same writer says regarding it : " Only one specimen shot on the Assiniboine, September, 1881."

It is mentioned by Dr. Bell as occurring at Hudson's Bay; in what number is not stated.

The male Gadwall is a very handsome bird, much sought after by collectors, the price charged in their lists showing its comparative scarcity.

——— -   —   ———

SUBGENUS MARECA Stephens.

ANAS AMERICANA Gmel.

40. Baldpate. (137)

Bill and feet, grayish-blue; top of head, white, or nearly so, plain or speckled; its sides and the neck, more or less speckled; a broad green patch on sides of head; fore breast, light-brownish; belly, pure white; crissum, abruptly black; middle and greater coverts, white, the latter black-tipped; speculum, green, black bordered. Length, 20-22; wing, 11; tail, 5; tarsus, 2; bill, 1⅛-1¼; female known by the wing markings.

HAB.—North America, from the Arctic Ocean south to Guatemala and Cuba.

Nest, on the ground in the marsh; it is composed of grass and weeds. It is neatly formed and lined with feathers and down from the breast of the bird.

Eggs, eight to twelve, pale buff.

Resembles the Gadwall in appearance, but can always be distinguished by the creamy white crown which has suggested for the species the familiar name of "Baldpate." It is also more abundant than the Gadwall, being often seen in flocks of fifty to one hundred during the season of migration. It has a wide breeding range throughout the United States and British America. At the St. Clair Flats it has often been seen at midsummer, but so far I have no record of its nest or eggs having been found there. It seems rather tender, and is one of the first to retire to the south in the fall.

In the "Birds of Manitoba," we read regarding this species : "Abundant summer resident, breeding at Lake Manitoba and in all the sloughs in this vicinity. This is the last duck to arrive in the spring and the first to leave in the fall. In 1884, first seen April 16th."—NASH.

It has been observed in Alaska but only in limited numbers, arriving there early in May and leaving early in October.

SUBGENUS NETTION KAUP.

ANAS CAROLINENSIS GMELIN.

41. Green-winged Teal. (139)

Head and upper neck, chestnut, with a broad glossy green band on each side, uniting and blackening on the nape; under parts, white or whitish, the fore breast with circular black spots ; upper parts and flanks closely waved with blackish and white; a white crescent in front of the wing; crissum, black, varied with white or creamy; speculum, rich green, bordered in front with buffy tips of the greater coverts, behind with light tips of secondaries; no blue on the wing; bill, black; feet, gray. Female differs in the head markings, but those of the wing are the same. Small; length, 14-15; wing, 7½; tail, 3½; bill, 1½; tarsus, 1¼.

HAB.—North America, chiefly breeding north of the United States, and migrating south to Honduras and Cuba.

Nest, on the ground, built of dried grass, and lined with feathers.

Eggs, usually eight, pale dull green or buff.

In Southern Ontario this dainty little duck is known only as a passing migrant in spring and fall. During the latter season it is much sought after at the shooting stations on account of the delicacy of its flesh. It breeds abundantly throughout Manitoba and the great North-West, and is mentioned among the birds found by Dr. Bell at Hudson's Bay.

Of the two teals common in Ontario, this seems the hardier, breeding farther north, and remaining later with us in the fall. In the spring it appears as soon as there is open water in the marshes, and at this season, being very properly protected by the Game Act, it passes on in peace.

Of its occurrence in Alaska, Mr. Nelson says : " It is found widely spread and rather common over the mainland, and it is resident throughout the entire length of the Aleutian Chain.

"They are the least suspicious of the ducks, probably because the Eskimo usually consider them too small to waste a charge of powder and shot upon."

SUBGENUS QUERQUEDULA STEPHENS.

ANAS DISCORS LINN.

42. Blue-winged Teal. (140)

Head and neck of the male, blackish plumbeous, darkest on the crown, usually with purplish iridescence ; a white crescent in front of the eye ; under parts thickly dark spotted ; wing coverts, sky blue, the greater white-tipped ; speculum, green, white-tipped ; axillars and most under wing coverts, white ; scapulars striped with tawny and blue, or dark green ; fore back, barred ; rump and tail, dark, plain ; crissum, black ; bill, black ; feet, dusky yellow. *Female:*— With head and neck altogether different ; under parts much paler and obscurely spotted, but known by the wing marks ; size, same as *carolinensis*.

HAB. —North America in general, but chiefly the Eastern Province ; north to Alaska, and south to the West Indies and northern South America ; breeds from the northern United States northward.

Nest, composed of dry grass and weeds, lined with feathers.

Eggs, eight to ten, dull greenish or buff.

At Hamilton very few of this species are seen in spring, but in the fall they often appear in flocks of considerable size, and during their short stay afford good sport to the gunners, who wait for them in the evening near their feeding ground.

At St. Clair I have seen them in June, evidently mated, and have been told that a few pairs still breed there, though the number of summer residents is small compared with former years.

Dr. Macallum reports the same particulars regarding the occurrence of the species on Mohawk Island. Several couples still nest there, and are seen during the summer with their young, but, compared with former years, the numbers are greatly reduced. Early in the fall these are joined by flocks which have bred farther north, and all retire to the south before the Green-winged Teal arrives.

The species is very abundant throughout the North-West, where Mr. Macoun found it exceedingly plentiful during the fall of 1880. At Hudson's Bay it is mentioned as rare by Dr. Bell; and Mr. Nelson says that during the four years he passed at St. Michael's he failed to see a single individual of this species. In his list of Alaskan birds, Mr. Dall reports it being found sparingly at Fort Yukon and the Yukon mouth.

Genus SPATULA Boie.

SPATULA CLYPEATA (Linn.).

43. **Shoveller.** (142)

Bill, much longer than head or foot, widening rapidly to the end, where it is twice as wide as at the base, with very numerous and prominent laminæ; head and neck of male, green; fore breast, white; belly, purplish-chestnut; wing coverts, blue; speculum, green, bordered with black and white; some scapulars blue, others green, all white-striped; bill, blackish; feet, red. Female known by bill and wings. Length, 17-21; wing, 9½; tail, 3; bill, 2¾.

Hab.—Northern hemisphere. In North America, breeding from Alaska to Texas; not abundant on the Atlantic coast.

Nest, on the ground.

Eggs, eight to ten, greenish-gray.

An adult male Shoveller procured in the month of May makes a handsome specimen for the cabinet, for there are few of our waterfowl so gaily attired. The large spoonbill somewhat spoils his beauty of proportion, but it serves as a distinguishing mark for individuals of the species, of any age or sex.

It is not common in Ontario, but is occasionally found by the gunners steering up some sluggish creek, or sifting the mud along its shores. Its flesh being held in high estimation for the table, it is never allowed to get away when it can be stopped.

Dr. Macallum has observed it leading out its young within half a mile of the town of Dunnville. It is reported occurring at all the points of observation in Southern Ontario during the migratory season. In the North-West, Macoun found it breeding abundantly on the prairie ponds and about Pleasant Hills. It is also mentioned by Dr. Bell as breeding in large numbers on Lake Winnipeg. One or two specimens have been found by Turner in Alaska.

This species is said to have the widest distribution of any of the duck tribe, being more or less common in every portion of Europe and Asia, except in the extreme north. It occurs in northern and central Africa, is somewhat rare in England, but more common in Scotland. They all leave Ontario in the fall, but in spring return regularly to their old haunts.

Genus DAFILA Stephens.

DAFILA ACUTA (Linn.).

44. Pintail. (143)

Tail cuneate, when fully developed the central feathers projecting and nearly equalling the wing; much shorter and not so narrow in the female and young, four to nine inches long; wing, 11; total length, about 24. Bill, black and blue; feet, grayish-blue; head and upper neck, dark brown, with green and purple gloss; sides of neck, with a long white stripe; lower neck and under parts, white; dorsal line of neck, black, passing into the gray of the back, which, like the sides, is vermiculated with black; speculum, greenish-purple anteriorly bordered by buff tips of the greater coverts, elsewhere by black and white; tertials and scapulars, black and silvery. *Female and young:*—With the whole head and neck speckled or finely streaked with dark brown, and grayish or yellowish-brown; below, dusky freckled; above blackish, all the feathers pale-edged; only a trace of the speculum between the white or whitish tips of the greater coverts and secondaries.

HAB.—Northern hemisphere. In North America breeds from the northern parts of the United States northward, and migrates south to Panama and Cuba.

Nest, on the ground, in a bunch of tall grass near the water.

Eggs, eight to twelve, dull grayish-olive.

This is another very handsome member of the duck family which is common in Southern Ontario in spring and fall. According to Mr. Saunders, a few spend the summer and raise their young on St. Clair Flats.

The Pintail is quite common throughout the North-West. Dr. Bell mentions it having been found breeding near Norway House, and Mr. Nelson says : " It is one of the most common, if not *the* most common, of the ducks which breed along the Alaskan shore of Behring Sea. It is about the first of the water-fowl to commence nesting. The date when the first eggs are laid varies from May 18th to 25th, according to the season. The nest, which is placed in a tussock of grass, is lined with grass, stems and feathers, and is pretty well concealed. The eggs are rather small for the size of the bird, and are pale olive green when fresh. When the young are hatched the parents lead them to the adjacent pool, and they keep in the most secluded parts of the marsh until able to take wing. In the fall the Pintails feed upon the various berries growing on the hill-side till they become extremely fat, and a young bird at this season is the most delicious of the water-fowl found in the north. Toward the end of August they unite in flocks of from five to fifty, and the end of September finds but few remaining of the large numbers seen a few weeks previous."

GENUS AIX BOIE.

AIX SPONSA (LINN.).

45. Wood Duck. (144)

Male:—Head crested, metallic green and purple; line above and behind the eye, white; throat, white; above, coppery black with a gloss of green and purple; beneath, white; upper part of the breast, chestnut; sides, buffy, very finely variegated with black; the shoulder bordered also with black; covert and quills with more or fewer tips and shades of white and purple. *Female:*— Chestnut of the neck detached and dull; sides, not striped; head and neck, dull; bill, reddish, edges dusky; legs and feet, yellowish; iris, red. Length, 19; extent, 27½; wing, 9; tarsus, 1½.

HAB. —Temperate North America, breeding throughout its range.

Nest, in a hole in a tree.

Eggs, about twelve in number, pale buff slightly tinged with green.

This, the most beautiful of all our water-fowl, is very generally distributed throughout the country, arriving from the south about the time the ice disappears from our lakes and rivers, and retiring early in the fall. Owing to the great beauty of the male, these birds are much sought after by all classes of sportsmen, and are now seldom seen except near the retired ponds and marshes where they breed. Twenty-five years ago I used to see them leading out their young from one of the inlets of the Dundas marsh. They were known at that time to breed near Gage's inlet also, but of late years they have been observed only as passing migrants in spring and fall. The Wood Duck has frequently been domesticated, and adds greatly to the interest and beauty of an artificial pond in a pleasure ground.

The fact of its nesting in a hole in a tree is one of the interesting points in the history of the Wood Duck, although it is not the only duck with this habit.

The hole selected is a natural cavity, a woodpecker's or squirrel's hole, or the decayed end of a broken branch. The nest is warmly lined with feathers, and there the female rests in peace during incubation, her lord having for the time deserted her society for that of his own sex. If the nest is placed over the water, the young soon after being hatched, scramble up to the edge, spread their little wings and feet and courageously take their first leap in life toward the water. If it is a short distance off, the mother takes the tiny youngsters gently in her bill and drops them carefully on the surface, where for many days she stays with them, directing all their movements.

The Wood Duck, though found in all parts of Ontario, except perhaps in the extreme north, is nowhere abundant throughout the Province. At St. Clair Flats it used to breed in considerable numbers, but of late years has not been so often observed.

Throughout the North-West it is mentioned as a rare summer resident, and among the birds of Alaska it does not appear at all.

In some of the States to the west of us, in the interior, these ducks are said to be abundant, keeping by themselves in large flocks in the fall, as the Redheads and Bluebills do. Dr. Hatch, in the " Birds of Minnesota," says of them : " Arriving simultaneously with the other early species, none other braves the last rigors of the departing winter in the closing days of a Minnesota March with greater spirit, and when they come, like the rains in the tropics, they pour in until every pool in the woodlands has been deluged with them. This may sound strangely and exaggerated to ears unfamiliar with the history of bird-life on the borders of civilization, yet such has heretofore been my personal observation at the very location of our city " (Minneapolis).

Genus AYTHYA Boie.

AYTHYA AMERICANA (Eyt.).

46. Redhead. (146)

Bill, dull blue with a black belt at end, broad and depressed, shorter than head (two or less). the nostrils within its basal half; color of head, rich, pure chestnut, with bronzy or red reflections; in the female, plain brown; body anteriorly. rump and tail coverts, black; in the female, dark brown; back, scapulars and sides, plumbeous-white. finely waved with unbroken black lines, less distinct in the female; speculum, bluish-ash. Length, about 20; wing, 9-10; tarsus, 1⅜-1¾.

Hab.—North America, breeding from California and Maine northward.

Nest, like that of a coot, composed of broken bits of rushes on a clump of bog, often afloat.

Eggs, seven or eight, dull buff or creamy white.

The Redhead is one of the most abundant species which visits Lake Ontario, and, judging by the numbers which are sent down from the shooting stations farther west, it seems to be equally so at other points. They are strong, hardy birds, and a heavy charge, skilfully aimed, is necessary to stop them when on the wing. During the past two seasons a flock of 100 to 150 remained in Lake Ontario all winter, about half a mile from the shore, opposite the village of Burlington. The birds spent most of their time at one particular place, sometimes diving, sometimes sitting at rest on the water, and always close together, as if for greater warmth. When the weather moderated in March they shifted about for a few days, and then went off to the north-west, the direction taken by most water-fowl when leaving this part of Ontario in spring.

These large flocks of Redheads are somewhat capricious in their movements. They are seldom found more than one or two seasons at one place, having perhaps to "shift the pasture." They are among the most abundant species at all the shooting stations in Southern Ontario in the fall, and are said to breed abundantly all through the North-West, but are not named as being found at Alaska nor at Hudson's Bay.

They are greatly prized for the table, and are often sold for Canvas-backs, which they resemble in taste and color.

AYTHYA VALLISNERIA (Wils.).

47. Canvas-back. (147)

Similar to the preceding, but bill blackish, high at the base and narrow throughout, not shorter than head (two and a half or more), the nostrils at its middle; head, much obscured with dusky; black waved lines of the back sparse and broken up into dots, the whitish thus predominating.

Hab.—Nearly all of North America, breeding from the north-western States northward to Alaska. Breeds in the North-West.

Nest, on the ground, of grass and weeds lined with feathers.

Eggs, six to ten, pale greenish-buff.

The Canvas-back occurs with us, occasionally, in limited numbers. It resembles the Redhead in many respects, but can readily be distinguished by its low forehead and by the sooty color of the head and upper part of the neck. Its mode of diving is also peculiar. Before going under the water it throws itself upward and forward, describing a curve as if seeking to gain impetus in the descent, just as boys sometimes do when taking a header off a point not much above the water level.

Its reputation as a table duck is very high, but the excellence is attained only when the birds have for some time been feeding on wild celery, of which they are very fond. When that fare is not available they are no better for the table than Redheads or Bluebills.

The Canvas-back has been taken at Ottawa, Toronto, Hamilton, St. Clair Flats, and other points in Southern Ontario, but only as a straggler. It is rare throughout the North-West, but is mentioned by Mr. Dall as breeding abundantly at Fort Yukon, though it had not been observed at any other point in Alaska. Being generally distributed throughout the marshes in the interior through the summer, it gathers in immense flocks along the Atlantic sea coast in winter, especially on Chesapeake Bay, where the much-prized wild celery abounds.

Subgenus FULIGULA Stephens.

AYTHYA MARILA NEARCTICA Stejn.

48. American Scaup Duck. (148)

Male:—With the head, neck and body anteriorly, black, the former with a green gloss; back and sides, whitish, finely waved in zigzag with black; below, and speculum of wing, white; bill, dull blue with black nail; legs, plumbeous.

Female: With the head and anterior parts brown, and other black parts of the male, rather brown; face, pure white. Length, about 20; wing, 9.

Hab.—North America, breeding far north.

Nest, of weeds and dry grass, lined with down, placed on the ground.

Eggs, six to ten, grayish-green.

This and the next species, which are nearly allied, are the ducks most frequently met in Southern Ontario, where they are known as Bluebills. In the fall they remain in Hamilton Bay till they are frozen out, and in spring, even before the bay is open, they appear outside on Lake Ontario, and make frequent excursions inward to watch for the moving of ice. In spring many remain in the bay till about the first of May, by which time they seem all to be paired, but I have no record of their having been found breeding, and think it likely that nearly all spend the summer in the north of the Province.

This is the larger bird of the two species, and is considered to be somewhat hardier than its little brother. It breeds commonly throughout the North-West and in Alaska. Mr. Nelson says regarding it: "Everywhere in suitable locations over all the mainland portions of Alaska this is a common and frequently an abundant summer resident. In the north, as in the south, these birds show a predilection for the larger bodies of water, and at once, after the young are hatched, they are marshalled off to the largest pond in the vicinity."

The big Bluebill is common at all the shooting stations in Southern Ontario, where its large size makes the game-bag heavy, a fact which is duly appreciated by the hunters.

AYTHYA AFFINIS (Eyt.).

49. **Lesser Scaup Duck.** (149)

Similar to the preceding, but smaller, about 16; wing, 8; gloss of head chiefly purple; flanks and scapulars less closely waved with black (?) It is very difficult to define this bird specifically, and it may be simply a small southern form; but it appears to preserve its characteristics though constantly associated with the last.

Hab.—North America in general, breeding chiefly north of the United States, migrating south to Guatemala and the West Indies.

Closely resembles the preceding, except in being considerably less in size. Nesting habits and eggs are the same.

According to Dr. Coues, this is a southerly bird, not breeding so far north as the American Scaup Duck, and going farther south in winter.

In Southern Ontario it is about equal in abundance with the preceding, with which it is often associated, but it does not leave Hamilton Bay till about the middle of May, which would lead us to suppose that it does not go so far north to breed as some of the others.

In Southern Ontario it is the more abundant of the two during migrations. Mr. Saunders mentions that a few breed on the St. Clair Flats, and Dr. Macallum states that some still breed in the marsh near Dunnville. Throughout the North-West it is spoken of by all the observers as an exceedingly abundant summer resident. Macoun says: "Breeding more commonly than the preceding."

In Alaska, Mr. Nelson had heard of it having been obtained at the mouth of the Yukon, and says regarding it: "This record is extremely doubtful, since during my visit to the Yukon mouth in the spring of 1879, and my long residence only sixty miles north of there, not a single example of this bird was obtained or seen, nor did any of my collectors in the various parts of the territory observe or secure it."

It remains in Southern Ontario till the waters are frozen over, when it moves to the south to spend the winter.

AYTHYA COLLARIS (Donov.).

50. Ring-necked Duck. (150)

Similar to the foregoing, but an orange-brown ring around the neck; speculum, gray; back, nearly uniform blackish; bill, black, pale at base and near tip. *Female:-* With head and neck brown, and no collar, but loral space and chin, whitish, as is a ring around eye; bill, plain dusky. In size, between the two foregoing.

HAB.—North America, breeding far north, and migrating south to Guatemala and the West Indies.

Nest, on the ground, composed of grass, lined with feathers.

Eggs, eight to ten, varying from grayish to buff.

This handsome little duck is not so common as either of the preceding. While here it resembles the Teal in its habits, preferring marsh to open water, on account of which the gunners have given it the name of Pond Bluebill.

In distribution its range is about the same as that of the Blue-

6

bills. It has not been observed in summer in Southern Ontario, but breeds commonly throughout the North-West. In Alaska, Turner mentions having seen the species at St. Michael's and on the Aleutian Islands, but in small numbers, and so shy that he was unable to secure a specimen.

GENUS GLAUCIONETTA STEJNEGER.

GLAUCIONETTA CLANGULA AMERICANA (BONAP.).

51. American Golden-eye. (151)

Male:—With the head and upper neck, glossy green, and a white oval or rounded loral spot, not touching the base of the bill-throughout; lower neck all round, lower parts, including sides, most of the scapulars, wing coverts and secondaries, white; the white of outer surface of wings, continuous; lining of wings and axillars, dark; most of upper parts, black; no waving on the back or sides; bill, black with pale or yellow end, with nostrils in anterior half; feet, orange; webs, dusky; eyes, yellow; head, uniformly puffy. *Female:*—With head snuff-brown, and no white patch in front of the eye, and white of wings not always continuous. Length, 16-19; wing, 8-9.

HAB.—North America, breeding from Maine and the British Provinces northward; in winter, south to Cuba.

Nest, in hollow trees; it is made of grass, leaves and moss, lined with down. Eggs, eight to ten, ashy-green.

This species is a regular visitor at Hamilton Bay during the spring and fall migrations. While here they do not keep by themselves, but seek the society of whatever species may be at hand. They are very watchful, and difficult of approach. If any one of my readers has ever tried to scull up behind the rushes towards a bunch of Bluebills, among which were one or two Golden-eyes, and succeeded in getting a shot, he has had much better luck than I have had. Frequently, before getting within one hundred yards, I would hear the whistling of the Golden-eyes' wings, and looking up, see them going off, with the others following. Like many others which are known in Southern Ontario only as visitors in spring and fall, the Golden-eyes breed in suitable places throughout the North-West Territory.

Dr. Bell mentions their breeding near Norway House, on Hudson's Bay, and Mr. Nelson says: "In the Aleutian Islands this bird is a winter resident, as observed by Mr. Dall. The same author also reports it as being always one of the first arrivals along the Yukon." At the shooting stations in Southern Ontario it is not abundant, but at certain points, where the favorite mussels are obtained, it is more common. Dr. Macallum mentions the mouth of the Grand River, near Dunnville, as one of the resorts where these birds assemble in great numbers, and remain till frozen out.

GLAUCIONETTA ISLANDICA (Gm.).

52. Barrow's Golden-eye. (152)

Very similar to the preceding, differing chiefly in being larger in size; gloss of the head, purple and violet; loral spot, larger; white on the wing divided by a dark bar; feathers on the hind head lengthened into a crest; bill blotched with red. Length, 19-22; wing, 9-10. The female can probably not be distinguished from the preceding.

Hab.- Northern North America, south in winter to New York, Illinois and Utah; breeding from the Gulf of St. Lawrence northward, and south in the Rocky Mountains to Colorado.

Nest and eggs, like the preceding.

Dr. Garnier, who resides at Lucknow, a little to the east of Lake Huron, reports the finding of this species occasionally in winter in the inlets along the lake shore. The Doctor, who is not entirely in harmony with the modern school of ornithologists, thinks this a case of unnecessary subdivision. At all events, he claims to have found both forms, and he is likely correct, for the present species is found on Lake Michigan, which is within easy reach of the point to which the Doctor refers. It was also taken at Toronto, by Mr. C. Pickering, on the 18th of April, 1885; and at Hamilton I am aware of three being obtained, one of which came into my possession. They may, however, be more common than we are aware, for the hunters do not trouble the Whistlewings if anything more suitable for the table is in view.

In the "Birds of the North-West," Dr. Coues says: "Barrow's Golden-eye, upon which some doubt has been cast by myself among others, appears, nevertheless, to be a valid species, the differences pointed out in the Key and in other works being apparently constant as well as appreciable; and there being, moreover, certain anatomical peculiarities in the form of the skull, of which I have only lately become aware."

"Originally described in 1788, it was subsequently re-named and figured in 1831. The species was ignored by Audubon, who mistook it for the summer plumage of the common Golden-eye. It is the most northerly species of the genus, having apparently a circumpolar dis-

tribution, breeding only (?) in high latitudes, and penetrating but a
limited distance south in winter. Its claim to a place in the present
connection rests upon its occurrence in the Rocky Mountains as far
south as Utah, where it was procured by Mr. H. W. Henshaw; on
the eastern coast it occurs as far south in winter as New York."

Mr. Edwin Carter was the first to find the nest and eggs of this
species. He met with them in the mountains of Colorado in 1876.

Dr. Bell mentions its occurrence at Hudson's Bay, and it has been
obtained in Manitoba, but only as a rare straggler.

Genus CHARITONETTA Stejneger.

CHARITONETTA ALBEOLA (Linn.).

53. Buffle-headed Duck. (153)

Somewhat similar to the *clangula americana* in color, but *male* with the head
particularly puffy, of varied rich iridescence, with a large white auricular patch
confluent with its fellow on the nape; small. Length, 14-16; wing, 6-7; bill, 1,
with nostrils in its basal half. *Female;* Still smaller, an insignificant looking
duck, with head scarcely puffy, dark gray, with traces of the white auricular
patch.

Hab. North America, south in winter to Cuba and Mexico. Breeds from
Maine northward, through the fur countries and Alaska.

Dr. Cones (Birds N.-W., 575) describes the nest of this duck placed in the
hollow of a dead tree, and composed of feathers.

The eggs, from six to fourteen in number, are described as varying from buff
to a creamy-white or grayish-olive color.

The Buffle-heads are common at all the shooting stations in Southern Ontario in spring and fall, but owing to their small size they are
not much sought after. The male in full spring dress is a very handsome little fellow, and, like many other animals of diminutive proportions, seems to feel himself as big as any of those about him. I have
in my collection a young male of this species of a uniform cream
color, which was shot in Hamilton Bay a few years ago.

Mr. Saunders mentions that a few pairs breed at St. Clair Flats.
Throughout the North-West their distribution seems to be somewhat
irregular. Macoun says of them: "Abundant in the ponds in the
autumn, not seen in the prairie regions." While Thompson, writing
from Carberry, says: "Common summer resident, breeding also at
west slope of Duck Mountain, Portage la Prairie."

From Alaska, the reports are similar. Nelson says: " Bischoff
found it at Sitka, and at the Yukon mouth Dall notes them as

abundant, and probably breeding. During my visit to the latter point, extending through the latter half of May and the first half of June, not a single individual of this species was seen, nor was it found by me along the coast farther to the north."

Dr. Hatch, in his report on the birds of Minnesota, says of the Buffle-head: "Such has been my confidence that to a limited extent they breed here, that I have left no opportunity unimproved to discover the final proof by the finding of a veritable nest. But for this testimony I must still wait, notwithstanding the oft-repeated assurances of several persons that they have found them. In one instance my hopes had been nearly realized, when I found the nest to be that of the Wood Duck."

In Southern Ontario they are among the first to arrive in spring and the last to leave in the fall, being apparently quite hardy and able to stand the cold.

GENUS CLANGULA LEACH.

CLANGULA HYEMALIS (LINN.).

54. Old Squaw; Long-tailed Duck. (154)

Tail, of fourteen narrow pointed feathers, in the male in summer the central ones very slender and much elongated, nearly or quite equalling the wing; nail of bill occupying the whole tip: seasonal changes remarkable. *Male, in summer:*—With the back and the long narrowly lanceolate scapulars varied with reddish-brown, wanting in winter, when this color is exchanged for pearly-gray or white; general color, blackish or very dark brown; below from the breast abruptly, white; no white on the wing; sides of head, plumbeous-gray; in winter, the head, neck and body anteriorly, white, but the gray cheek patch persistent, and a large dark patch below this; bill, at all seasons, black, broadly orange barred. *Female:*—Without lengthened scapulars or tail feathers; the bill, dusky greenish, and otherwise different; but recognized by presence of head and neck patches, and absence of white on the wing. Length, 15-20 or more, according to tail; wing, 8-9.

Hab.—Northern hemisphere, in North America south to the Potomac and the Ohio; breeds far northward.

Nest, on the ground.

Eggs, six or seven, drab color, shaded with green.

Vast numbers of "cowheens" (as these birds are called here) spend the winter in Lake Ontario, out on the deep water away from the shore. Even there they are not free from danger, for great numbers get entangled in the gill nets. Passing along the beach in winter, strings of drowned, draggled cowheens may be seen dangling from the clothes lines about the fishermen's outhouses. I have frequently heard the fishermen, when trying to force a sale, declare positively, that if buried in the earth for twenty-four hours before being prepared for the table, these birds are excellent eating. Notwithstanding this assertion, the supply still keeps ahead of the demand, and numbers are turned over to the pigs, a sorrowful end for the beautiful, lively *Clangula hyemalis.*

This species frequents the northern shores of both continents, making its summer home in the Arctic regions, where, among the tall grass by the margins of retired lakes and ponds, the nests are found in great numbers. Nelson says, regarding its place among the birds of Alaska: "The Old Squaw is the first duck to reach high northern latitudes in spring, and along the Alaskan coast of Behring Sea is one of the most abundant species during the summer. The seal hunters find them in open spaces in the ice off St. Michael's, from the 1st to the 20th April, but the first open water near the shore is sure to attract them. In the fall they retreat before the ice, and by the 15th or 20th October they are either on their way south or well out to sea."

"During the pairing season the males have a rich, musical note, frequently repeated in deep, reed-like tones. Amid the general chorus of water-fowl which is heard at this season, the notes of the Old Squaw are so harmonious that the fur-traders of the upper Yukon have christened him the Organ Duck, a well-merited name. I have frequently stopped and listened with deep pleasure to these harmonious tones while traversing the broad marshes in the dim twilight at midnight, and while passing a lonely month on the dreary banks of the Yukon delta I lay in my blankets many hours at night, and listened to these rhythmical sounds which, with a few exceptions, were the only ones to break the silence." This cry is very familiar to all who have occasion to be near the shores of Hamilton Bay in spring. Here the birds assemble in large flocks, before leaving for the north, and when this cry is started and kept up with spirit by each member of the flock, the concert is heard a long way off, and is a subject of wonder to all who hear it for the first time. The note consists of five syllables often repeated, and is variously translated in different regions. Along the shores of the north of Scotland, where large bands of the birds spend the winter, it is said to call for two articles which are indispensable during the long, dark nights of this dreary season : "Coal and can'le licht, coal and can'le licht."

GENUS HISTRIONICUS LESSON.

HISTRIONICUS HISTRIONICUS (LINN.).

55. Harlequin Duck. (155)

Bill, very small and short, tapering to the tip, which is wholly occupied by the nail, and with a membraneous lobe at its base; tertiaries, curly; plumage, singularly patched with different colors. *Male :*—Deep bluish lead color, browner below ; sides of the head and of the body posteriorly, chestnut ; coronal stripe and tail, black ; a white patch at the base of the bill and another on the side of the occiput, of breast and of tail, two transverse ones on side of neck forming a nearly complete ring, and several on the wings ; a white jugular collar; speculum, violet and purple. *Female :*—Dark brown, paler below, a white patch on auriculars and before the eye. Length, 15-18 inches; wing, 8 ; bill, 1.

HAB.—Northern North America, breeding from Newfoundland, the Rocky Mountains, and the Sierra Nevada northward; south in winter to the Middle States and California.

Nest, composed of weeds and grass, lined with down from the breast of the owner; it is sometimes placed in a hollow tree or stump, more frequently on the ground, not far from the water.

Eggs, six to eight, pale green shaded with buff.

The Harlequin is found on the northern shores of Europe, Asia and North America. On the last named continent, it breeds sparingly in Maine and in the North-West as far as Alaska. It has also been found in the northern Rocky Mountains and in the Sierra Nevada. In winter it descends to the Middle States and California.

With these facts before us, we naturally expect to hear of the species having been seen occasionally in Ontario, but of such occurences the records are very few.

William Loane, of Toronto, reports having killed a pair near that city in the spring of 1865, and in the fall of 1881 he killed another, a female, which is now in the rooms of the Toronto Gun Club.

One of the residents on the beach, near Hamilton, told me some years ago that he had seen a pair there in spring. The male in full plumage was correctly described by my informant, and spoken of as the most "dapper little drake" he had ever seen. The name Harlequin is suggested by the peculiar markings on the head of the male, which are supposed to resemble those often assumed by the clown in a circus.

In the eighth volume of the bulletin of the Nuttal Club, Dr. Merriam gives the following summary of the bird's breeding range : " In Siberia it is known to breed about Lake Baikal and in the Bureza Mountains (Radde), in Mantchuria, and at various points in the great Stanowi Range (von Niddendorf), about the upper Amoor (von Schreuck) and in Kamtschatka. On the American continent it has been found breeding along the tributaries of the Yukon in Alaska (Dall), in the interior of the fur countries and about Hudson's Bay (Richardson), on the fresh-water ponds in Labrador (Audubon), and in the Rocky Mountains within the limits of the United States (Coues). It also nests in Greenland, Iceland and Newfoundland."

In all these places, and wherever else it appears, it is regarded as very rare.

GENUS SOMATERIA LEACH.

SUBGENUS SOMATERIA.

SOMATERIA DRESSERI SHARPE.

56. American Eider. (160)

Bill, with long club-shaped frontal processes extending in a line with the culmen upon the sides of the forehead, divided by a broad feathered interspace. *Male :* In breeding attire, white, creamy-tinted on breast and washed with green on the head ; under parts from the breast, lower back, rump, tail, quills, and large forked patch on the crown, black. *Female :* With the bill less developed, general plumage an extremely variable shade of reddish-brown or ochrey-brown, speckled, mottled and barred with darker ; male in certain stages resembling female. Length, about 2 feet ; wing, 11-12 inches.

HAB.—Atlantic coast of North America, from Maine to northern Labrador, south in winter to the Delaware.

Nest, on the ground, composed of dry grass, moss and sea weed, lined with down and feathers.

Eggs, six to ten, drab, tinged with green.

The Eider Duck is essentially a bird of the sea coast, breeding abundantly along the shores of Newfoundland and Labrador. Its visits to our inland waters are made during the season of migration, when the movements of all migratory birds are considerably affected by the prevailing winds. On Lake Ontario it is a casual visitor in winter, but is seldom, if ever, seen there in mature plumage.

The one in my collection is a young male in the garb of the female. I shot it from the pier of the canal at the entrance to Hamilton Bay a few years ago. They were seen occasionally all that winter, but they were known to be "fishy," and as there is nothing attractive in their dress they were not much disturbed, although they allowed a nearer approach than other water-fowl are disposed to do.

On the 7th November, 1889, Mr. George R. White captured a young male of this species on the river near Ottawa. It was in the plumage of the female.

Mr. William Cross obtained a specimen which was shot at the island near Toronto on December 6th, 1890.

Mr. Frazer found the Eider Duck breeding on the small islands along the coast of Labrador. The nest was built in a hollow among the soft short grass, or at the foot of a rock where it was sheltered from the wind. It was composed of grass and lined with slate-colored down from the breast of the bird.

SOMATERIA SPECTABILIS (Linn.).

57. King Eider. (162)

Adult, male:—Bill, pale yellow; at the base of the upper mandible is a compressed gibbous substance of a bright orange color, the front covered with short black feathers, the sides margined with the same color, the feathers extending back in a point nearly to the eye; head, bluish-gray, darkest behind; cheeks, shaded with sea green, a spot of black below the eye; on the throat, two lines of black forming an inverted **V**; middle of neck, white; lower neck and forepart of the breast, buff; lower plumage, blackish; a large spot of white on either side of the rump; posteriorly, black; wings and tail, brown, the former broadly marked with white. When in full plumage the secondaries curve over the primaries. Length, 25 inches. *Young*:—Dark brown, many feathers on the neck margined with white; gibbous substance on the bill scarcely perceptible. *Female*:—Much like the common Eider, the shape of the bill being the principal point of difference. Of circumpolar distribution, breeding abundantly on the shores of the Arctic seas; in winter, south on the west coast to the Aleutian Islands in great numbers. On the east, south irregularly as far as New York.

Nest, a depression in the ground lined entirely with down.

Eggs, usually six, grayish-green.

This is a royal visitor from the north whom we are pleased to welcome, though he rarely comes in his royal robes. Nearly all of those found in Ontario are in immature plumage. In this garb Eider Ducks have occasionally been seen near Toronto and Hamilton during the winter, but they all looked so much alike that it was only after a close examination that many of them were found to be the young of the King Eider.

On the 25th November, 1889, Mr. Cross reports having obtained a fine male in summer plumage which was shot in Toronto Bay. There must have been something irregular about this specimen, for Mr. Murdoch and others, who have seen them in the summer, report that as soon as the breeding season is over the male loses the gay, light-colored plumage, and assumes a plain brown dress similar to that of the female, the change in the majority of cases taking place in September.

The species is reported from Lake Erie by Dr. Wheaton, of Columbus, and also by Dr. Bergtold, of Buffalo, but neither mentions in what dress it appeared.

Mr. Murdoch names this as the most abundant bird at Point Barrow, where it appears in enormous numbers during the season of migration. Very few remain there to breed, the great bulk of the species going along the coast to the eastward of the Point, where they settle down early in May.

GENUS OIDEMIA FLEMING.

SUBGENUS OIDEMIA.

OIDEMIA AMERICANA (Sw. & Rich.).

58. American Scoter. (163)

Plumage of *male*, entirely black; bill, black; the gibbosity, orange. *Female:*—Sooty-brown, paler below; on the belly, grayish-white, speckled with dusky; waved with dusky on the sides and flanks; throat and sides of the head, mostly whitish; feet, livid-olive; webs, black. Length, 22 to 24 inches; female, 18 to 20 inches.

HAB.—Coasts and larger lakes of northern North America. Breeds in Labrador and the northern interior. South in winter to New Jersey, the Great Lakes and California.

Nest, in a hollow in the ground near the water. It is lined with coarse grass, feathers and down.

Eggs, six to eight, pale brownish-buff.

This is one of the sea ducks whose home is in the north, and its line of migration being mostly along the sea coast, its visits to the inland waters are only accidental. The specimen in my collection was obtained at the west end of Lake Ontario, where the species is often seen in the fall, in company with others of its class.

Being undesirable either for use or ornament, it is allowed to spend the time of its visit here in peace.

Mr. White reports it as a regular visitor at Ottawa, where it appears singly, or in small numbers, in the fall.

Dr. Bergtold also mentions its being found in Lake Erie near Buffalo.

Mr. E. W. Nelson, speaking of these birds in Alaska, where they breed in great numbers, says:

"At St. Michael's these ducks are never seen in spring until the ice begins to break off shore and the marshes are dotted with pools of open water. May 16th is the earliest date of arrival I recorded. Toward the end of this month they leave the leads in the ice and are found in abundance among the salt and fresh water ponds in the great marshes, from the Yukon mouth north and south. The mating is quickly accomplished, and a nesting site chosen on the border of some pond. The spot is artfully hidden in the standing grass, and the eggs, if left by the parent, are carefully covered with grass and moss. As the set of eggs is completed, the male gradually loses interest in the female, and soon deserts her to join great flocks of his kind along the sea shore, usually keeping in the vicinity of a bay, inlet, or the mouth of some large stream. These flocks are formed early in June and continue to grow larger until the fall migration occurs. The numbers gradually decrease until the 10th to the 15th of October, when all have gone south. Until the young are about half grown, the female usually keeps them in some large pond near the nesting place, but as August passes they gradually work their way to the coast, and are found about the shores and inner bays until able to fly. They do not at any time ascend rivers, preferring to keep near the sea shore."

SUBGENUS MELANITTA Boie.

OIDEMIA DEGLANDI Bonap.

59. White-winged Scoter. (165)

Male: With a large patch of white on the wing and another under the eye; feet, orange-red, with dusky webs; bill, black, broadly tipped with orange. *Female:* Sooty-brown, grayish below; whitish about the head; speculum, white. Length, 24 to 26 inches; female, 20 to 22 inches.

HAB.—Northern North America; breeding in Labrador and the fur countries; south in winter to the Middle States, southern Illinois and southern California.

Audubon found this species breeding in Labrador. The nests were built by the side of small lakes, two or three miles distant from the sea, and usually placed under low bushes. They were formed of twigs, mosses and various plants matted together, and were large and almost flat, several inches thick, and lined with feathers.

Eggs, usually six, pale buff, clouded with green.

This is the most numerous of the three Scoters which are found in Ontario, for it seems more partial to the inland waters than either of the others.

In spring and fall it is common on all the large lakes, and it is reported at Buffalo, Ottawa, Kingston, Toronto, Hamilton, etc.

They are large, heavy birds, and their jet-black color makes them look larger than they really are.

When moving about from one place to another, they fly heavily, at no great height above the water. They have not the restless habits of some other species, and if left undisturbed, will remain for days together feeding near the same spot.

At Hamilton Bay they are regular visitors, appearing toward the end of April, and remaining for two or three weeks. Very soon after their arrival, they are affected by some malady which results in many of their number being washed up dead on the shore. These birds are in fine plumage and excellent condition, but that does not save them. Whether they bring the seeds of disease with them when they come, or whether the emptying of the city sewage and refuse from the oil refineries into the bay has anything to do with their trouble, has not yet been determined.

In the spring of the present year (1893), this disease prevailed to as great an extent as formerly, so many as six bodies being observed in a walk of half a mile along the shore.

Mr. Nelson saw very few of this species in Alaska, but they breed abundantly on the lower Anderson River. They have also been observed in summer on Lake Winnipeg and other lakes in Manitoba, where they were supposed to be breeding.

SUBGENUS PELIONETTA KAUP.

OIDEMIA PERSPICILLATA (LINN.).

60. Surf Scoter. (166)

Bill, narrowly encroached upon by the frontal feathers, on the culmen, nearly or quite to the nostril, but not at all upon its sides; about as long as the head, with nail narrowed anteriorly. The swelling lateral as well as superior, nostrils beyond its middle. Bill of *male*, orange-red, whitish on the sides, with a large circular black base; plumage, black, with a patch of white on the forehead, and another on the nape; none on the wing. *Female.*—Smaller; bill, black; feet, dark, tinged with reddish; webs, black; plumage, sooty-brown; below, silvery-gray; whitish patches on each side of the head. Length, 22 to 24 inches; female, 20 to 22 inches.

HAB.—Coasts and larger inland waters of northern North America. South in winter to the Carolinas, Ohio River and Lower California.

Audubon observed this species breeding in Labrador. He found a nest in a fresh-water marsh, among the tall grass and weeds. It was composed entirely of withered weeds, lined with the down of the birds, and contained five eggs of a pale yellowish or cream color.

This handsome Scoter visits the waters of Ontario in spring and fall, where it is observed in limited numbers in company with the white-winged species, which it resembles in its habits, the clear white patches in marked contrast to the deep black of the plumage serving, even at a distance, to mark its presence in a flock. It is never numerous, though more frequently seen than the black Scoter; and has been observed at Toronto, Hamilton, Ottawa and along the sea coast. It breeds in immense numbers in the north, and judging from the fact of so few being seen elsewhere, it is probable that the bulk of the species remain there over the winter.

Mr. Nelson, in his "Birds of Alaska," says regarding this species: "During the summer of 1881 I found them common about the head of Norton Sound, on both shores of Behring Strait, and in Kotzebue Sound. Although I did not find these birds nesting commonly near St. Michael's, yet from about the first of July until autumn, immense flocks of males frequented the shores of St. Michael's and the adjoining Stewart Island. The seaward shores formed the ordinary haunts of these birds until the approach of a gale forced them to seek the lee of the islands or the sheltering bays. From the fact that these flocks are formed exclusively of males, it is evident that the females assume the duties of incubating the eggs and rearing the young.

"The main breeding ground of this species remains unknown to me, for, although females and young were not rare in summer, they were never numerous enough to account for the vast numbers of males to be found.

"On August 23rd, 1878, I visited Stewart Island, about ten miles to the seaward of St. Michael's. As I neared the island in my kyak, I found the water literally black with the males of this species, which were united in an enormous flock, forming a continuous band around the outer end of the island for a distance of about ten miles in length, and from one-half to three-fourths of a mile in width. As the boat approached them, those nearest began to rise heavily by aid of wings and feet from the glassy surface of the undulating but calm water. The first to rise communicated the alarm to those beyond, until as far as could be seen the water was covered with flapping wings, and the air filled with a roar like that of a cataract. The rapid vibration produced in the air by tens of thousands of wings could be plainly felt.

"In all my northern experience among the water-fowl which flock there in summer, I never saw any approach to the number of large birds gathered here in one flock, nor shall I soon forget the grand effect produced by this enormous body of birds as they took wing and swept out to sea in a great black cloud, and settled again a mile or so away."

GENUS ERISMATURA BONAP.
ERISMATURA RUBIDA (WILS.).

61. **Ruddy Duck.** (167)

Male in full plumage:—Bill, slaty-blue; the nail, black; neck all round and the upper parts, bright chestnut; the lower parts, silky white, watered with dusky; chin and sides of the head, white; the crown and nape, black. *Female:*—Brown above, finely dotted and waved with dusky; paler and duller below, with sometimes a slight tawny tinge, which also occurs on the side of the head. Length, 14-17; wing, 5-6; tarsus, 1¼.

HAB.—Northern North America, generally breeding throughout its range.

Nest, on the bog near the water.

Eggs, five to ten; grayish-white.

Mr. Shields writes from Los Angeles, California, that it breeds there abundantly, preferring the abandoned nest of a coot to one of its own making.

The Ruddy Duck is very generally distributed throughout Ontario, but, except near its breeding places, it appears only as a visitor in spring and fall. At the St. Clair Flats, I have seen it in summer, and have been told that a few pairs breed there every season. The greatest number I ever saw at one place was in a fisherman's wagon in the Hamilton market. It was early in May. A large flock, composed chiefly of males and numbering about 150, had become entangled in the gill nets, and been drowned in Lake Ontario, where they had tarried for rest and refreshment. The fishermen, maintaining that all is fish which comes into the net, tried to make the most of their "haul." But, although the birds attracted a good deal of notice on account of their being strangers and richly dressed, they did not meet with a ready sale.

Throughout Ontario they are found at all suitable places, and Dr. Bell, of the Geological Survey, reports one being shot at York Factory, where it was considered rare, because their breeding places are usually farther south.

Dr. Coues found them nesting in Dakota and Montana, and they have even been known to breed in Cuba, West Indies.

As divers, they have no equal among the ducks; and they also have the power of suddenly sinking backward, forward, or sideways, after the manner of the Grebes, without disturbing the surface to any extent. They can remain a long time under water, and a chase after a wounded one is a hopeless task.

The eye is placed high in the head, the lower plumage is stiff and glossy, and the tail is black, short and rounded, the feathers being very stiff and narrow. When not in use it is carried erect, giving the bird a very spirited look when seen amongst others which carry their tails horizontally.

SUBFAMILY ANSERINÆ GEESE.

GENUS CHEN BOIE.

CHEN HYPERBOREA NIVALIS (FORST.).

62. Greater Snow Goose. (169a)

Bill, with laminæ very prominent, owing to arching of the edges of the bill; adult plumage, pure white, the head often washed with rusty red; primaries, broadly tipped with black; bill, lake-red, with white nail; feet, the same; claws, dark. "*Young:*—Dull bluish or pale lead color on the head and upper part of the body."—CASSIN. Length, about 30 inches; wing, 17 to 19; tail, 5½ to 6; bill, 2½; tarsus, 3¼.

HAB.—North America; breeding far north, and migrating south in winter, chiefly along the Pacific coast, reaching Cuba.

Eggs, five to eight, yellowish-white.

In Ontario the Snow Goose can only be regarded as a casual visitor during the season of migration, for its summer home, according to Dr. Bell, is "still to the north of the regions known to the Eskimo," whence it retires to the south at the approach of winter. It is seen at different points in this province during spring and fall, but as there are seldom more than two or three together, they are regarded as stragglers from the main body, whose line of migration is along the Mississippi or the Pacific coast. The specimen in my collection was killed at the Beach, in the month of December, a few years ago, while making its way toward the open water in Lake Ontario.

Dr. Macallum, speaking of the occurrence of this species at Dunnville, says: "The Greater Snow Goose is often seen here in small flocks in the fall, when they frequent fall-wheat fields near the lake, often in company with the Canada Geese. I have three specimens in my collection taken here, two of which came up to farmers' barns with the tame ducks and were domesticated."

Sir John Richardson, in the *Fauna Boreali Americana* after noting the abundance of these birds, continues: "The eggs, of a yellowish-white color, and regularly ovate form, are a little larger than those of an Eider Duck, their length being three inches and their greatest breadth two. The young fly in August, and by the middle of September all have departed southward.

"The Snow Goose feeds on rushes, insects and, in autumn, on berries, particularly those of the *empetrum nigrum*. When well fed it is a very excellent bird, far superior to the Canada Goose in juiciness and flavor. It is said that the young do not attain their full plumage before the fourth year, and until that period they appear to keep in separate flocks. They are numerous at Albany Fort, in the southern part of Hudson's Bay, where the old birds are rarely seen, and, on the other hand, the old birds in their migrations visit York Factory in great abundance, but are seldom accompanied by the young."

Samples of the eggs in the Smithsonian Institute agree exactly with the above description, but show the usual variation in size, some being noticeably less than three inches in length by over two in breadth.

The occurrence of *one specimen* of the *Lesser* Snow Goose (*Chen hyperborea*) is reported by Mr. W. E. Saunders.

Genus ANSER Brisson.

ANSER ALBIFRONS GAMBELI (Hartl.).

63. American White-fronted Goose. (171*a*)

Tail, normally of sixteen feathers; bill, smooth, the laminæ moderately exposed. *Adult:*—Bill, pink, pale lake or carmine; nails, white; feet, yellow; claws, white. A white band along base of upper mandible, bordered behind by blackish; upper tail coverts, white; under parts, whitish, blotched with black; sides of the rump and crissum, white; head and neck, grayish-brown, shading lighter as it joins the breast; back, dark gray, the feathers tipped with brown; greater coverts and secondaries, bordered with whitish; primaries and coverts, edged and tipped with white; shaft of quills, white. *Young:*—Prevailing color, brown; no white on the forehead, which is darker than the rest of the head; no black on under parts. Length, about 27 inches; wing, 16; tail, 5; tarsus, 2.75.

Hab.—North America; south to Cuba and Mexico.

Nest, a depression in the sand, lined with hay, feathers and down.

Eggs, six or seven, smooth, dull yellowish, with an olive shade, marked in places with a darker tint.

This, like the preceding species, is only a casual visitor in Ontario, the vast flocks which annually leave their breeding grounds in the north at the approach of winter evidently preferring to make their journey along the western coast rather than by the Atlantic or through the interior. Stragglers have been observed at the various shooting stations, where they are considered rare. The specimen in my collection was shot at the St. Clair Flats, and is an immature male.

Speaking of this species in the "Birds of Alaska," Nelson says: "In early seasons the first White-fronted Goose reaches St. Michael's about April 27th, but the usual time is from the 5th to the 8th of May. From about the 10th of May they are very common, and remain to breed in considerable numbers all along the Alaskan shore of Behring Sea and on the Arctic coast of Point Barrow, where they are plentiful, arriving the latter half of May."

"During the summer of 1881, a number were found breeding upon St. Lawrence Island, and they also nest on the Siberian shore in the vicinity of Behring Strait.

"During the migrations, they occur at various points along the Aleutian chain, but are not known to breed there. Dall found their eggs all along the Yukon, from Fort Yukon to the sea, and it is well known as a widely spread species, breeding all around the Arctic mainland portions of America."

7

"All through September, old and young, which have been on the
wing since August, gather in larger flocks, and as the sharp frosts
toward the end of September warn them of approaching winter, commence moving south. The marshes resound with their cries, and
after some days of chattering, flying back and forth, and a general
bustle, they suddenly start off in considerable flocks, and the few
laggards which remain get away by the 7th or 8th of October," to
return again late in April or early in May, according to the season.

Genus BRANTA Scopoli.

BRANTA CANADENSIS (Linn.).

64. Canada Goose; Wild Goose. (172)

Tail, normally, eighteen feathered. Grayish-brown, below paler or whitish-
gray, bleaching on the crissum, all the feathers with lighter edges ; head and
neck, black, with a broad white patch on the throat mounting each side of the
head ; tail, black, with white upper coverts. Length, about 36 ; wing, 18-20 ;
tail, 6½-7½ ; bill, 1⅔-2 ; tarsus, usually over 3.

Hab.—The whole of North America, breeding in the United States, as well
as further north. Accidental in Europe.

Nests, usually a hollow in the sand, lined with down, and a few sticks
round the outer edge. In the "Birds of the North-West," Dr. Coues makes
mention of their breeding in trees in the upper Missouri and Yellowstone
regions, carrying their young to the water in their bill.

Eggs, five, pale dull green.

The Canada Goose is the most abundant and best known of its
class in Ontario. Early in April the Λ-shaped flocks are seen passing
on to their breeding grounds in the North-West, led by an experienced
gander, whose well-known call is welcomed by the Canadian people
as the harbinger of spring. It is associated with the return of
warmer days, and the passing away of the ice which for so many
months has held everything under control.

In former years the flocks used sometimes to settle in Hamilton
Bay, and similar places on their route, for rest and refreshments ; but
now their haunts have been invaded by trolley cars, electric lights,
telegraph wires and other innovations, which cause them to fly high
and to pass on with fewer stoppages.

Their return in the fall is eagerly watched for at the shooting
stations ; but so acute is their sense of hearing, and so careful are

they in the selection of a resting place, that only a very few are obtained.

Dr. Bell, of the Geological Survey of Canada, who is well acquainted with our native birds, and has furnished much valuable information regarding their habits in the different regions he has visited, says of this species: "The southern limit of the ordinary breeding ground of the Canada Goose runs north-westward across the continent from the Maritime Provinces to the valley of the McKenzie. I have met with them breeding in considerable numbers in the interior of Newfoundland, but in the same latitude, between the Great Lakes and James' Bay, only chance pairs lag behind in their northward flight to hatch their broods. They also breed on the islands along the east coast of Hudson's Bay. To the westward of the bay they are first met with, raising their young on the lower part of Churchill River. To the eastward it is said that very few Canada Geese breed northward of Hudson's Strait."

In these remote regions they no doubt enjoy the quiet which is necessary to the raising of their young. Farther south, while on their migratory journeys, they are subjected to continual persecution, which has trained them to be extremely vigilant; and when feeding or reposing on the water, sentinels are placed on the outskirts of the flock, who at once spread the alarm on the slightest appearance of danger. "So acute," says Audubon, "is their sense of hearing, that they are able to distinguish the different sounds or footsteps of their friends or foes with astonishing accuracy. Thus the breaking of a stick by a deer is distinguished from the same accident occasioned by a man. If a dozen large turtles drop into the water making a great noise in their fall, or if the same effect has been produced by an alligator, the wild goose pays no regard to it; but however faint and distant may be the sound of an Indian paddle that may by accident have struck the side of the canoe, it is at once marked. Every individual raises its head and looks intently towards the place from which the noise has proceeded, and in silence all watch the movements of the enemy."

BRANTA CANADENSIS HUTCHINSII Sw. & Rich.

65. Hutchin's Goose. (172*a*)

Tail, of sixteen feathers; colors, exactly as in the Canada Goose but size less. Length, about 30 inches; wing, 15-17; tail, 5-6; bill, 1¼-1⅜.

Hab. North America. Breeds in the Arctic regions, south in winter to Mexico.

Nest, usually a depression in the sandy beach, lined with leaves, grass, feathers and down. In the Anderson River region, the eggs of this species have been taken from the deserted nests of crows and hawks.

Eggs, white, four to six, laid in June or July.

This is apparently a small race of the preceding, from which it differs slightly in plumage, and it has been raised to the rank of a separate subspecies, in which position it is as easily considered as in any other. Small geese are occasionally seen in company with the last groups of the others which pass in spring, but they are fewer in number and are less frequently obtained.

I once saw a fine pair of these birds in the hands of a local taxidermist, with whom they had been left to be "stuffed," and with such vigor was the operation performed, that when finished it was a hard matter for anyone to tell to which species the birds originally belonged.

The differences between Hutchin's and the Canada Goose seem to be constant, and most writers are satisfied to treat them as now placed.

Mr. E. W. Nelson, who was familiar with the appearance of this species as it occurred at St. Michael's and the mouth of the Yukon, says regarding it in the " Birds of Alaska " : " From my observations I should decide the centre of abundance of this species to be along the lower Yukon and thence south to the Kuskoquim."

" The main difference between this form and *canadensis* is the smaller size of the former. In *hutchinsii*, the black of the head and neck tends to assume a glossier black, and the dark color very commonly encroaches upon the white cheek patches, frequently separating them by a broad black throat-band. The main distinction, however, besides the smaller size, is in the much lighter color of the lower surface. The white abdominal area extends forward and almost encloses the thigh in some cases, and almost invariably there is no definite line of demarcation between the white and brown areas. In addition, the grayish-brown of the breast is very light, and the encroachment of the white upon its posterior border gives a mottled gray and white surface."

I have given the above details in full, so that anyone meeting the species may be able to identify it.

BRANTA BERNICLA (Linn.).

66. **Brant.** (173)

Bill, feet and claws, black; head, neck all round and a little of the forepart of the body, glossy black; on each side of the neck a small patch of white streaks, also some touches of white on the eyelids and chin; breast, ashy-gray, beginning abruptly from the black, fading on the belly and crissum into white; tail feathers, wing quills and primary coverts, blackish, the inner quills whitish toward the base. Length, 24 inches; tail, 4-5.

Hab.—Northern portions of northern hemisphere, partial to salt water, rare in the interior. Breeds only within the Arctic circle.

Nest, a hollow on a sandy beach, lined with feathers and down.

Eggs, four to six, grayish or dirty white.

This is another casual visitor to the waters of Ontario, where it is less frequently seen than any other of the geese. It is by no means a scarce species, but seems partial to the sea coast. In Mr. Saunders' "List of Birds of Western Ontario," it is mentioned as a rather rare migrant. I have only seen it once, flying past out of range.

The Brant is of almost cosmopolitan distribution, being found on the sea coast of Europe and eastern North America, breeding only within the Arctic circle. It is said to breed in immense numbers in Spitzbergen and on the islands along the coast. On the rocky shores of Greenland, where it also breeds, the nests are often placed on the ledges of the rocks.

Subfamily CYGNINÆ. Swans.

Genus OLOR Wagler.

OLOR COLUMBIANUS (Ord.).

67. **Whistling Swan.** (180)

Pure white; head often stained with rusty; bill, black, usually with small yellow spot; iris, dark brown; feet, black. Length, 4½ feet; wing, 21-22 inches.

Hab.—The whole of North America, breeding far north.

Nest, in a tussock of grass near the water, often surrounded by it so closely that the bird while sitting on the eggs has her feet submerged.

Eggs, two to five, white, often stained with brown.

These beautiful birds, never at any point abundant, are seldom seen in Ontario, because they breed in the far north, and generally make their migratory journey along the sea coast, east or west, where they spend the winter.

In the interior they are sometimes seen singly or in pairs at the shooting stations, where rifle bullets and buck shot at once come into demand.

I once saw four, in full adult plumage, come up Lake Ontario on a stormy afternoon toward the end of March. They alighted for a short time on the open water near the canal, but had a wild, restless look, evidently feeling themselves off their route, and they soon wheeled around and went off east again.

On another occasion a family of four visited Hamilton Bay in the fall. They were not allowed to remain long undisturbed, and one young bird was so disabled by a pellet of shot in the wing, that it was prevented from leaving with the others. It could still take care of itself, however, and remained till the bay was frozen over, when it walked ashore, and was captured in an exhausted condition by one of the fishermen.

Writing from Dunnville, in August, 1893, Dr. Macallum says: "On the 24th of April last, a fine young female Whistling Swan was shot here, which had been frequenting the river for about six weeks. One or more of this species visit us every spring."

During such visits they sometimes get bewildered by fog, and so fall into the hands of the enemy.

Dr. Bell tells us: "This species breeds near Churchill, and on the islands toward the eastern side of Hudson's Bay. Their skins constitute an article of trade, but only a small number of them are collected annually."

For further particulars of their habits we turn again to Mr. Nelson, who is one of the few who have been privileged to see the birds in their northern home.

He says: "The last of June or first of July the young are hatched, and soon after the parents lead them to the vicinity of some large lake or stream, and there the old birds moult their quill feathers, and are unable to fly. They are pursued by the natives at this season, and many are speared from canoes and kyaks. Although unable to fly, it is no easy task, single-handed, to capture them alive. The young men among the Eskimo consider it a remarkable exhibition of fleetness and endurance for one of their number to capture a bird by running it down.

"About twenty miles from St. Michael's, toward the Yukon mouth, is a small, shallow lake, about one-fourth of a mile in diameter, which is grown up with 'horse-tails' (*Equisitum*). This lakelet forms a general rendezvous for all the swans of that vicinity during the summer and fall. During the breeding season they gather there to feed, and the males make it their home. In autumn, as the old birds regain their wing feathers and the young are able to fly, all congregate here, so that I have rarely passed this place without seeing from one hundred to five hundred swans gathered in this small area.

"I have frequently sat and listened with the keenest pleasure to the organ-like swell and fall in their notes, as they were wafted on in rich, full harmony, then sank to a faint murmur, not unlike that of running water. A series of low hillocks afforded a cover by which the lake could be approached, and it was a majestic sight to lie there on a mossy knoll, and gaze on the unsuspecting groups of these graceful birds as they swam back and forth, within rifle shot, not suspecting our presence. Their snowy bodies and beautiful forms, as I last saw them in that far away spot, will linger long in my mind as one of the most unique and interesting sights of my experience in the north. The report of a rifle is sufficient to change the scene into wild confusion. A chorus of confused cries and the heavy beating of hundreds of mighty wings is heard. A cloud of white rises, breaks into numerous fragments, and the birds scatter over the wide flats on every side."

"Toward the end of September these birds begin to gather into flocks, preparatory to migrating, and from the last of this month to the 6th or 8th of October all leave for the south, the exact date varying with the season."

OLOR BUCCINATOR (Rich.).

68. Trumpeter Swan. (181)

Adult:—Plumage, entirely white; younger, the head and neck washed with a rusty brown; still younger, gray or ashy; bill and feet, black. Length, 4-5 feet; tail (normally), of twenty-four feathers; no yellow spots on bill, which is rather longer than the head, the nostrils fairly in its basal half.

HAB.—Chiefly the interior of North America, from the Gulf coast to the fur countries, breeding from Iowa and Dakota northward, west to the Pacific coast, but rare or casual on the Atlantic.

Nest, on dry, high ground, near the beach, a mixture of grass, down and feathers.

Eggs, two to five, dull white, stained with brown, shell rough.

Swans are seen nearly every spring and fall at one or other of the shooting stations in western Ontario, but the points of specific distinction are so inconspicuous that unless the birds are secured, it is difficult to tell to which species they belong. Dr. Garnier reports having taken one at Mitchell's Bay. There was one in the collection sent from Toronto to Paris in 1867, and I have seen two which were killed at Long Point, in Lake Erie.

The highway of this species from north to south is evidently by the Mississippi Valley, where it is quite common during the period of migration, those we see here being merely stragglers off the route.

The history of this swan is not so well known as that of the other. Nelson mentions one specimen with its eggs having been secured at Fort Yukon, which renders it an Alaskan species, though there is no further evidence of its presence in the territory. The lack of observations may be owing to the fact that the interior of Alaska remains almost unexplored, so far as its summer birds are concerned. Hearne speaks of both species breeding on the islands in the lakes to the north-west of Hudson's Bay, and Sir John Richardson gives the Trumpeter a breeding range of from 61° north to well within the Arctic circle.

ORDER HERODIONES. HERONS, STORKS, IBISES, ETC.

SUBORDER IBIDES. SPOONBILLS AND IBISES.

FAMILY IBIDIDÆ. IBISES.

GENUS PLEGADIS KAUP.

PLEGADIS AUTUMNALIS (HASSELQ.).

69. Glossy Ibis. (186)

Plumage, rich dark chestnut, changing to glossy dark green, with purplish reflections on the head, wings and elsewhere; bill, dark. *Young:*—Similar, much duller, or grayish brown, especially on the head and neck, which are white streaked. Claws, slender, nearly straight; head, bare only about the eyes and between the forks of the jaw. Length, about 2 feet; wing, 10-11; tail, 4; bill, 4½; tarsus, 3¼; middle toe and claw, 3.

HAB.—Northern Old World, West Indies, and eastern United States. Only locally abundant, and of irregular distribution in America.

Nest, among the reeds, built of dead and withered reeds, attached to the living ones, well and firmly built, not far above the water.

The eggs of the Glossy Ibis measure from 1·90 by 1·45 to 2·10 by 1·50, and are of a dull greenish-blue color, without markings. The number usually deposited is believed to be three.

About the end of May, 1857, Mr. John Bates, whose farm adjoins the creek near the Hamilton waterworks, saw two tired-looking birds which he took to be curlews, circling round with the evident intention of alighting near the creek. Mr. Bates' gun was always in order, and none in the neighborhood at that time knew better than he how to use it. In a few minutes he picked up a pair of Glossy Ibises, the only birds of the kind which have been observed in Ontario. This pair, which subsequently came into my possession, were male and female in fine adult plumage. They are not common anywhere on the American continent. Wilson knew nothing of the species, nor was it known to naturalists till after his death.

This bird is widely distributed and well known in Europe, and has also been observed in Africa. A few breed in the West Indies, and also in the more southern parts of the eastern United States, but the habitat is local, and the numbers small. In the west is a closely allied species, named the White-faced Glossy Ibis, which is very common along the coast of California to Oregon.

SUBORDER HERODII. HERONS, EGRETS, BITTERNS, ETC.

FAMILY ARDEIDÆ. HERONS, BITTERNS, ETC.

SUBFAMILY BOTAURINÆ. BITTERNS.

GENUS BOTAURUS HERMANN.

SUBGENUS BOTAURUS.

BOTAURUS LENTIGINOSUS (MONTAG.).

70. American Bittern. (190)

Plumage of upper part, singularly freckled with brown of various shades, blackish, tawny and whitish; neck and under parts, ochrey or tawny-white; each feather marked with a brown dark-edged stripe; the throat line, white, with brown streaks; a velvety-black patch on each side of the neck above; crown, dull brown, with buff superciliary stripe; tail, brown; quills, greenish-black, with a glaucous shade, brown tipped; bill, black and yellowish; legs, greenish; soles, yellow. Length, 23-28; wing, 10-13; tail, 4½; bill, about 3; tarsus, about 3½.

HAB.   Temperate North America, south to Guatemala and the West Indies.
The nest of the Bittern is placed on the ground.

The eggs, three to five in number, are brownish-drab, measuring about 2.00
by 1.50.

The American Bittern is a common summer resident, found in all
suitable places throughout Ontario, where, during the early summer,
may be heard the peculiar clunking sound which has gained for the
species the not inappropriate name of "Stake Driver." It seldom
leaves the marsh, where it makes its home and finds its favorite fare
of fish, frogs and lizards. It drops readily to a light charge of shot,
but when wounded makes a fierce resistance, raising the feathers of
the head and neck and striking straight at the eye of a dog with its
sharp-pointed bill. It arrives as soon as the flags begin to show
green, about the end of April, and leaves again for the south toward
the end of September, or later, according to the weather.

During the breeding season it has a wide distribution, being quite
common in the Northern States, and Dr. Bell tells us it is found on
both sides of Hudson's Bay. It does not frequent the clear, running
stream, but the wide, stagnant marsh, where its cry is one of the
characteristic sounds heard the summer through. The cry is very
peculiar, and I do not think it well understood just how it is produced. Many people believe that it is caused by water being taken
into the throat and quickly thrown out again, but so far as I have
noticed, no water is used in the operation.

Once or twice, while hidden among the reeds watching for ducks,
I have seen a bittern uttering his love notes, and the impression
made upon me was, that however pleasing they might be to the ears
for which they were intended, their production must be painful to
the operator, for he looked as if he had recently taken a violent
emetic, and was suffering the usual results.

There are two distinct calls used by these birds, but whether or
not any one bird can utter both is not at present known. These
calls consist of three syllables each, with the accent strongest on the
first and weakest on the middle syllable. They are heard with about
equal frequency. One is soft and windy, like "pumph-ah-gah," while
the other is harder and more decided, like "chunk-a-lunk." Both are
repeated a good many times, and in the still evening, when they are
most frequently observed, they can be heard a long way off. This
call, though loud and deep, is not equal in that respect to the call of
the British Bittern, which is said to resemble the bellowing of a bull,
and is the origin of the term *Botaurus*, by which the bird is now

known. In the "Old Country," it is very generally believed that when the Bittern booms, the whole floating bog on which he is standing at the time vibrates with the sound.

Burns knew this, and refers to it in his writings. We often wish that he had said more about the birds, for the incidental references he makes show that he was a close observer, and well acquainted with their habits.

This will be noticed in the passage in which the Bittern is referred to. He is calling on the feathered tribes to join him in mourning the loss of his friend, Capt. Matthew Henderson :

> " Mourn, ye wee songsters o' the wood ;
> Ye grouse that crap the heather-bud ;
> Ye curlews calling thro' a clud ;
> Ye whistling plover ;
> And mourn, ye whirring paitrick brood ;
> He's gane for ever !
>
> .   .   .   .   .   .
>
> " Mourn, sooty coots, and speckled teals,
> Ye fisher herons, watching eels ;
> Ye duck and drake, wi' airy wheels
> Circling the lake ;
> Ye bitterns, till the quagmire reels,
> Rair for his sake.
>
> " Mourn, clam'ring craiks, at close o' day,
> 'Mang fields o' flow'ring clover gay ;
> And when ye wing your annual way
> Frae our cauld shore,
> Tell thae far warlds, wha lies in clay,
> Wham we deplore.
>
> " Ye houlets, frae your ivy bow'r
> In some auld tree, or eldritch tow'r,
> What time the moon, wi' silent glow'r,
> Sets up her horn,
> Wail thro' the dreary midnight hour
> Till waukrife morn !"
>
> .   .   .   .   .   .

In the spring the plumage of the American Bittern often looks bleached and faded, but in the fall their shades of brown and yellow are exceedingly rich. When wishing to escape notice, I have seen them standing perfectly still among the bulrushes, with the neck extended and the bill pointing straight upwards. On these occasions, their colors were in such perfect harmony with the surroundings that so long as they kept still they were rarely observed.

SUBGENUS ARDETTA GRAY.

BOTAURUS EXILIS (GMEL.).

71. Least Bittern. (191)

No peculiar feathers, but those of the lower neck, long and loose, as in the Bittern; size, very small; 11-14 inches long; wing, 4-5; tail, 2 or less; bill, 2 or less; tarsus, about 1⅜. *Male:*—With the slightly crested crown, back and tail, glossy greenish-black; neck behind, most of the wing coverts, and outer edges of inner quills, rich chestnut, other wing coverts, brownish-yellow; front and sides of neck and under parts, brownish-yellow varied with white along the throat line, the sides of the breast with a blackish-brown patch; bill and lores mostly pale yellow, the culmen blackish; eyes and soles, yellow; legs, greenish-yellow. *Female:*- With the black of the back entirely, that of the crown mostly or wholly replaced by rich purplish-chestnut; the edges of the scapulars forming a brownish-white stripe on either side.   Length, 11-14; wing, 4-5; tail, bill, tarsus, 2 each.

HAB.--Temperate North America, from the British Provinces to the West Indies and Brazil.

Nest, among the rushes.

Eggs, three to five, white with a bluish tinge.

This diminutive Bittern, though seemingly slender and tender, is not only generally distributed in Southern Ontario, but has been reported by Professor Macoun "common throughout the country" in the North-West, and Dr. Bell has specimens from Manitoba and from York Factory. At Hamilton Bay it is a regular summer resident, raising its young in the most retired parts of the marsh. The nest is large for the size of the bird, a platform being made for its support by bending down the flags till they cross each other a foot or more above the water level. The whole affair is very loose and readily falls asunder at the close of the season. The Least Bittern is not supposed to be so plentiful as its big brother, but from its retiring habits may be more so than we are aware. It is seldom seen except by those who invade its favorite haunts, and when disturbed it rises without note or noise of any kind, and with a wavering, uncertain flight passes off for a short distance, again to drop among the rushes.

At other times it has been noticed by the hunter to drop at a point which he marks and goes to as quickly as possible, but can see nothing of the bird. He may find, however, that these little birds breed in communities, often associated with the Rails, and that the portion of the bog which they occupy is interlaced with a series of covered runs, like rat roads, among the flags, along which the little birds travel at a rate which neither dog nor man can emulate, and they are thus enabled to get quickly out of danger without being seen.

In Manitoba the Least Bittern is mentioned as an accidental visitor, only one specimen having been obtained in ten years.

At Hamilton Bay he arrives about the middle of May and leaves early in September.

- - - - - -

BOTAURUS NEOXENUS (Cory).

72. Cory's Bittern. (191*a*)

" Top of the head, back and tail, dark greenish-black, showing a green gloss when held to the light; sides of the head and throat, rufous chestnut; the feathers on the back of the neck showing greenish-black tips; breast and under parts nearly uniform rufous chestnut, shading into dull black on the sides; wing coverts, dark rufous chestnut; under wing coverts, paler chestnut; all the remiges entirely slaty plumbeous; under tail coverts uniform, dull black. Total length, 10.80; wing, 4.30; tarsus, 1.40; culmen, 1.80.

" Hab.—Florida, Okeechobee region.

"In the specimen above described, two of the flank feathers on one side are white, but this may be attributed to albinism. There is no trace of a stripe on the sides of the back as in *A. exilis*. The bird in question is claimed to have been shot in south-west Florida, and was brought to Tampa with a number of other species, including *A. exilis*, *Anas fulvigula* and *Ajaja ajaja*. It is without doubt perfectly distinct from any other known species."

The above is copied from the *Auk*, Vol. III., page 262, which is the first published notice of the species. The writer is Mr. Charles B. Cory.

In the *Auk*, Vol. VIII., page 309, is a notice of another specimen of this bird being secured. There is also given an account of the nest, which was discovered on the borders of the small lake near which Mr. Cory's specimens were found. It was quite similar to the nest of the Least Bittern which occurs in the same region, the one being known as the *brown*, and the other as the *black* bittern. The nest contained four young birds about two-thirds grown, the female allowing herself to be taken in the hand rather than leave the nest. The male was also within three or four feet during the examination.

I read the first account of this little bittern as given above in the *Auk*, for 1886, but thought no more about it till the summer of 1890, when Mr. Wm. Cross, taxidermist, of Toronto, sent me for identi-

fication a mounted specimen of a bird which had been shot in the
marsh near Toronto. It corresponded exactly with the description
given of Cory's Bittern, but to make sure I sent the specimen to Mr.
Ridgway, who at once pronounced it a genuine specimen of *neoxenus.*
Mr. Cross got the specimen in good condition and mounted it very
nicely. It is now in the collection of the Canadian Institute in Toronto. So the record stood for Canada until the summer of the
present year, when notice was received of a second specimen having
been captured at Toronto. On this occasion a female in perfect plumage was shot by a fisherman named Ramsden, in the Ashbridge
marsh, near where the first was found, and within two days of the
same date, the first being taken on the 18th of May, 1890, and the
second on the 20th May, 1893. This makes a total of eight specimens now known to be in collections.

This is evidently a southern bird, and it has not been found anywhere away from Florida, where it was first discovered, except these
two specimens which have been obtained so near the same place at
Toronto. No doubt this species associates with our common little
bitterns, many of which spend the winter in Florida, and it is just
possible that some gallant *exilis* has in his own way painted the
beauties of Ashbridge's marsh in such glowing colors as to induce this
little brown lady to accompany him to the north, when he started on
his annual journey in spring. Pity she did not fare better, as her
report might have induced others to follow the route.

SUBFAMILY ARDEINÆ. HERONS AND EGRETS.

GENUS ARDEA LINN.

SUBGENUS ARDEA.

ARDEA HERODIAS LINN.

73. Great Blue Heron. (194)

Back, without peculiar plumes at any season, but scapulars lengthened and
lanceolate; an occipital crest, two feathers of which are long and filamentous;
long loose feathers on the lower neck. Length, about 4 feet; extent, 6; bill, 5½
inches; tarsus, 6½; middle toe and claw, 5; wing, 18-20; tail, 7. *Female:*—
Much smaller than male. Adult of both sexes grayish-blue above, the neck
pale purplish-brown with a white throat line; the head, black, with a white
frontal patch; the under parts mostly black, streaked with white; tibia, edge
of wing and some of the lower neck feathers, orange-brown; bill and eyes,

yellow; culmen, dusky; lores and legs, greenish. The young differ considerably, but are never white, and cannot be confounded with any of the succeeding.

HAB.—North America, from the Arctic regions southward to the West Indies and northern South America.

Nest, usually in trees, sometimes on rocks.

Eggs, two or three, elliptical light, dull greenish-blue.

As the Great Blue Heron breeds in communities, it is not often seen during the summer, except in the vicinity of the heronry. In the fall, when the young birds are able to shift for themselves, they disperse over the country, their tall, gaunt figures being often seen standing motionless watching for eels by the shore of some muddy creek. In the report of the Ornithological Branch of the Ottawa Field Naturalists' Club, for 1883, is a most interesting account of a visit paid by a number of members of the club to a heronry situated on the bank of the river about twenty-five miles from the city. Limited space will admit of only a short extract: "The heronry is located in the centre of a thick swamp which, on the occasion of our first visit, was so deeply submerged as to bar all ingress. On the 19th of July, however, the water was but knee deep. After proceeding about half a mile into the swamp, our attention was arrested by a peculiar sound which we at first thought proceeded from some distant saw-mill or steamer on the river. As we advanced, however, the sound resolved itself into the most extraordinary noises, some of which resembled the yelping of dogs or foxes. On penetrating still deeper into the swamp, we discovered that the noises proceeded from immense numbers of herons, some perched on branches of trees, some sitting on the nests, and others flying overhead. The uproar was almost deafening, and the odor arising from the filth with which the trees and ground were covered was extremely disagreeable. We tramped all through the heronry, and calculated that it must extend about half a mile in each direction. The nests were all of the same pattern, great cumbersome piles of sticks, about a foot thick, with but a very shallow cavity and no lining.

"The birds were very tame, making no attempt to fly until we began to climb the trees on which they were; and even then they moved lazily off and manifested little or no alarm at our near approach to their young."

Usually the adult Heron is an exceedingly wary bird, and is seldom obtained except when he happens to fly above some hunter who is concealed among the rushes watching for ducks.

When thus brought down from above with neck, wings and legs all mixed up, he presents a most ragged appearance, but when seen alive at shooting distance, the graceful movements of the long, lithe neck, with its pointed plumes, present a sight we all like to look upon.

In Southern Ontario, Herons are seen occasionally wherever their favorite fish is to be found. Throughout the North-West, they are not common, though Mr. Thompson tells of being one of a party who accidentally found a heronry where it was little expected, in the Poplar Woods, at the head waters of Bird's-tail Creek.

SUBGENUS HERODIAS BOIE.

ARDEA EGRETTA GMEL.

74. American Egret. (196)

No obviously lengthened feathers on the head at any time; in the breeding season, back with very long plumes of decomposed feathers drooping far beyond the tail; neck, closely feathered; plumage, entirely white at all seasons; bill, lores and eyes, yellow; legs and feet, black. Length, 36-42 inches (not including the dorsal train); wing, 16-17; bill, nearly 5; tarsus, nearly 6.

HAB.—United States, southerly, straggling northward to Nova Scotia. Massachusetts, Canada West and Minnesota. West Indies, Mexico, Central and South America.

Nest, in trees or bushes.

Eggs, three or four, pale greenish-blue.

This species has a wide distribution in the south, but Canada seems to be its northern boundary. It is only an accidental visitor here, and, strange to say, nearly all of those obtained have been young birds. There is a record in the *Auk*, Vol. II., page 110, of a pair seen at Rockcliffe, on the Ottawa River, in the spring of 1883. The male was obtained, and is now in the Museum of the Geological Survey at Ottawa. These were adults, but the specimen in my collection, which was obtained at Rond Eau, near the west end of Lake Erie, and others which I have heard of along our southern border, were all young birds.

Dr. Wheaton gives the same account of those found in Ohio, and Dr. Coues in his "Birds of the North-west," page 521, says: "I may here observe that a certain northward migration of some southerly birds at this season is nowhere more noticeable than among the Herons and their allies, the migrants consisting chiefly of birds hatched that year, which unaccountably stray in what seems to us to be the wrong direction."

ARDEA CANDIDISSIMA GMEL.

75. Snowy Heron. (197)

Adult. With a long occipital crest of decomposed feathers and similar dorsal plumes. latter *recurved* when perfect; similar, but not recurved plumes on the lower neck, which is bare behind; lores, eyes and toes, yellow; bill and legs, black, former yellow at base. latter yellow at lower part behind. Plumage always entirely white. Length. 24; wing, 11-12; bill, 3; tarsus, 3½-4.

HAB.--Temperate and tropical America, from New Jersey, Minnesota and Oregon. south to Patagonia: casually on the Atlantic coast to Nova Scotia.

Nest. a platform of sticks. usually in top of a tall tree, sometimes in a bush above the water.

Eggs. three to five. pale bluish-green.

The young of the Snowy Heron inherit the family peculiarity of making their first journey in the wrong direction, and it is to this fact that we are indebted for the visits we occasionally receive from them along our southern frontier during the early fall. I have had them sent to me from Long Point, on Lake Erie, and have heard of their being captured at other places, but all were young birds, and I have no record of the species being found breeding in Ontario.

8

On the east coast, it is found breeding as far north as Long Island, and in the interior it occurs in Oregon, but the true home of these little herons is farther south. In all suitable places throughout South and Central America, the West Indies and Mexico, they breed in colonies in immense numbers, though I grieve to say that of late years they have been almost exterminated by plume hunters.

— ┼

SUBGENUS BUTORIDES BLYTH.

ARDEA VIRESCENS LINN.

76. Green Heron. (201)

Adult:—In the breeding season with the crown, long soft occipital crest, and lengthened narrow feathers of the back, lustrous dark green, sometimes with a bronzy iridescence, and on the back often with a glaucous cast; wing coverts, green, with conspicuous tawny edgings; neck, purplish-chestnut, the throat-line variegated with dusky or whitish; under parts, mostly dark brownish-ash, belly variegated with white; quills and tail, greenish-dusky, with a glaucous shade, edge of the wing white; some of the quills, usually white-tipped; bill, greenish-black, much of the under mandible, yellow; lores and iris, yellow; legs, greenish-yellow; lower neck, with lengthened feathers in front, a bare space behind. *Young:*—With the head less crested, the back without long plumes, but glossy greenish; neck, merely reddish-brown, and whole under parts, white, variegated with tawny and dark brown. Length, 16-18; wing, about, 7; bill, 2½; tarsus, 2; middle toe and claw, about the same; tibia, bare, 1 or less.

HAB.—Canada and Oregon, southward to northern South America and West Indies; rare or absent in the Middle Provinces.

Nest, composed of twigs, placed in a bush or low tree in a swamp or by the bank of a stream.

Eggs, three to six, pale greenish-blue.

This handsome little Heron finds its northern limit along the southern border of Ontario. According to Dr. Macallum, it breeds regularly on the banks of the Grand River, near Dunnville, and has also been observed, occasionally, near Hamilton and at the St. Clair Flats. Like the others of its class, the Green Heron feeds mostly at night, and is seldom seen abroad by day, except by those who have occasion to invade its marshy haunts. On this account it may be more numerous than it is supposed to be. It arrives about the end of April, and leaves for the south again in September.

In the North-West it has been taken in the Assiniboine, and is said to breed in the sloughs around Turtle Mountain, but is nowhere common.

Genus NYCTICORAX Stephens.

Subgenus NYCTICORAX.

NYCTICORAX NYCTICORAX NÆVIUS (Bodd.).

77. Black-crowned Night Heron. (202)

No peculiar feathers, excepting two or three very long filamentous plumes springing from the occiput, generally imbricated in one bundle; bill, very stout; tarsi, reticulate below in front. Length, about 2 feet; wing, 12-14 inches; bill, tarsus and middle toe, about 3. Crown, scapulars and inter-scapulars, very dark glossy green; general plumage, bluish-gray, more or less tinged with lilac; forehead, throat-line and most under parts, whitish; occipital plumes, white; bill, black; lores, greenish; eyes, red; feet, yellow. Young:— Very different; lacking the plumes; grayish-brown, paler below, extensively speckled with white; quills, chocolate-brown, white-tipped.

Hab.—America, from the British possessions southward to the Falkland Islands, including part of the West Indies. Breeds in communities, returning to the same place year after year.

Nest, a large loose platform of sticks and twigs, placed well up in a tall tree. Eggs, four to six, pale greenish-blue.

In Ontario the Night Heron, or "Quawk," as it is commonly called, is not generally distributed, though stragglers are occasionally seen at different points throughout the Province. Their breeding places are by no means common, the vicinity of the sea being evidently preferred to the interior.

Along the banks of the lower St. Lawrence they breed in immense numbers, every tree in certain districts having several nests among its boughs. When viewed from a distance these trees have the appearance of being heavily coated with dirty whitewash, and the entire vegetation underneath them is killed by the accumulated droppings of the birds.

Though somewhat untidy in their surroundings at home, the birds themselves when seen in spring plumage are very handsome, the fiery-red eyes and long, flowing plumes giving them quite an interesting appearance.

In the North-West they occur in limited numbers, but are not regularly distributed. There the nests are made in the marsh, and fixed to the reeds, eight or ten inches above the water.

Night Herons have been found throughout the greater portion of South America, and in some sections of the United States they have large heronries, where thousands breed together. They usually select a clump of tall trees, not easy of access, and have four

or five nests in each tree. They all go wandering when the young
are able to fly, but return again to their breeding place with the
return of the season.

Order PALUDICOLÆ. Cranes, Rails, etc.
Suborder GRUES. Cranes.
Family GRUIDÆ. Cranes.
Genus GRUS Pallas.
GRUS AMERICANA (Linn.).

78. Whooping Crane. (204)

Adult:—With the bare part of head extending in a point on the occiput
above, on each side below the eyes, and very hairy; bill, very stout, convex,
ascending, that part of the under mandible as deep as the upper opposite to it.
Adult plumage, pure white, with black primaries, primary coverts and alula:
bill, dusky greenish; legs, black; head, carmine, the hair-like feathers blackish.
Young:—With the head feathered; general plumage, gray (?), varied with
brown. Length, about 50 inches; wing, 24; tail, 9; tarsus, 12; middle toe, 5;
bill, 6.

Hab.—Interior of North America, from the fur countries to Florida, Texas
and Mexico, and from Ohio to Colorado. Formerly on the Atlantic coast, at
least casually, to New England.

Nest, on the ground among weeds or rank grass, built of fine, tough grass,
firmly put together and neatly formed.

Eggs, two or three, light brownish-drab, marked with large, irregular spots
of dull chocolate-brown and obscure shell-markings.

Mr. John Ewart, of the village of Yarker, in the county of
Addington, has a mounted specimen of the Whooping Crane in his
collection, which is the only record I have of the species in Ontario.
In the fall of 1871, it was observed frequenting the borders of a
small, shallow lake in the township of Camden, and for a week or
ten days the local gunners were on its track. It was very shy,
but finally fell before the gun of Wesley Potter on the 27th of
September.

Wilson speaks of this species as an occasional visitor in the
marshes at Cape May during its migration, but now it is hardly
known in the East, its line of migration being along the Mississippi
valley. It breeds in Manitoba, and is said to be found also in the
fur countries.

When wounded, it is a dangerous bird to approach, for it drives its sharp bill with great force and precision at its antagonist.

For many years the Sandhill Crane was believed to be the young of this species, but in all stages of plumage they can now be readily identified.

GRUS MEXICANA (Müll.).

79. Sandhill Crane. (206)

Adult:—With the bare part of head forking behind to receive a pointed extension of the occipital feathers, not reaching on the sides below the eyes, and sparsely hairy; bill, moderately stout, with nearly straight and scarcely ascending gonys, that part of the under mandible not so deep as the upper at the same place; adult plumage, plumbeous-gray, never whitening; primaries, their coverts and alula, blackish. *Young:*—With head feathered, and plumage varied with rusty brown. Rather smaller than the last.

HAB.—Southern half of North America: now rare near the Atlantic coast, except in Georgia and Florida.

Eggs, two, light brownish-drab, marked, except at the greater end, with blotches of dull chocolate-brown; shell, rough, with numerous warty elevations.

I am indebted to Dr. Garnier, of Lucknow, for the first record I have had of the occurrence of the Sandhill Crane in Ontario. Writing, under date December 6th, 1884, he says : "About twenty-two years ago a pair of these birds spent the summer in the marshes near Murphy's Landing, County Kent. Later in the season they were seen stalking about, accompanied by two young, and finally all disappeared as the weather grew cold."

" In 1881 a pair spent the summer near Mud Creek, in the same locality, and were often seen by the people residing there. On the 1st of November, Mr. Joseph Martin, while out shooting in his canoe, suddenly came upon them at short distance. He killed one, and the other, being hard hit, dropped on a shaking bog close by. Mr. Martin brought me the dead one, and next day I went with him in search of its mate. We saw it lying quite dead on the bog, but, though my partner and I tried hard to force our way to where it was, we were compelled to give it up, to my very great regret."

In the *Auk*, Vol. V., page 205, is a notice by Mr. W. E. Saunders, of London, stating that he had received from Mr. M. J. Dodds, of St. Thomas, a Sandhill Crane, which was killed at Rond Eau in 1869, by Mr. John Oxford.

These are the only well authenticated instances of the occurrence

of the Sandhill Crane in Ontario that I know of, but as they are
known to breed in Michigan, we cannot be surprised at their sometimes being found on the east side of the Detroit River. The species
is most abundant in the Mississippi Valley, west to the Pacific coast.
It is irregularly distributed, having been found breeding in suitable
places outside of its ordinary range. It was seen by Dr. Bell near
Norway House, on Hudson's Bay. In the North-West it is reported
as tolerably common, and is often taken young and domesticated,
making an interesting pet. Mr. Thompson says of it : " As a game
bird, I am inclined to place the present species as first on the game
list in Manitoba. An average specimen weighs about ten pounds,
and the quality of the flesh is unsurpassed by any of our ordinary
birds, unless it be the partridge. I should strongly advocate the
protection of this bird by the game law were it not that it is so
thoroughly able to take care of itself that legislation in its favor
seems altogether unnecessary."

SUBORDER RALLI. RAILS, GALLINULES, COOTS, ETC.

FAMILY RALLIDÆ. RAILS, GALLINULES, COOTS, ETC.

SUBFAMILY RALLINÆ. RAILS.

GENUS RALLUS LINNÆUS.

RALLUS ELEGANS (AUD.).

80. King Rail. (208)

Above, brownish-black, variegated with olive-brown, becoming rich chestnut
on the wing coverts ; under parts, rich rufous or cinnamon-brown, usually paler
on the middle of the belly and whitening on the throat ; flanks and axillars,
blackish, white-barred. Length, about 16 ; wing, 5-6 ; tail, 2-2½ ; bill, 2½ ;
tarsus, 2 ; middle toe and claw, 2¼. Female.—Smaller.

HAB.—Fresh-water marshes of the eastern portion of the United States,
from the Middle States, northern Illinois, Wisconsin and Kansas southward.
Casually north to Massachusetts, Maine and Ontario.

Nest, a rude mass of reeds and grass, on marshy ground close to the water.

Eggs, six to twelve, buff or cream color, speckled and blotched with reddish-
brown.

This large and handsome Rail, which, until recently, was considered
to be only a casual visitor to Ontario, is now known to breed plentifully in the marshes all along the River St. Clair. It has also been

found at other points in Southern Ontario, but the St. Clair Flats seem to be its favorite breeding place. The extent of the marsh and the almost stagnant water appear to suit the taste of these birds, and here they spend the summer and raise their young without being disturbed.

They are seldom seen on the wing, but become very noisy and excited before rain, keeping up an incessant cackling, which, better than anything else, gives an idea of the number that are moving about under cover of the rushes.

This seems to be the northern boundary of their habitat in Ontario, for they are not found in the North-West, nor anywhere else to the north of us. They arrive in May and leave in September.

RALLUS VIRGINIANUS (Linn.).

81. **Virginia Rail.** (212)

Coloration, exactly as in *elegans*, of which it is a perfect miniature. Length, 8½-10½; wing, about 4; tail, about 1½; bill, 1½-1⅞; tarsus, 1½-1½; middle toe, 1½-1¾.

Hab.—North America, from British Provinces south to Guatemala and Cuba.

Nest, in a tuft of reeds or rushes, some of them bent down to assist in forming the structure, which is usually placed close to the water.

Eggs, six to nine, buff or creamy, speckled and blotched with reddish-brown and obscure lilac.

Although this cannot be said to be a numerous species, it is very generally distributed, being found in all suitable places throughout the Province. When not disturbed, it may be seen quietly wading in the shallow ponds in search of its food, which consists of aquatic insects, snails, worms, and the seeds of such grasses as grow near its haunts. If alarmed, it at once takes to the rushes, and passes with such swiftness along the covered runways which interlace the rush beds, that it will elude the pursuit of an active dog, and avoid exposing itself to the aim of the sportsman.

It is found in the North-West, but is not abundant.

In Southern Ontario it arrives early in May and leaves late in September.

GENUS PORZANA VIEILLOT.

SUBGENUS PORZANA.

PORZANA CAROLINA (LINN.).

82. Sora. (214)

Above, olive-brown, varied with black, with numerous sharp white streaks and specks; flanks, axillars and lining of wings, barred with white and blackish; belly, whitish; crissum, rufescent. *Adult:*—With the face and central line of the throat black, the rest of the throat, line over eye, and especially the breast more or less intensely slate-gray, the sides of the breast usually with some obsolete whitish barring and speckling. *Young:*—Without the black, the throat whitish, the breast brown. Length, 8-9; wing, 4-4½; tail, about 2; bill, ⅞-¾; tarsus, 1¼; middle toe and claw, 1⅞.

HAB.—Temperate North America, but most common in the Eastern Province, breeding chiefly northward. South to West Indies and northern South America.

Builds a rude nest of grass and rushes on the ground near the water.

Eggs, eight to ten, dull drab, marked with reddish-brown.

Here, as elsewhere, the Sora is the most numerous of the Rail family, and is found breeding in all suitable places throughout the country. Many also pass up north, and when they return in the fall, accompanied by their young, they linger in the marshes along the southern border till they are found swarming everywhere. They are very sensitive to cold, and a sportsman may have good rail shooting till late in the evening, but should a sharp frost set in during the night, he may return in the morning and find that the birds have all left.

Many spend the summer in the North-West, but they are most abundant in the Middle Atlantic States, where great numbers are killed for the table in the fall.

In Southern Ontario they arrive in May, and leave in September at the first touch of frost.

SUBGENUS COTURNICOPS BONAPARTE.

PORZANA NOVEBORACENSIS (GM.).

83. Yellow Rail. (215)

Above, varied with blackish and ochrey-brown, and thickly marked with narrow white semicircles and transverse bars; below, pale ochrey-brown, fading on the belly, deepest on the breast where many of the feathers are tipped with dark brown; flanks, rufous with many white bars; lining of the wing, white; a brownish-yellow streak over the eye. Length, about 6 inches; wing, $3\frac{1}{2}$; tail, $1\frac{1}{2}$; bill, $\frac{1}{2}$.

HAB.—Eastern North America, from Nova Scotia and Hudson's Bay west to Utah and Nevada. No extra-limital record except Cuba and the Bermudas.

Nest, like that of the other rails.

Eggs, six to eight, dark buff color, marked with reddish spots at the greater end.

We know little of this bird, partly because it belongs to a class much given to keeping out of sight, but chiefly because it is a rare species everywhere. During the present year, I saw a fine mounted specimen in the store of Mr. Cross, taxidermist, Toronto. It was obtained in the marsh near that city, and I have heard of another which a few years ago was shot near the same place, and is now in the public museum at Ottawa. The greater number of specimens of the Yellow Rail now in existence have been found in New England, but that may be owing to the greater number of collectors there. It would be well for our Canadian sportsmen to look out for the species when visiting its haunts, because from its general resemblance to the Sora, it may readily be overlooked.

One observer reports it as a tolerably common summer resident near Winnipeg, and it has also been noticed at Fort George by Dr. Bell. It has, therefore, a wide distribution, but is nowhere abundant.

SUBGENUS CRECISCUS CABONIS.

PORZANA JAMAICENSIS (GMEL.).

84. Black Rail. (216)

Upper parts, blackish, finely speckled and barred with white; the hind neck and fore neck, dark chestnut; head and under parts, dark slate color, paler or whitening on the throat; the lower belly, flanks and under wing and tail covert barred with white; quills and tail feathers with white spots, very small. Length, about 5.50; wing, 2.75-3.00; tail, 1.35; tarsus, 0.75.

HAB.—South and Central America and West Indies.

Nest, in a deep cup-shaped depression resembling that of the Meadow Lark, only deeper in proportion to its width; the outer portion composed of grass-stems and blades, the inner portion of soft blades of grass arranged in a circular manner and loosely interwoven.

Eggs, ten, clear white, thinly sprinkled with reddish-brown dots which become more numerous toward the greater end.

This is one of the rarest of American birds, or, rather, it is one of those least frequently found. From its small size and the fact that

its life is spent mostly among the rank vegetation of the marsh,
where it cannot be seen, it may not be so rare as we suppose.

I mention it here on the authority of the late Dr. Cottle, of
Woodstock, who, in an article in the *Canadian Journal* for September, 1859, claims to have found a bird of this species near Inger-
soll in 1857, which at the time of his writing was in the collection of
Wm. Poole, jun.

I have not seen this specimen, but I knew Dr. Cottle, and feel
sure that no mistake would be made in the identification.   There is
no reason why the Black Rail should not be found in Ontario, for it
occurs to the east and west of us, and will yet, I expect, be found
in one or more of the many suitable haunts which occur throughout
the Province.

The Dundas marsh is exactly the sort of place where one might
expect to meet with this species.   The marsh extends from the
Hamilton city limits about four miles westward to the town of
Dundas, and has an average width of a mile, with many inlets winding inland.   The banks are generally steep and wooded, and the
water in the summer is covered with aquatic plants and clumps of
floating bog.   Malaria reigns there, and there are mosquitoes in
abundance.

About 1873-74, I heard that a young man, named Nash, was
diligently exploring this uninviting locality, to find out what birds
were there during the summer, and that he had been successful in his
researches; but he left for the North-West about that time, and the
matter was forgotten.

On learning that Mr. Nash had quite recently returned to reside
in Toronto, I wrote to him on the subject, and promptly received the
following reply, which came to me while the first part of this article
was passing through the hands of the printer :

"TORONTO, January 12th, 1894.
"THOS. McILWRAITH, ESQ.

"DEAR SIR,—My original note as to the capture of the Black Rails is as
follows:

"'August 18th, 1874.—Shot four of these birds this evening at the upper
end of the Dundas marsh. My dogs put them up where the rushes had been
mowed. This is the first time I ever noticed them here.'

"After this date I saw several others about the same place, during this
same year (1874).   I also shot a few Yellow Rails, and saw many.

"Yours, etc.,
"C. W. NASH."

SUBFAMILY GALLINULINÆ.

GENUS IONORNIS REICHENBACH.

IONORNIS MARTINICA (LINN.).

85. Purple Gallinule. (218)

Head, neck and under parts beautiful purplish-blue, blackening on the belly, the crissum white: above olivaceous-green, the cervix and wing coverts tinted with blue; frontal shield blue; bill red, tipped with yellow: legs yellowish. *Young:*—With the head, neck and lower back brownish, the under parts mostly white, mixed with ochrey. Length, 10-12; wing, 6½-7; tail. 2½-3; bill from gape. about 1¼; tarsus. about 2¼: middle-toe and claw, about 3.

HAB.—South Atlantic and Gulf States, north casually to New England (Maine, Nova Scotia) and Ontario.

Nest. built among rushes over the water; the taller rushes are bent down and woven together as a support.

Eggs, eight or nine, cream color, finely dotted with chestnut-brown and umber.

In April, 1892, I received a letter from Pickering. describing a bird which had recently been shot by Mr. James Cowan, at the mouth of the Rouge in that township, and asking me to identify it. The description was so perfectly correct that I had no difficulty in deciding it to be the Purple Gallinule, though I had not before heard of its occurrence in Ontario, and I replied to that effect.

I have since learned that the specimen is now mounted and located somewhere in Toronto.

This beautiful Gallinule is a resident of the South Atlantic and Gulf States, but occasionally strays as far north as Ohio, Wisconsin, Maine and New York.

It is a very handsome bird, whose presence in our marshes would add to their interest, but we are too far north to expect it. save as a very rare visitor.

GENUS GALLINULA BRISSON.

GALLINULA GALEATA (LICHT.).

86. Florida Gallinule. (219)

Head, neck and under parts grayish-black, darkest on the former, paler or whitening on the belly; back, brownish-olive; wings and tail dusky; crissum edge of wing, and stripes on the flank, white; bill, frontal plate. and ring around tibiæ red, the former tipped with yellow; tarsi and toes, greenish: 12-15 long; wing, 6½-7½; tail, 3½; gape of bill, about 1½; tarsus, about 2.

HAB. -Temperate and tropical America from Canada to Brazil and Chili.

Nest, a mass of broken, rotten reeds and rushes, with a slight hollow in the centre: it is seldom much above water level, and often afloat, but is fastened to the sedges.

Eggs, ten to twelve, brownish-buff, thickly spotted with reddish-brown.

This is a common summer resident, breeding in suitable places throughout Southern Ontario. Near Hamilton it is quite common, a few pairs generally spending the summer in the Waterdown creek, and also in the Dundas marsh. Its retired haunts are seldom invaded during the summer months, for the mosquitoes form a bar to the intrusion of visitors, and its flesh not being in demand for the table, it is not much disturbed.

Southern Ontario seems to be the northern limit of its habitat. It is not mentioned among the birds of Manitoba, neither have I heard of it appearing elsewhere in the north.

Like most of its class, it arrives in May and leaves in September.

SUBFAMILY FULICINÆ.

GENUS FULICA LINNÆUS.

FULICA AMERICANA GMEL.

87. American Coot. (221)

Dark slate, paler or grayish below, blackening on the head and neck, tinged with olive on the back; crissum, whole edge of wing, and top of the secondaries white; bill, white or flesh-colored, marked with reddish-black near the end; feet, dull olivaceous. Young:—Similar, paler and duller. Length, about 14; wing, 7-8; tail, 2; bill, from the gape, 1½-1½; tarsus, about 2; middle toe and claw, about 3.

HAB.—North America, from Greenland and Alaska southward to West Indies and Central America.

Nest, of vegetable rubbish from the marsh, often afloat and fastened to the rushes like the Grebes, but sometimes on dry ground back from the water.

Eggs, ten to twelve, clear clay color, dotted minutely with dark brown.

This species is very generally distributed in suitable places throughout Ontario, and also in the North-West. It breeds abundantly at St. Clair, but at Hamilton is only a migratory visitor in spring and fall. They are hardy birds, often arriving in spring before the ice is quite away, and again lingering late in the fall, as if unwilling to depart. They are sometimes mistaken for ducks by amateur gunners, and in this way a few lose their lives, but except in such cases they are not molested, "Mud-hens" not being generally looked upon as game.

The Coots are strong of wing, good swimmers, and capable of enduring both cold and fatigue. They are very abundant throughout the North-West, their haunts being in the marshes, for which their lobed feet are admirably adapted. There they spend the summer and find the enjoyment peculiar to their race. In the fall they assemble in vast flocks and generally all disappear at once during the night. In Alaska, only one wanderer of the species is reported. There is a similar report from Greenland, which is its most northerly record.

Order LIMICOLÆ. Shore Birds.

Family PHALAROPODIDÆ.

Genus CRYMOPHILUS Vieillot.

CRYMOPHILUS FULICARIUS (Linn.).

88. Red Phalarope. (222)

Adult.—With the under parts, purplish chestnut of variable intensity, white in the young; above, variegated with blackish and tawny. Length, 7-8 inches; wing, 5; tail, 2¾; bill, 1, yellowish, black-tipped; tarsus, ¾, greenish.

Hab.—Northern parts of the northern hemisphere, breeding in the Arctic regions, and migrating south in winter; in the United States south to the Middle State, Ohio, Illinois and Cape St. Lucas; chiefly maritime.

Nest, a hollow in the ground lined with dry grass.

Eggs, three or four, variable in color, usually brownish-olive, spotted or blotched with dark chocolate-brown.

Vast numbers of Phalaropes breed in Spitzbergen, and on the shores of the Polar Sea. At the approach of winter they retire to the south, but in these migratory journeys they follow the line of the sea coast, so that the stragglers we see inland are most likely bewildered by fog, or driven by storm away from their associates and their regular course.

Dr. Garnier saw a flock of six, one of which he secured, at Mitchell's Bay, near St. Clair, in the fall of 1880; and on the 17th of November, 1882, Mr. Brooks, of Milton, shot a single bird, which he found swimming alone on Hamilton Bay, a little way out from Dynes' place. On the 21st of October, 1886, Mr. White shot one on the Rideau River, and on the 1st September, 1888, he got a second specimen on the River Ottawa, which completes the record for Ontario, so far as I know at present.

Turning to our usual authority for information about northern species, we find Mr. Nelson saying: "This handsome Phalarope arrives at the Yukon mouth and adjacent parts of Behring Sea coast during the last few days of May or first of June, according to the season. Its preference is for the flat, wet lands bordering the coast and rivers, where it remains to breed. Very early in June the females have each paid their court, and won a shy and gentle male to share their coming cares. The eggs are laid in a slight depression, generally on the damp flats where the birds are found. There is rarely any lining to the nest." The eggs are hatched in July, and as soon as the young are able to fly, all leave the nesting ground and are found only at sea.

"They breed all along the Arctic shores of Alaska and Siberia, wherever suitable flats occur, and even reach those isolated islands, forever encircled by ice, which lie beyond. It is not rare in Spitzbergen, where its eggs have been found laid upon the bare ground.

"During the cruise of the *Corwin*, in the summer of 1881, we found this and the Northern Phalarope abundant wherever we went on the Alaskan or Siberian shores of the Arctic, and their pretty forms, as they flitted here and there over the surface of the smooth sea, now alighting a moment and gliding quickly to right and left, pecking at the minute animals in the water, then taking wing for an instant, appeared in ever-changing groups." "In winter these birds pass south and occur along the coasts of the Pacific on both shores, reaching the south coasts of India on the Asiatic side."

The foregoing are but brief extracts from Mr. Nelson's interesting account of the home habits of these little-known birds, my limited space having prevented me from making more lengthy quotations.

GENUS PHALAROPUS BRISSON.

SUBGENUS PHALAROPUS.

PHALAROPUS LOBATUS (LINN.).

89. Northern Phalarope. (223)

Adult:—Dark opaque ash or grayish-black, the back variegated with tawny: upper tail coverts and under parts, mostly white; side of the head and neck. with a broad stripe of rich chestnut, generally meeting on the jugulum; breast, otherwise with ashy-gray. *Young:*—Lacking the chestnut. Length, about 7 inches: wing, 4½; tail, 2; bill, tarsus and middle toe, each under 1, black.

HAB. Northern portions of northern hemisphere, breeding in Arctic latitudes; south in winter to the tropics.

Nest, a hollow in the ground, lined with dry grass.

Eggs, three or four, similar to those of the Red Phalarope, but smaller.

Like the preceding, this is a bird of the sea coast, but, singly or in pairs, it is sometimes seen inland during the season of migration. The two in my collection were found in the fall on one of the inlets of Hamilton Bay.

In the List of the "Birds of Western Ontario," mention is made of three having been taken in Middlesex, and one found dead at Mitchell's Bay in 1882.

While this was passing through the press, K. C. McIlwraith shot a young male of the species, as it rose from one of the inlets which run from the bay up to the Beach road near Hamilton.

Although a bird of the sea coast, the Northern Phalarope is found more frequently in the interior than the Red Phalarope. It has been observed at Ottawa, Toronto, Hamilton, London, and also at Dunnville, where Dr. Macallum says it may be counted on with tolerable certainty every season during October.

In Manitoba, most of the observers are mute regarding it, though Mr. Nash says: "Common autumn visitor to Portage la Prairie, and very abundant at the prairie sloughs near Winnipeg, where I saw immense flocks of them in August and September, 1886."

Of Alaska, Mr. Nelson says that the first arrivals reach St. Michael's in full plumage about the middle of May, and by the first of June they are in full force, and ready to begin the business of the season. The young are hatched during June, and by the 20th of July are fledged and on the wing. Soon they begin to gather in parties of from five, to one hundred or more, keeping by the large ponds and inlets till about the end of September, from which date they are seen no more for the season.

"They breed on all the islands of Behring Sea, the north coast of Siberia, and we saw them common about Herald and Wrangel islands in July and August, 1881. It is plentiful throughout the interior of Northern Alaska, as well as on the salt marshes of the coast."

SUBGENUS STEGANOPUS VIEILLOT.

PHALAROPUS TRICOLOR (VIEILL.).

90. Wilson's Phalarope. (224)

Adult: Ashy; upper tail coverts and under parts, white; a black stripe from the eye down the side of the neck, spreading into rich purplish-chestnut, which also variegates the back and shades the throat. *Young:*—Lacking these last colors. Length, 9-10; wing, 5; tail, 2; bill, tarsus and middle toe, each over 1, black.

HAB.—Temperate North America, chiefly the interior, breeding from northern Illinois and Utah northward to the Saskatchewan region, south in winter to Brazil and Patagonia.

Nest, in moist meadows.

Eggs, three or four, variable in pattern, usually brownish-drab, marked with splashes, spots and scratches of chocolate-brown.

This is the largest of the Phalaropes and the handsomest of all our waders. Unlike the others of its class, it is rare along the sea coast, but common inland, its line of migration being along the Mississippi Valley. Another peculiarity of the species is that the female is the larger and more gaily attired, and, from choice or necessity, the eggs are incubated by the male. In some other respects their domestic relations are not in accordance with the recognized rules of propriety.

The first record I had of it as an Ontario species was in Mr. Saunders' "List of Birds of Western Ontario," where mention is made of one having been taken at Mitchell's Bay in May, 1882. I did not hear of it again until I received the report of the Sub-section of the Canadian Institute, in which it is stated that, at a meeting held June 2nd, 1890, Mr. Wm. Cross reported having received on the 2nd inst. a female Wilson's Phalarope in full breeding plumage, which had been shot in Toronto marsh. At a subsequent meeting held September 23rd, it was stated that, while Mr. Bunker was watching for ducks off the sand bar at the west end of the Island, a Wilson's Phalarope pitched among his decoys and was secured. So the record of Ontario stands for the present.

It is more common in the interior than along the sea coasts, and is now known to breed in suitable places throughout the northern tier of States, and also from the Red River to the Rockies, along the boundary line.

On July 24th, 1880, Mr. Macoun reports finding it breeding around the ponds at Moose Mountain.

The prairie ponds seem to be the favorite resort of this beautiful species, and as these are not common in Ontario, we may not have the birds except as visitors.

FAMILY RECURVIROSTRIDÆ.

GENUS RECURVIROSTRA LINNÆUS.

RECURVIROSTRA AMERICANA GM.

91. American Avocet. (225)

White; back and wings, with much black; head and neck, cinnamon-brown in the adult, ashy in the young; bill, black, 3¼ to gape; legs, blue; eyes, red. Length, 16-18; wing, 7-8; tail, 3½; tarsus, 3½.

HAB.—Temperate North America, from the Saskatchewan and Great Slave Lake south, in winter, to Guatemala and the West Indies. Rare in the Eastern Province.

Eggs, three or four, variable in size and marking, usually brownish-drab, marked with spots of chocolate-brown.

This is another delicate inland wader, rare on the sea coast, but abundant in the Mississippi Valley. Stragglers appear occasionally at far distant points, and are at once identified by their peculiar markings and awl-shaped bill. I am aware of three individuals having been taken at different times at Rond Eau, on the north shore of Lake Erie, but these are all I have heard of in Ontario. In Manitoba it is spoken of as being exceedingly rare, but it is very abundant around the saline ponds and lakes in the North-West (Macoun).

According to Dr. Coues, "It is more abundant than elsewhere in the interior of the United States along the Mississippi Valley, and thence westward, in all suitable localities, to the Rocky Mountains."

Its preference for salt or brackish waters is indicated by its abundance at Great Salt Lake, in Utah, and about the alkaline waters of Dakota.

FAMILY SCOLOPACIDÆ. SNIPES, SANDPIPERS, ETC.

GENUS PHILOHELA GRAY.

PHILOHELA MINOR (GMEL.).

92. American Woodcock. (228)

Above, variegated and harmoniously blended black, brown, gray and russet; below, pale warm brown of variable shade. Length, *male*, 10-11; *female*, 11-12; extent, 16-18; wing, 4½-5; bill, 2½-3; tarsus, 1¼; middle toe and claw, 1½; weight, 5-9 ounces.

HAB.—Eastern Province of North America, north to the British Provinces,

9

west to Dakota, Kansas, etc.; breeding throughout its range. No extralimital records.

The nest, which is composed of a few dead leaves, is usually placed at the root of a tree, or in a clump of weeds.

Eggs, three or four, grayish-brown, marked with spots and blotches of lilac and chocolate.

The Woodcock is a summer resident in Southern Ontario in uncertain numbers, appearing about the time the snow is going out of sight. In the fall it is much sought after by sportsmen, with varying success. Occasionally good bags are made, but in this respect no two seasons are alike.

The birds seem to be paired on their arrival in spring, and at once select a site for the nest, which is usually placed in dense woods or swampy thickets. When the breeding season is over, they change their places of resort and are often found in corn fields, orchards and moist places, where they feed mostly during the night. They remain as long as the ground is soft enough for them to probe, after which they retire to the south.

Writing from Hamilton, I may say that this species seems to be better known to the south and east of us than it is to the north and west. Dr. Bell says: "I saw one specimen of the Woodcock in August last at York Factory. This bird is not uncommon in Manitoba, though the fact is not generally known." Other observers in Manitoba report single birds having been procured at long intervals, so that it must either be very scarce or seldom seen.

Throughout the Eastern States it is more common, but is so highly prized as a game bird that it is persecuted wherever it is known to exist.

Genus GALLINAGO Leach.

GALLINAGO DELICATA (Ord.).

93. Wilson's Snipe. (230)

Crown, black with a pale middle stripe; back, varied with black, bright bay and tawny, the latter forming two lengthwise stripes on the scapulars; neck and breast speckled with brown and dusky; lining of wings, barred with black and white; tail, usually of sixteen feathers, barred with black, white and chestnut; sides, waved with dusky; belly, dull white; quills, blackish, the outer, white edged. Length, 9-11; wing, $4\frac{1}{2}$-$5\frac{1}{4}$; bill, about $2\frac{1}{2}$: whole naked portion of leg and foot, about 3.

Hab.—North and middle America, breeding from northern United States northward; south in winter to West Indies and northern South America.

Nest, usually a depression in a grassy meadow.

Eggs, three or four; grayish-olive, heavily marked with umber-brown and irregular lines of black.

This is *the* Snipe of America, although the name is often erroneously applied to other species. It is sometimes called English Snipe, owing to the close resemblance it bears to the British bird, but those who have compared the two species state positively that they are different in their markings.

In Southern Ontario this species is known only as a migrant in spring and fall. During the former season they are more protected by the Game Act, so that the short visit they pay us in April must really be to them a time of enjoyment. In the fall it is quite different, for every nominal sportsman wants to go snipe shooting, and the birds are so frequently fired at, they are kept continually on the move from the time of their arrival till they take their final departure for the season.

In former years, the breeding ground of the Snipe was a matter of speculation. It is now known to breed along the northern border of the northern tier of States, and is also common during the summer season in suitable places throughout Manitoba and the North-West.

Here again, Nelson's report from Alaska is quite interesting. He says: "This is a rather uncommon but widely spread species in Alaska along the mainland shore of Behring Sea. I found it both at St. Michael's and on the lower Yukon in small numbers, making its presence known in spring-time by its peculiar whistling noise as it flew high overhead. It nests wherever found in the north, and is a rather common species along the entire course of the Yukon, extending thence north to within the Arctic Circle, but its limit in this direction is not definitely known.

"It was found at Sitka and Kadiak by the Western Union Telegraph explorers, but is not known on any of the Behring Sea islands nor on the coast of Siberia, but it is to be looked for from the latter region, at least."

GENUS MACRORHAMPHUS LEACH.

MACRORHAMPHUS GRISEUS (GMEL.).

94. Dowitcher. (231)

Tail and its coverts, at all seasons, conspicuously barred with black and white (or tawny), lining of the wings and axillars the same; quills, dusky; shaft of first primary, and tips of the secondaries, except long inner ones, white; bill and feet, greenish-black. In summer, brownish-black above, variegated with bay; below, brownish-red, variegated with dusky; a tawny superciliary stripe, and a dark one from the bill to the eye. In winter, plain gray above, and on the breast, with few or no traces of black and bay; the belly, line over eye and under eyelid, white. Length, 10-11; wing, 5-5½; tail, 2½; bill, about 2½; tarsus, 1½; middle toe and claw, 1¼.

HAB.—Atlantic coast of North America, breeding far north.

Nest, a hollow near the borders of marshy lakes or ponds, lined with a few leaves and grass.

Eggs, three or four; identical in appearance with those of the common snipe.

Although this species is abundant along the sea coast during the season of migration, it can only be regarded as an accidental traveller in Ontario. The specimen in my collection is the only one I have ever found near Hamilton. In the " List of Birds of Western Ontario " it is spoken of as rare, and Dr. Wheaton, in his exhaustive " List of the Birds of Ohio," says he never saw it in that State, but has had it reported as a rare spring and fall migrant.

Geo. R. White has found it on one or two occasions at Ottawa, but it is restricted to the Atlantic coast, and those found inland are only stragglers from the ranks during the season of migration.

They are gentle, unsuspicious birds, allowing a near approach, and, as they fly in compact flocks and gather very closely together when alighting, there is great opportunity for unlimited slaughter among them by anyone bent on filling the " bag."

They are a very abundant species and must breed in great numbers somewhere, though exactly where I have not found on record.

Dr. Coues says that it breeds in high latitudes, and Ridgway describes it as " breeding far northward, Nushagak River, Alaska (straggler)." It is spoken of by Dr. Richardson as having " an extensive breeding range throughout the fur countries, from the borders of Lake Superior to the Arctic Ocean." Which of the two species he found on the borders of Lake Superior is not apparent.

These birds are seen in greatest numbers along the shores of the Atlantic States in fall and winter. They are highly esteemed for the table, and are slaughtered in great numbers for the market.

In the North-West, including Alaska, the class is represented by the Long-billed Dowitcher, a bird very similar in habit and appearance but of larger size, the bill, especially, being longer than in the present species.

GENUS MICROPALAMA BAIRD.

MICROPALAMA HIMANTOPUS (BONAP.).

95. Stilt Sandpiper. (233)

Adult in summer:—Above, blackish, each feather edged and tipped with white and tawny or bay, which on the scapulars becomes scalloped; auriculars, chestnut; a dusky line from bill to eye, and a light reddish superciliary line; upper tail coverts, white with dusky bars; primaries, dusky with blackish tips; tail feathers, ashy-gray, their edge and a central field white; under parts mixed, reddish, black and whitish, in streaks on the jugulum, elsewhere in bars; bill and feet, greenish-black. *Young and adult in winter:*—Ashy-gray above, with or without traces of black and bay, the feathers usually with white edging; line over the eye and under parts white; the jugulum and sides suffused with the color of the back, and streaked with dusky; legs, usually pale. Length, 8-9 inches; wing, 5; tail, 2¼; bill and tarsus, both 1½-1⅜; middle toe, 1.

HAB.—Eastern Province of North America, breeding north of the United States, and migrating in winter to the West Indies, Central and South America.

Nest, a depression in the ground, lined with grass and leaves.

Eggs, three or four, light-drab or grayish-white, with bold spots and markings of chestnut-brown.

I have some scruples about including this species in my list, for I have no record of its having been taken within the Province; but, when we consider that it breeds to the north of us, and winters far to the south, there can be no reasonable doubt that it passes through Ontario. Being rather a scarce species, it may have escaped the notice of sportsmen, or it may have been taken and no record made of the occurrence. I anticipate that when this list is made public, I shall learn of birds having been found in Ontario which are not included here, for the simple reason that I had not heard of them. There is no convenient way of placing such records before the public, and they drop out of sight and are forgotten.

It is to be hoped that the writer of the next list of the birds of Ontario will, for this reason, have many additions to make to the present one.

While this article is in the hands of the printer, Mr. Cross, taxidermist, of Toronto, sends me a bird for identification, which proves to be this species. It is one of two which were shot near Toronto about the 25th of June last, by Mr. Heinrich. Mr. Cross has made a happy hit in mounting them. They look like a pair of miniature curlews.

The above was written eight years ago, and since then there has been but little to add to our acquaintance with this species in Ontario. That little comes from Toronto, where the birds seem to have found a place to suit them, but I fear they are not to be allowed peaceful possession of the same. The first record appears in the report of the Ornithological Sub-section of the Canadian Institute for 1889, where it is stated: "On September 26th, we secured three of these rare Sandpipers, all shot at Toronto." At a meeting of the same sub-section, held on the 23rd September, 1890, it was stated by Mr. T. Hannar, that on the 28th July he shot a fine Stilt Sandpiper on Ashbridge's Bar. The fact of this specimen having been obtained in July would indicate that the birds are breeding in that neighborhood, but so far we have no account of their nests.

GENUS TRINGA LINNÆUS.

SUBGENUS TRINGA.

TRINGA CANUTUS LINN.

96. **Knot**. (234)

Bill, equalling or rather exceeding the head, comparatively stout. *Adult in summer:*—Above, brownish-black, each feather tipped with ashy-white, and tinged with reddish on scapulars; below, uniform brownish-red, much as in the robin, fading into white on the flanks and crissum; upper tail coverts white with dusky bars, tail feathers and secondaries grayish-ash with white edges; quills, blackish; gray on the inner webs and with white shafts; bill and feet, blackish. *Young:*—Above, clear ash, with numerous black and white semicircles; below white, more or less tinged with reddish, dusky speckled on breast, wavy barred on sides. Length. 10-11; wing, 6-6½; tail, 2⅓, nearly square; bill about 1⅓ (very variable).

HAB.—Nearly cosmopolitan. Breeds in high northern latitudes, but visits the southern hemisphere during its migration.

Nest, a depression in the sand.

Eggs, light pea-green.

This is the largest and handsomest of the Sandpipers, and though common along the sea coast, it is only an occasional visitor inland. The specimen in my collection I killed many years ago on the muddy shore of one of the inlets of the Bay. I did not see the Knot again till May, 1884, when K. C. McIlwraith killed four very fine specimens in a moist vegetable garden on the beach. Dr. Wheaton met with it only once in Ohio, and it is not mentioned in the "List of the Birds of Western Ontario," from which it may be inferred that we are not on the line of its migrations.

Mr. White reports its occurrence at Ottawa but once. He says: "On the 4th June, 1890, E. White obtained eight out of a flock of about seventy birds. They were all in full adult plumage, but, strange to say, we have not seen a single specimen since."

In Manitoba, it occurs occasionally during migrations, but irregularly and not in large numbers. One observer (Hunter) says: "I have never seen the Knot along the Red River, but have seen large flocks west of Brandon."

In the North-West, Prof. Macoun says that it is frequently found along the borders of salt marshes.

In Alaska, Mr. Nelson secured a single specimen, which was the only one seen during his residence there.

Along the shores of New England, during spring and fall, it is still abundant, though the numbers have during past years been greatly reduced in comparison with what they once were.

It is observed, occasionally, along the coasts of Scotland, and in England it is sometimes seen in very large flocks, all of which are migrants.

On the west coast of the Pacific, it migrates as far south as Australia and New Zealand to spend the winter, and at that season has been found in Damara Land, Africa, and also in Brazil.

In the *Auk* for January, 1893, page 25, Mr. Geo. H. Mackay gives a most interesting and exhaustive history of the haunts and habits of this species, from which I should like to quote at length, did my limits permit. Mr. Mackay says: "This bird, which formerly sojourned on these shores in great abundance, and occurs now to a limited extent during its migrations, has been the subject of considerable inquiry as to the cause of its appearing now in such reduced numbers. As each contribution to the subject may add something in assisting to correct conclusions, I have to present the following *résumé*, especially of the habits and movements of this bird during its short stay in Massachusetts, while on migration." Mr. Mackay then speaks of the clouds of these birds which visited the coast of New England thirty or forty years ago, and how they were slaughtered wholesale by a most barbarous practice called "firelighting." He continues :

"I have it directly from an excellent authority, that he has seen, in the spring, six barrels of these birds (all of which had been taken in this manner) at one time on the deck of the Cape Cod packet for Boston. He had also seen barrels of them, which had spoiled during the voyage, thrown overboard in Boston Harbor on the arrival of the packet. The price of these birds at that time was ten cents per dozen; mixed with them would be Turnstones and Black-bellied Plover. Not one of these birds had been shot, all had been taken with the aid of a 'firelight.'

" Besides those destroyed on Cape Cod in this way, I have reasons for believing that they have been shot also in large numbers on the coast of Virginia in the spring, on their way north to their breeding grounds; one such place shipping to New York city in a single spring, from April 1st to June 3rd, upwards of 6,000 Plover, a large share of which were Knots.

" It is not my intention to convey the impression that the Knots are nearly exterminated, but they are much reduced in numbers, and are in great danger of extinction, and comparatively few can now be seen in Massachusetts, where formerly there were twenty to twenty-five thousand a year, which I consider a reasonable estimate of its former abundance."

For many years the great *desiderata* among oölogists were the eggs of the Knot. Even now there are very few in existence, and it is only a few years since the first authenticated specimen was procured. The members of every expedition which visited the lands where the Knot was known to breed had instructions to search for these eggs, but one after another returned without success. Major N. W. Fielding, naturalist to the Nares' Arctic Expedition of 1875-76, says: "I was not so fortunate as to obtain the eggs of the Knot during our stay in the polar regions, though it breeds in some numbers along the shores of Smith Sound and the north coast of Grinnell Land. During the month of July my companions and I often endeavored to discover the nest of this bird, but none of us were successful.

" However, on July 30th, 1876, the day before we broke up our winter-quarters, where we had been frozen in for eleven months, three of our seamen, walking along the border of a small lake not far from the ship, came upon an old bird accompanied by three nestlings which they brought to me. These young ones I have since seen in the British Museum at South Kensington, where, in company with a pair of the old birds, they constitute one of the most attractive of the many natural groups which adorn Mr. Sharpe's department."

Finally, Lieut. A. W. Greely, commander of the late expedition to Lady Franklin Sound, succeeded in obtaining the long sought-for egg, and with great propriety requested Dr. C. H. Merriam to accept the honor of being the first to publish the account of it.

The specimens of bird and egg were obtained in the vicinity of Fort Conger, latitude 81° 44′ N.

In color the egg was light pea-green, closely spotted with brown in small specks about the size of a pin-head.

Subgenus ARQUATELLA Baird.

TRINGA MARITIMA (Brünn.).

97. Purple Sandpiper. (235)

Bill, little longer than the head, much longer than the tarsus, straight or nearly so; tibial feathers, long, reaching to the joints; though the legs are rarely bare a little way above. *Adult:*—Above, ashy-black, with purplish and violet reflections, most of the feathers with pale or white edgings; secondaries, mostly white; line over eye, eyelids and under parts, white; the breast and jugulum, a pale cast of the color of the back; and sides marked with the same. In winter, and most immature birds, the colors are similar but much duller; very young birds have tawny edgings above, and are mottled with ashy and dusky below. Length, 8-9 inches; wing, 5; tail, 2¾, rounded; bill, 1¼; tarsus, ¾; middle toe, 1, or a little more.

Hab.—Northern portions of the northern hemisphere; in North America, chiefly the northern portions, breeding in the high north, migrating in winter to the Eastern and Middle States, the Great Lakes and the shores of the larger streams in the Mississippi Valley.

Nest, a mere depression in the ground with a scant lining of grass.

The eggs are said to be four in number, clay color, shaded with olive, and marked with rich umber-brown.

This, like the preceding species, is common to both continents, and is of circumpolar distribution. If it has been in the habit of passing this way, it did so without being observed till the 31st of October, 1885, when one individual was killed at Hamilton Beach, by Dr. K. C. McIlwraith. This is the only record we have of it in the Province.

As its name (*maritima*) implies, it is a bird of the sea coast, but though a Sandpiper, it is not so fond of the sandy shores as it is of the rocky ledges covered with sea weeds, where it no doubt finds something to suit its taste. The name *purple* might lead a stranger to expect this to be a bird of showy colors, but in general appearance it is perhaps the least so of its class, and might be described as about the size and build of the Black-heart, dull slaty-blue above, belly and vent white. When in full plumage, the feathers feel soft and silky for a bird of this class, and in certain rays of light seem slightly glossed with purple.

Since the above was written, eight years ago, one or two more specimens have been found at Hamilton. Mr. White got one at Ottawa, and Mr. Cross had one brought to him at Toronto. This tends to show that the bird is a rare straggler so far from the sea.

It does not appear among the "Birds of Manitoba," nor those of Alaska, but breeds abundantly on the shores of Hudson's Bay and Melville Peninsula.

SUBGENUS ACTODROMAS KAUP.

TRINGA MACULATA VIEILL.

98. Pectoral Sandpiper. (239)

Coloration much as in Baird's Sandpiper, but crown noticeably different from cervix; chestnut edgings of scapulars, straight-edged; chin, whitish, definitely contrasted with the heavily ashy-shaded and sharply dusky-streaked jugulum. Large. Length, 8½-9 inches; wing, 5-5¼; bill, tarsus and middle toe with claw, about 1½: bill and feet, greenish.

HAB.—The whole of North America, the West Indies, and the greater part of South America. Breeds in the Arctic Regions. Of frequent occurrence in Europe.

Nest, in a tuft of grass.

Eggs, four, pale grayish-buff, varying to pale olive-green, blotched and spotted with vandyke-brown.

While on their extended migratory journey in spring and fall, these birds rest and refresh themselves on the marshes and lake shores of Ontario, where they are frequently observed by sportsmen, in flocks of considerable size.

Near Hamilton they are not of regular occurrence, though they occasionally appear in the fall in goodly numbers, and if the weather keeps soft, remain till October.

While here they frequent the grassy meadows and muddy inlets near the Bay, being very seldom noticed on the sand.

Like several others of the same class, this species has a wide geographical distribution, being found in Iceland, Europe and Asia.

Those who have seen this species only in Ontario can have but little idea of its appearance during the breeding season, as seen by Mr. Nelson in Alaska, and I regret that his description is too long to be copied in full. He was under his tent on a lonely island near the mouth of the Yukon. He says : " My eyelids began to droop and the scene to become indistinct, when suddenly a low, hollow, booming note struck my ear, and sent my thoughts back to a spring morning in Northern Illinois, and to the loud, vibrating tones of the prairie chickens. A few seconds passed and again arose the note : a moment later and, gun in hand, I stood outside the tent. Once again the note was repeated close by, and a glance revealed its author. Standing in the thin grass, ten or fifteen yards from me, with its throat inflated until it was as large as the rest of the bird, was a male Pectoral Sandpiper.

" The succeeding days afforded opportunity to observe the bird as it uttered its singular notes, under a variety of situations, and at various hours of the day or during the light Arctic night. Before the bird utters these notes, it fills its œsophagus with air to such an extent that the breast and throat is inflated to more than twice its natural size, and the great air sac thus formed gives the peculiar resonant quality to the note. Whenever the Pectoral pursues his love-making, his rather low but pervading note swells and dies in musical cadences, which form a striking part of the great bird chorus heard at this season in the north."

TRINGA FUSCICOLLIS Vieill.

99. White-rumped Sandpiper. (240)

Size. medium; upper tail coverts, white; feet, black; bill, black, light-colored at base below; coloration otherwise much as in Baird's Sandpiper. An ashy wash on the jugulum is hardly perceptible except in young birds, and then it is slight : the streaks are very numerous, broad and distinct, extending as specks nearly or quite to the bill, and as shaft lines along the sides.

Hab. Eastern Province of North America, breeding in the high north. In winter, the West Indies, Central and South America. south to the Falkland Islands. Occasional in Europe.

Nest, a depression in the ground, lined with grass and a few withered leaves. Eggs. three or four, light olive-brown, spotted with deep dark chestnut.

Several of our Sandpipers resemble each other so much in general appearance that by the gunner they are considered as all of one sort and treated alike—that is, they are tied in bunches by the neck or legs and handed over to be prepared for the table. With the collector it is different : every individual is carefully examined as to species, sex, age and condition, so that nothing may be lost that is worth preserving. In the present species, the white rump is always a distinguishing mark, most conspicuous while the birds are on the wing. Inland it is not very common, but a few are usually seen associating with the others during the season of migration. The pair in my collection I found on the sandy shore of Lake Ontario near the Burlington Canal.

It is an eastern species, with a long range north and south. It is very abundant on the shores of New England during the season of migration.

Inland it is found in Ontario, Manitoba and the North-West, but only in small flocks while migrating. It does not appear west of the Rocky Mountains, and only one or two stragglers have been found in Alaska. It is said to breed abundantly in the Mackenzie River region, and it was found by McFarlane breeding on the shores of the Arctic Sea. As winter approaches it moves south through the United States to the West Indies, Central and South America and the Falkland Islands. Stragglers have also been found in Europe.

TRINGA BAIRDII (Coues).

100. Baird's Sandpiper. (241)

Adult male:—Bill, wholly black, small and slender, slightly shorter than the head, just as long as the tarsus, or as the middle toe and claw, slightly expanded or lancet-shaped at the end, the point acute; grooves, long, narrow, deep; feathers on the side of lower mandible evidently reaching further than those on upper. Upper parts, brownish-black (deepest on the rump and middle upper tail coverts, and lightest on the neck behind), each feather bordered and tipped with pale brownish-yellow, the tipping of the scapulars broadest and nearly white, their marginings broad and brightest in tint, making several deep scallops toward the shafts of the feathers; only the outer series black, the others plain gray, with paler margins: jugulum, tinged with light, dull yellowish-brown, spotted and streaked with ill-defined blackish markings, as are also the sides under the wings; throat and other under parts, white, unmarked; feet, black, like the bill. Length, 7.25; extent, 15.25; wing, 4.90; bill, 0.85; tarsus, middle toe and claw, the same. The *female* is entirely similar, but slightly larger. The *young* have the upper parts wholly light brownish-ash, darker on the rump, and all the feathers with a dark field, and pale or whitish edging;

waves of brownish-black on the scapulars; jugulum and breast, suffused with
dull, light reddish-brown, the spotting small, sparse and very indistinct.

HAB.--The whole of North and South America, but chiefly the interior of
North and the western portions of South America. Rare along the Atlantic
coast, and not yet recorded from the Pacific coast. Known to breed only in the
Arctic Regions.

Nest, a slight depression, lined with grass, usually shaded by a tuft of grass.
Eggs, three or four, clay color, spotted with rich umber-brown.

Dr. Coues, in his new "Key to North American Birds," says that
this is the most abundant small Sandpiper in some parts of the west
during migrations, but it has not been found on the Pacific coast
and is quite rare on the Atlantic. The only record we have of its
occurrence in Ontario is that of a fine specimen now in my collection,
which was shot at Hamilton Beach on the 25th of August, 1885, by
K. C. McIlwraith. It was singled out among a flock of small sandpipers by its peculiar erratic snipe-like flight, and on being secured,
its dainty little body was picked up with feelings which only the
enthusiastic collector can understand.

It is named after S. F. Baird, of the Smithsonian Institute, and,
so far as known, is peculiar to the American continent.

On the 23rd of August, 1886, while this article was in the
printer's hands, the locality where the specimen herein referred to
was obtained was again visited, and, strange to say, another individual of the species was secured at the same place, under similar
circumstances. On the 1st of September the place was again visited,
and two more were obtained, but on two subsequent visits, made
within a day or two, no more were seen. Those who are observant
of the migratory movements of the birds must often have been
astonished to see with what persistent regularity certain birds appear
at certain places at a given time. In the present instance these
are the only birds of the kind we have ever seen or heard of in
Ontario, but they were all found within a few yards of the same
spot, and within ten days of the same date in different years.

This species seems to prefer travelling inland, as it is rare on
the Atlantic coast, and has not yet been observed on the Pacific.
Mr. Nelson found only one, an immature bird, during his residence
in Alaska; but it is known to breed in the barren lands in the
Arctic Regions.

It was not until 1861, that Dr. Coues disentangled this species
from the general crowd, and gave it the name and rank by which it
has since been called. It is so little known, that we have yet much
to learn of its peculiarities.

TRINGA MINUTILLA Vieill.

101. Least Sandpiper. (242)

Upper parts in summer, with each feather blackish centrally, edged with bright bay, and tipped with ashy or white; in winter and in the young, simply ashy; tail feathers, gray, with whitish edges, the central blackish, usually with reddish edges; crown, not conspicuously different from hind neck; chestnut edgings of scapulars usually scalloped; below, white, the jugulum with dusky streaks and an ashy or brownish suffusion; bill, black; legs, dusky greenish. Smallest of the sandpipers. Length, 5½-6 inches; wing, 3¼-3½; tail, 2 or less; bill, tarsus and middle toe with claw, about ¾.

HAB.—Whole of North and South America, breeding north of the United States. Accidental in Europe.

Nest, a depression in the ground, lined with grass and leaves.

Eggs, three or four, light drab, thickly sprinkled with reddish-brown spots.

The appearance of this, the smallest of the Sandpipers, always excites a feeling of pity as he is seen hurrying along the sand in rear of his big brothers, uttering his feeble "peep" as if begging them to leave a little for him.

In Ontario it is a common species, found in all suitable places in spring and fall, but its breeding ground is far north, and little, if anything, is known of its nest or eggs. Some might consider that a matter of no consequence, but here is what Dr. Coues says about it in his "Birds of the North-West": "Fogs hang low and heavy over rock-girdled Labrador. Angry waves, palled with rage, exhaust themselves to encroach upon the stern shores, and, baffled, sink back howling into the depths. Winds shriek as they course from crag to crag in mad career, till the humble mosses that clothe the rocks crouch lower still in fear. Overhead the Sea Gulls scream as they winnow, and the Murres, all silent, ply eager oars to escape the blast. What is here to entice the steps of the delicate birds? Yet they have come, urged by resistless impulse, and have made a nest on the ground in some half-sheltered nook. The material was ready at hand, in the mossy covering of the earth, and little care or thought was needed to fashion a little bunch into a little home.

"Four eggs are laid (they are buffy-yellow, thickly spotted over with brown and drab), with the points together, that they may take up less room and be more warmly covered; there is need of this, such large eggs belonging to so small a bird. As we draw near the mother sees us, and nestles closer still over her treasures, quite hiding them in the covering of her breast, and watches us with timid eyes, all anxiety for the safety of what is dearer to her than her own life.

Her mate stands motionless, but not unmoved, hard by, not venturing even to chirp the note of encouragement and sympathy she loves to hear.

"Alas! hope fades, and dies out, leaving only fear; there is no further concealment—we are almost upon the nest almost trodden upon, she springs up with a piteous cry and flies a little distance, re-alighting, almost beside herself with grief : for she knows only too well what is to be feared at such a time. If there were hope for her that her nest were undiscovered, she might dissimulate, and try to entice us away by those touching deceits which maternal love inspires. But we are actually bending over her treasures, and deception would be in vain : her grief is too great to be witnessed unmoved, still less portrayed ; nor can we, deaf to her beseeching, change it to despair. We have seen and admired the home —there is no excuse for making it desolate ; we have not so much as touched one of the precious eggs, and will leave them to her renewed and patient care."

It is found as a migrant in Ontario and the North-West as far as Alaska. In the latter region it is rare, none having been found on the islands of Behring Sea.

SUBGENUS PELIDNA CUVIER.

TRINGA ALPINA PACIFICA (COUES).

102. **Red-backed Sandpiper.** (243a)

Adult in summer:—Above, chestnut, each feather with a central black field, and most of them whitish-tipped; rump and upper tail coverts, blackish; tail feathers and wing coverts, ashy-gray; quills, dusky with pale shafts; secondaries, mostly white; and inner primaries, edged with the same; under parts, white; belly, with a broad jet black area; breast and jugulum, thickly streaked with dusky; bill and feet, black. *Adult in winter, and young:*—Above, plain ashy-gray, with dark shaft-lines, with or without red or black traces; below, white; little or no trace of black on the belly; jugulum, with a few dusky streaks and an ashy suffusion. Length, 8-9 inches; wing, $4\frac{1}{2}$-5; tail, 2-$2\frac{1}{2}$; bill, $1\frac{1}{2}$-$1\frac{3}{4}$, longer than head, compressed at base, rather depressed at the end; tibia, bare about $\frac{1}{2}$; tarsus, 1, or rather less.

HAB.—North America in general, breeding far north, and straggling to eastern coast of Asia.

Nest, in the vicinity of lakes and ponds, a hollow in the ground lined with a few withered leaves.

Eggs, three or four, clay color, spotted, stained and blotched with chestnut.

This is the Black-heart Plover of sportsmen. It is a regular visitor in Ontario in the season of migration, appearing on the shores of Lake Ontario with wonderful regularity on the Queen's birthday (May 24th), as if to afford sport to our gunners on that Canadian holiday. It is much in favor with those who are fond of killing a great number of birds at once, for it usually appears in large, compact flocks and is not very difficult of approach. I once saw seventy-six killed or wounded with the discharge of two barrels. They had just arrived on the shore, and, seeming tired after a long flight, settled on a partially submerged log near the water's edge, from which they were unwilling to rise, and allowed the gunner to do as stated, to his extreme delight. It did not occur to one, looking at so large a number of dead and wounded birds, that any very commendable feat had been accomplished, but so it was considered at the time, and so it will be again, I presume, with that class of sportsmen, but the like opportunity may not happen soon again, as the number of Black-hearts which now visit that locality is very small.

On leaving Southern Ontario in spring they pass on to the North-West, where they breed abundantly in Alaska and in the Arctic Regions.

They are again seen in the fall, when they spend a few days before leaving for winter-quarters in the south.

TRINGA FERRUGINEA Brünn.

103. Curlew Sandpiper. (244)

Adult:—Crown of the head and entire upper parts, greenish-black, each feather tipped and indented with yellowish-red; wing coverts, ashy-brown, each feather with dusky shaft line and reddish edging; upper tail coverts, white, with broad dusky bars, tinged at their extremities with reddish; tail, pale gray, with greenish reflection; sides of the neck and entire under parts uniform, deep brownish-red; under tail coverts, barred with dusky; axillars and under wing coverts, white; bill and feet, greenish-black. Length, $8\frac{1}{2}$; wing, about 5; bill, $1\frac{1}{2}$.

Hab.—Old World in general; occasional in eastern North America.

Nest, by the margin of lakes and rivers, a slight hollow lined with leaves and grass.

Eggs, three or four, pale-greenish buff, spotted and blotched with chestnut-brown.

So far as at present known, the Curlew Sandpiper is only a straggler on the American continent, about ten or a dozen being all the recorded captures. It is quite a common British species, and

10

like others peculiar to those eastern lands, may occasionally be
wafted westward against its inclinations, but no nest of the species
has yet been found on this side of the Atlantic.

In 1867, the Board of Arts of Western Canada prepared a "Catalogue of Birds Observed in the Country," in connection with the
collection which, during that year, was sent to the Paris Exposition.
The Curlew Sandpiper is named in the catalogue, but no specimen
was available for the collection. I have mentioned it here, chiefly
with the view of placing the technical description in the hands of
those interested, so that they may be able to identify the species
should they at any time fall in with it.

Genus EREUNETES Illiger.

EREUNETES PUSILLUS (Linn.).

104. Semipalmated Sandpiper. (246)

Adult in summer:—Above, variegated with black, bay and ashy or white,
each feather with a black field, reddish edge and whitish tip; rump and upper
tail coverts, except the lateral ones, blackish; tail feathers, ashy-gray, the
central darker; primaries dusky, the shaft of the first white; a dusky line from
the bill to the eye, and a white superciliary line; below, pure white, usually
rufescent on the breast, and with more or less dusky speckling on the throat,
breast and sides, in young birds usually wanting; in winter the upper parts
mostly plain ashy-gray; but in any plumage or under any variation the species
is known by its small size and semipalmated feet. Length, 5½-6½ inches;
wing 4¼-4⅞; tarsus and middle toe and claw, about 1; bill, variable from ⅝ to
1¼, averaging ⅞.

HAB.—Eastern Province of North America, breeding north of the United
States; south in winter to the West Indies and South America.

Nest, a depression in the ground, in or near some moist place, lined with
withered grass.

Eggs, three or four, variable in color, usually clay color, blotched or spotted
with umber-brown.

This is a very abundant species during the season of migration,
thronging alike the shores of the Atlantic and those of our inland
lakes and marshes.

They visit the borders of Hamilton Bay in spring and fall in considerable numbers, but are so much disturbed by amateur gunners
that they soon seek for more retired feeding grounds elsewhere.

They are usually found associating with the Least Sandpiper,
which they much resemble in general appearance, but the semipal-

mated toes of the present species are always a reliable distinguishing mark.

This is an eastern species, which prefers the shores of the Atlantic as its line of travel, passing up north in spring. It is tolerably common in Manitoba and the North-West, but is not seen in Alaska. Its centre of abundance seems to be the shores of Labrador and Hudson's Bay, where it breeds in great numbers. They can raise only one brood, for they return from the north before August is out, and soon pass on south, where they are said to winter from the Carolinas southward.

Genus CALIDRIS Cuvier.

CALIDRIS ARENARIA (Linn.).

105. Sanderling. (248)

Adult in summer:—Head, neck and upper parts varied with black, ashy and bright reddish; below, from the breast, pure white; tail, except central feathers, light-ash, nearly white; primaries, gray with blackish edges and tips, the shafts of all and bases of most, white; secondaries, white, except a space at the end, and greater coverts broadly white tipped; bill and feet, black. *Adult in winter and young:*—No reddish; speckled with black and white; sometimes tawny tinged on the jugulum. Length, 7½-8; wing, 4½-5; tail, 2½; bill, about 1; tarsus, 1, or rather less; middle toe and claw, ¾.

HAB.—Nearly cosmopolitan, breeding in the Arctic and subarctic regions, migrating in America, south to Chili and Patagonia.

Nest, a hollow in the ground, lined with grass and dead leaves.

Eggs, two to four, greenish-brown, spotted and blotched with brown of different shades.

The Sanderling is a species of very wide geographical distribution, being found in suitable places nearly all over the world.

It visits the shores of the great lakes in Ontario during the season of migration, and appears in different dress according to age or the season of the year. In spring the breast and foreneck are tinged with pale rufous, but in autumn the whole lower parts are as white as snow. It is a very active species, and, when feeding along the shore, shows great celerity in following the receding wave, or keeping clear of the next one that rolls up on the beach. When wounded in the wing, it will run with great swiftness, and even take to the water and swim well. In spring their visits to Hamilton Bay are uncertain and of short duration, but on the return trip they appear about the end of August, and are found all through the fall.

The flocks of Sanderlings which come over the boundary into Southern Ontario in May, soon pass on by easy stages up north, being noticed at various points on the way. They have been found breeding on the islands along the coast of Greenland, also in Grinnell Land, and on the shores and islands of Hudson's Bay. Mr. Nelson did not see the species at St. Michael's, but he says that it breeds along the barren Arctic shore of the north Alaskan coast, east of Point Barrow, but not in any numbers south of this point.

It is found wintering in low latitudes on both shores of the Pacific, but is rather rare on the coast of India.

GENUS LIMOSA BRISSON.

LIMOSA FEDOA (LINN.).

106. Marbled Godwit. (249)

Tail, barred throughout with black and rufous; rump and upper tail coverts like the back; no pure white anywhere. General plumage, rufous, or cinnamon-brown; below, nearly unmarked, and of very variable shade, usually deepest on the lining of the wing; above, variegated with black and brown or gray; quills, rufous and black; bill, flesh-colored, largely tipped with black; feet, dark. Large. Length, 16-22; wing, about 9; tail, about 3½; bill, 4-5; tibia bare, 1-1½; tarsus, 2½-3¼; toes, 1¼, stout.

HAB.—North America, breeding in the interior (Missouri region and northward), migrating in winter southward to Central America and Cuba.

Nest, on the prairie, not far from water.

Eggs, three or four; olive-drab, spotted with various shades of umber-brown.

The Marbled Godwit is occasionally seen singly, or in pairs, on the lake shores of Ontario during the season of migration; but these can only be regarded as stragglers, for we learn that in spring it passes up the Mississippi Valley in flocks of considerable size, and has been found nesting in Iowa, Minnesota and Dakota. It was also found by Prof. Macoun "feeding in large flocks along the salt marshes at Old Wives Lakes and other points " in Manitoba and the North-West.

It is a handsome bird, in general appearance resembling the Curlews, from which, however, it can readily be distinguished by its straight bill.

LIMOSA HÆMASTICA (Linn.).

107. **Hudsonian Godwit.** (251)

Tail, black, largely white at base, its coverts mostly white; rump, blackish; lining of wings extensively blackish; under parts in the breeding season intense rufous (chiefly barred) with dusky; head, neck and upper parts brownish-black, variegated with gray, reddish and usually some whitish speckling; quills, blackish, more or less white at the base. *Young* and apparently winter specimens, much paler, tawny-whitish below, more gray above. Considerably smaller than the foregoing; about 15; wing, 8 or less; bill, 3½ or less; tarsus, 2½ or less.

Hab.—Eastern North America, and the whole of Middle and South America. Breeds only in the high north.

Nest, a hollow, lined with leaves and grass.

Eggs, four, olive-drab with dark spots.

This species is less abundant than the preceding. It seems to prefer the line of the Atlantic for its migrations, but is also noticed inland in smaller numbers. I have seen it in spring at St. Clair Flats, and also on the shores of Hamilton Bay, where the specimen in my collection was obtained.

It is not known to breed anywhere within the limits of the United States, and Prof. Macoun, in recording its presence in the North-West, speaks of it as "less abundant than the preceding and more to the north."

In spring the prevailing color of the plumage is rich chestnut-red, crossed with wavy lines of black. In the fall, it is less attractive, being mostly ashy-gray.

Though not abundant, this species is generally distributed east of the Rocky Mountains. It breeds abundantly on the barren lands of the Arctic Ocean, and on the lower Anderson River region. It associates with the Marbled Godwit, and has the same habits and characteristics.

GENUS TOTANUS BECHSTEIN.

SUBGENUS GLOTTIS KOCH.

TOTANUS MELANOLEUCUS (GMEL.).

108. Greater Yellow-legs. (254)

Bill, straight or slightly bent upwards, very slender, grooved half its length or less, black; legs, long and slender, yellow. In summer, ashy-brown; above varied with black and speckled with whitish; below, white; jugulum, streaked; breast, sides and crissum, speckled or barred with blackish; these latter marks fewer or wanting in winter and in the young; upper tail coverts, white, with dark bars; tail feathers, marbled or barred with ashy or white; quills, blackish. Large. Length, over 12 inches; wing, over 7; tail, 3 or more; bill, 2 or more; tarsus, about 2¼; middle toe and claw, 1½; tibia, bare, 1½.

HAB.—America in general, breeding in the cold, temperate and subarctic portions of North America, and migrating south to Buenos Ayres.

Nest, a hollow, lined with grass and leaves.

Eggs, three or four, grayish-white, marked with spots of dark brown and lilac.

In spring, even before the ice is quite gone from the lakes and rivers of Ontario, the shrill, piercing cry of this bird may be heard overhead, as it circles round in search of some quiet, marshy inlet as a temporary resting place. It is a very watchful species, sure to observe the stealthy approach of the gunner, and give the alarm to the neighborhood, on account of which it has been dubbed the *Tell-tale.*

But a short stay is made in spring, for it passes quickly on to its breeding place in the far north. As early as the end of August the birds again appear, toned down in dress and manners, accompanied by their families. Many of them become the victims of misplaced confidence by exposing themselves within reach of the ever-ready breech-loader, which at that season of the year seems omnipresent in the marshes.

Like others of its kind, this species is an occasional visitor at the Beach, near Hamilton, but the visits of all this class of birds at that point are now of less frequent occurrence, and of shorter duration, than in former years.

In Manitoba and the North-West, it is observed migrating in spring and fall, but is not known to breed. Mr. Nelson thinks it breeds in Alaska, but he has not found the eggs. Audubon found it breeding in Labrador. Its eggs are still scarce in collections.

TOTANUS FLAVIPES (Gmel.).

109. **Yellow-legs.** (255)

A miniature of the last; colors precisely the same; legs comparatively longer; bill grooved rather farther. Length, under 12; wing, under 7; tail, under 3; bill, under, 2; tarsus, about 2; middle toe and claw, and bare tibia, each 1¼.

HAB.—America in general, breeding in the cold, temperate and subarctic districts, and migrating south in winter to southern South America. Less common in the Western than in the Eastern Province of North America.

Nest, a slight depression in the ground, lined with dried grass or leaves.

Eggs, three or four, variable in color, usually clay color, blotched or spotted with umber-brown.

In color, haunts and habits, this species closely resembles the preceding, but the difference in size serves at all times to distinguish one from the other. Both are esteemed for the table, and they are therefore sought for by sportsmen, and often exposed for sale in the market. When one is wounded in a flock, the others raise a great outcry, and remain near it so long that their ranks are often still further thinned before they move off. Alone, or in company with the preceding, this species pays a passing visit to the shores of Hamilton Bay in spring and fall.

In some localities it is more numerous than the Greater Yellow-legs, but both follow the same route.

Coming into Canada from the Northern States in May, they pass up north through the British possessions, but do not stop to breed till they reach the far north.

The Yellow-legs has been found breeding abundantly in the Macfarlane and Anderson River regions, where its eggs were found early in June. It is also known to breed at the mouth of the Porcupine River, which empties into the Yukon in Alaska. In the fall it is again seen in noisy groups, that become reduced in numbers at the different points they visit on their way to their winter quarters, which are said to be in the West Indies and South America.

SUBGENUS RHYACOPHILUS KAUP.

TOTANUS SOLITARIUS (Wils.).

110. **Solitary Sandpiper.** (256)

Bill, perfectly straight, very slender, grooved little beyond its middle; dark lustrous olive-brown, streaked on the head and neck, elsewhere finely speckled with whitish; jugulum and sides of neck, with brownish suffusion and dusky

streaks; rump and upper tail coverts, like the back; tail, axillars and lining of
wings, beautifully barred with black and white; quills, entirely blackish; bill
and feet, very dark olive-green. *Young:*—Duller above, less speckled, jugulum
merely suffused with grayish-brown. Length, 8-9; wing, 5; tail, 2½; bill,
tarsus and middle toe, each about 1-1⅓; tibiæ, bare, ⅞.

HAB.—North America, breeding throughout the temperate portions (more
commonly northward), and migrating southward as far as Brazil and Peru.

Nest, a hollow in the ground, not far from water.

Eggs, clay color with a reddish tinge, thickly marked with reddish and
blackish brown.

As its name implies, this is a solitary bird, nowhere abundant,
yet widely distributed. It is seen during the summer months in
Southern Ontario. Prof. Macoun reports it as "of frequent occurrence on the plains" of the North-West, and it has been found in
Alaska.

In the "List of Birds of Western Ontario," published in the *Canadian Sportsman and Naturalist*, for November, 1882, it is stated that
"in the summer of 1879 this bird bred very commonly along the
streams in Middlesex, but has since then been quite rare." Most of
those I have seen near Hamilton have risen unexpectedly from some
pool by the roadside, frequently from places where cattle have been
in the habit of visiting to obtain water. I have not seen more than
two together. In their motions they are quiet and sedate, but have
the habit peculiar to others of this class, of nervously jerking their
hinder parts in a manner apparently satisfactory to themselves,
though what particular purpose is served by it, is not to us apparent.
From having seen this species in all the summer months, I have
placed it on the list as a rare summer resident here.

I have often fancied myself in close proximity to its nest, but
so far I have not found it. I may state, however, that I do not
search for eggs, and, therefore, do little collecting during June and
July.

There is no trouble in deciding the breeding range of a common
species, but the reverse is the case with a rare bird. In the "Birds
of Manitoba," several of the observers mention it as probably breeding, but Macoun states positively: "Frequent throughout the Winnipegosis region, along the route of 1881, breeding, for young were
observed with some down yet." It is reported as breeding at several
points in Alaska, but even there it is not abundant, and the eggs are
still rare in collections.

Genus SYMPHEMIA Rafinesque.

SYMPHEMIA SEMIPALMATA (Gmel.).

111. Willet. (258)

Bill, straight, comparatively stout, grooved little, if any, more than half its length. In summer, gray above, with numerous black marks; white, below; the jugulum, streaked; the breast, sides and crissum, barred, or with arrow-shaped marks of dusky (in winter, and in young birds, all these dark marks few or wanting, except on jugulum); upper tail coverts, most of the secondaries, and basal half of primaries, white; ends of primaries, their coverts, lining of wings and axillars, black; bill, bluish or dark; toes, with two conspicuous basal webs. Length, 12-16; wing, 7-8; tail, 2½-3; bill or tarsus, 2-2¾; tibia, bare, 1 or more; middle toe and claw, 1½-2.

Hab.—Temperate North America, south to the West Indies and Brazil.

Nest, in a tussock of grass in the marsh, just above water level.

Eggs, three or four, usually clay color, splashed or spotted with varying shades of umber-brown and purple.

Very little is known of this species in Ontario. On two occasions I have seen it brought in by gunners from the marsh, but have not met with it alive. That it passes this way in spring and fall is probable, for it breeds generally throughout the United States as far north as Dakota, and has also been observed in the North-West by Prof. Macoun. In general appearance it resembles the Greater Yellow-shanks, but in the present species the legs are bluish-lead color. The Willets are very wary birds, and along the sea coast, where they are more common and much sought after, decoys are used to attract them within range. In the fall they are said to get extremely fat, and are much prized for the table.

Although generally distributed throughout Ontario and across the interior to the Pacific coast, they are nowhere numerous, the centre of abundance being farther to the south. Cobb's Island, Virginia, is mentioned as one of the breeding places.

In Davie's "Nests and Eggs," it is stated that there is a tract of salt grass in Beaufort County, South Carolina, where it breeds in great numbers. A hundred pairs or more are commonly observed breeding in this locality at the same time. The eggs are very difficult to discover, but the crows find them out and feed upon them, the empty shells being strewed plentifully over the field.

GENUS PAVONCELLA LEACH.

PAVONCELLA PUGNAX LINN.

112. Ruff. (260)

Above, varied with black, rufous, and gray; the scapulars and tertials
exhibiting these colors in oblique bands; beneath, white, varied on the jugulum
and throat; primaries, dark brown, with greenish reflection above; the inner
webs finely mottled towards the base; outer three tail feathers plain, the
remainder transversely barred; bill, brown; sides of rump, white; legs, yellow.
Male in spring dress with the feathers of the neck greatly developed into a ruff;
the face covered with reddish papillæ. Length, about 10 inches; wing, 6.40;
tail, 2.60; bill, 1.25.

HAB.—Northern parts of the Old World, straying occasionally to Eastern
North America.

Nest, made by the females, in a dry tussock of grass in a wet swamp.

Eggs, four, grayish-green, blotched and spotted with reddish-brown.

A wanderer from the Old World, which has been occasionally
obtained on Long Island, on the coast of New England and in the
Middle States.

The fact of a specimen having been killed on the island near
Toronto, in the spring of 1882, gives me the privilege of recording
this species as a rare visitor to Ontario. This is farther inland than
any of the others occurred, and the probabilities are that it will not
often be found so far from the sea. The specimen referred to is
apparently a young male in nearly perfect plumage, and is now
mounted, and in the possession of Mr. Young, of Toronto.

Along the eastern shores of England and Scotland, the Ruffs are
migrants in spring and fall. In former years a few pairs used to
remain and raise their young, but now, owing to the clearing of the
land and the birds being more shot at than formerly, nearly all pass
on to Scandinavia, where they breed in great numbers. Ruffs are
polygamous, and the males have a curious habit of assembling on
bare knolls in the spring to fight for the females. There they erect
their long feathers and charge each other with a great deal of fuss
and flutter, but their differences are usually settled without blood-
shed, and soon afterwards the females retire, select the site, build
the nest, and raise their brood without receiving any further attention from the other sex.

In autumn they pass south to the Mediterranean, thence down
both coasts of Africa as far as Cape Colony.

GENUS BARTRAMIA LESSON.

BARTRAMIA LONGICAUDA (BECHST.).

113. Bartramian Sandpiper. (261)

Above, blackish, with a slight greenish reflection, variegated with tawny and whitish; below, pale tawny of varying shade, bleaching on throat and belly; jugulum with streaks, breast and sides with arrowheads and bars of blackish; axillars and lining of wings, pure white, black-barred; quills blackish, with white bars on the inner webs; tail, varied with tawny, black and white, chiefly in bars; bill and legs, pale, former black-tipped. Length, 11-13 inches; wing, 6-7; tail, 3-4; bill, 1-1¼; middle toe and claw about the same; tarsus, about 2.

HAB. --Eastern North America, north to Nova Scotia and Alaska, breeding throughout its North American range, migrating in winter southward as far even as southern South America. Occasional in Europe.

Nest, on the ground, a slight depression lined with grass often in an old pasture field.

Eggs, four, clay color, marked all over with small spots of umber-brown, most numerous at the larger end.

The Field Plover, as this species is frequently called, is now very seldom seen in Ontario, though the older sportsmen tell us that in former times it was often observed in the pasture fields in spring and fall. The few that I have noticed near Hamilton have always been in such places, but these can only be regarded as stragglers, bewildered by fog, or driven by adverse winds away from their regular habitat. In all the country between the Mississippi and the Rocky Mountains, this species is said to be exceedingly abundant during the seasons of migration, many remaining to raise their young in Illinois, Iowa, Minnesota and Dakota, while large flocks pass on for the same purpose, going as far north as the Yukon. According to Prof. Macoun, they are abundant on the prairies of the North-West, where they will afford good sport and a table delicacy to many a future settler in that promising country.

The only point in Southern Ontario at which I have heard of these birds being seen lately is on the Lake Erie shore not far from Dunnville, where Dr. Macallum is aware of at least two pairs having raised their broods during the two past summers. They have also been heard of on the lake shore farther west, but the increased cultivation of the land, and the increased number of people firing their guns at them, lead such birds to seek for greater retirement elsewhere.

GENUS TRYNGITES CABANIS.

TRYNGITES SUBRUFICOLLIS (VIEILL.).

114. Buff-breasted Sandpiper. (262)

Quills, largely white on the inner web, and with beautiful black marbling or mottling, best seen from below; tail, unbarred, gray, the central feathers darker, all with subterminal black edging and white tips; crown and upper parts blackish, the feathers with whitish or tawny edging, especially on the wings; sides of the head, neck all round and under parts, pale rufous or fawn color, speckled on the neck and breast with dusky; bill, black; feet, greenish-yellow. Length, 7-8; wing, 5-5½; tail, 2¼; tarsus, 1¼; middle toe and claw, and bill, under an inch.

HAB.—North America, especially in the interior; breeds in the Yukon district and the interior of British America, northward to the Arctic coast; South America in winter. Of frequent occurrence in Europe.

Nest, a depression in the ground, lined with dry grass or leaves.

Eggs, four; clay color, blotched or spotted with umber-brown.

In the early fall I have several times met with these interesting little birds, running among the short grass on the sandy knolls, north of the canal at the beach, but have not seen them elsewhere.

They are said to breed in high latitudes, a dozen sets of eggs in the Smithsonian Institute having all been collected by Mr. Macfarlane in the Anderson River region, and along the Arctic coast.

With this record before me, I was not a little surprised to receive from Dr. G. A. Macallum, of Dunnville, a notice of his having found a nest of the species near his home, a few miles back from the north shore of Lake Erie. In answer to my request for further particulars, I received a prompt and full reply, from which the following is an extract: "About the Buff-breasted Sandpiper—I find on turning up my notes that it was taken June 10th, 1879, when two of the eggs were hatched and the other one chipped, but of this, however, I was able to make a good specimen, and it is now in my cabinet.

"The female was shot, and with the two little fellows, stands in my collection. The young are fawn-colored, with black spots over the whole body; the egg measures 1.25 x .95, is pyriform in shape; color, ground, buff, thickly covered with dark blotches of two shades of brown, making the general appearance very dark—almost as dark as the eggs of Wilson's Snipe.

"The nest was placed between two tussocks of grass on the ground, a short distance from the bank of the river, where the ground is tolerably high, and where it is the custom to cut marsh hay. The nest was of a decided shape, and was composed of the fine moss or weed which grows between the tussocks of marsh grass. This is the only case of its breeding here to my knowledge."

This species not being common anywhere, there is not much opportunity for obtaining positive information regarding its distribution during the breeding season. It may be that the case referred to by Dr. Macallum is an isolated one, but it may yet be found, like its near relative, Bartram's Sandpiper, breeding occasionally in suitable places throughout the country. The Buff-breasted has a wide geographical range, and although many pairs breed in the far north, a few remain and raise their young in the middle districts. Those I obtained were got on the 5th of September, 1885, and, though evidently young birds, were in good plumage at that time.

In the "Birds of Manitoba," it is mentioned only as a rare transient visitor.

GENUS ACTITIS ILLIGER.

ACTITIS MACULARIA (LINN.).

115. **Spotted Sandpiper.** (263)

Above, olive (quaker-color, exactly as in the Cuckoo), with a coppery lustre, finely varied with black; line over eye, and entire under parts pure white, with numerous sharp circular black spots, larger and more crowded in the *female* than in the *male*, entirely wanting in very young birds; secondaries, broadly white-tipped, and inner primaries with a white spot; most of the tail feathers like the back, with sub-terminal black bar and white tip; bill, pale yellow, tipped with black; feet, flesh color. Length, 7-8; wing, about 4; tail, about 2; bill, tarsus and middle toe, each about 1.

HAB.—North and South America, south to Brazil. Breeds throughout temperate North America. Occasional in Europe.

Nest, on the ground, usually in shelter of high weeds, composed of dried grass. Eggs, four, clay color, blotched with blackish-brown.

No bird of its class is so well known throughout Ontario as the "Teeter Snipe." Merry bands of children, getting out to the woods to pick flowers in the early summer, listen with delight to its soft "peet-weet," as it flits from point to point along the margin of the stream, and find great amusement in watching the peculiar jerky, teetering motions which give rise to its common name. It thus becomes associated in the mind of the rising generation with the return of summer and its many outdoor enjoyments, and so is always

welcome.   About the middle of April, the Peet-weets cross our
southern boundary, and are soon dispersed in pairs all over the
country, where they are heard and seen by every brook side till
about the end of September, when they move off to spend the winter
in the Southern States.   In the fall they become quite numerous, and
many may be seen along the lake shore at one time, but they are not
gregarious, each individual choosing its own time to rise and place to
alight.   The female is rather larger and more heavily spotted than
the male.

The species is common throughout the North-West, and has been
found by Dr. Bell in the Hudson's Bay regions, but it rarely, if ever,
visits Alaska.

Genus NUMENIUS Brisson.

NUMENIUS LONGIROSTRIS Wils.

116. Long-billed Curlew   (264)

Bill, of extreme length and curvature, measuring from 5 to 8 or 9 inches;
total length, about 2 feet; wing, 1 foot or less; tail, about 4; tarsus, 2½ to 2¼.
Plumage very similar to that of the Godwit, prevailing tone, rufous, of varying
intensity in different birds and in different parts of the same bird, usually more
intense under the wing than elsewhere; below, the jugulum streaked, and the
breast and sides with arrow-heads and bars of dusky; above, variegated with
black, especially on the crown, back and wings; tail, barred throughout with
black and rufous; secondaries, rufous; primaries, blackish and rufous; no pure
white anywhere; bill, black, the under mandible flesh-colored for some distance; legs, dark.

Hab.--Temperate North America, migrating south to Guatemala and the
West Indies.   Breeds in the south Atlantic States, and in the interior through
most of its North American range.

Nest, on the ground, a slight hollow lined with grass.

Eggs, three or four, clay color, blotched or spotted with umber-brown.

The Long-billed Curlew is a bird of the prairie rather than the
coast, though it is often met with along the shores of the sea.   It is
said to breed in suitable places from Carolina to Minnesota, but is
spoken of by Prof. Macoun as rare in the North-West.   In Ontario,
it is occasionally seen along the shores of the lakes, but only as an
irregular visitor, and not in large numbers.   Among the veteran
sportsmen near Hamilton, it is spoken of as one of the kinds which
have been scared away by the railroads.   Whether or not the snorting
of the locomotive has anything to do with the disappearance of the
birds from their former haunts it is hard to say, but certain it is that

the number of waders and swimmers we now see is small compared
with former years.

In the "Birds of Minnesota," Dr. Hatch, at the commencement of
his article on the Long-billed Curlew, says: "This widely distributed
species is nowhere better represented than in Minnesota," but referring to this statement he says, in a foot-note: "The above was true
when written, but the curlews of this species, once so common, have
become less so within the last decade; and now, having been driven
back from both coasts by the advance of civilization, are found in
great numbers far inland on the dry plains, where they are killed
in scores and hundreds."

Dr. Bell remarks that it has not been seen near Hudson's Bay.
It does not migrate so far north as either of the other species with
which we are acquainted. It still breeds abundantly on the south
Atlantic coast, and is resident from the Carolinas south to Mexico.

NUMENIUS HUDSONICUS Lath.

117. **Hudsonian Curlew.** (265)

Bill, medium, 3 or 4 inches long; length, 16-18; wing, 9; tail, 3½; tarsus,
2½-2½. Plumage, as in the last species in pattern, but general tone much paler;
quills, barred.

Hab.—All of North and South America, including the West Indies; breeds
in the high north, and winters chiefly south of the United States.

Nest, a depression in the ground, lined with grass.

Eggs, ashy yellow, marked with chocolate and umber-brown.

According to Dr. Coues, *hudsonicus* is less abundant than either
of the other two Curlews, but at Hamilton it is, of the three, most
frequently observed. I was once on the Beach in May, when there
appeared to be a migratory movement of Hudsonian Curlews toward
the north. They flew high, in regular order, like geese, and showed
no inclination to alight till a boy, with a long shot, brought down
one, wing broken, from a passing flock.

Knowing the habits of the birds, he quickly tied it to a stake in a
moist meadow, and concealing himself close by, had good shooting
during the afternoon, for the loud outcry made by his prisoner
brought down every passing flock.

Of late years very few have been seen.

This is a truly northern species, for Mr. Nelson reports it as a
migrant in Alaska, only an isolated pair remaining here and there to
pass the summer, the main body going farther north, where they

have been found breeding on the barren land of the Arctic Regions. They seem to prefer the eastern route during migrations. Mr. Chamberlain reports them as abundant along the Atlantic, as far north, at least, as Anticosti. Dr. Bell found them in plenty near Fort Churchill, but in the "Birds of Manitoba," no mention is made of them. In Southern Ontario they are still rare. Occasionally a straggling flock of migrants is seen in spring, and Dr. Macallum reports them as regular visitors at Mohawk Island every June, but on such occasions they appear only in small numbers.

NUMENIUS BOREALIS (Forst.).

118. Eskimo Curlew. (266)

Bill, small, under 3 inches long; length, 12-15 inches; wing, under 9; tail. 3; tarsus, 2. Plumage, in tone and pattern almost exactly as in the last species, but averaging more rufous, especially under the wings, and primaries. not barred.

Hab.—Eastern Province of North America, breeding in the Arctic regions, and migrating south to the southern extremity of South America.

Nest, in open plains, a hollow lined with grass and leaves.

Eggs, olive-drab shading to green gray, or brown marked with dark chocolate.

The Curlews all resemble each other in plumage, but in size they vary considerably, this being the smallest of the three. It is very abundant in the remote regions which it frequents in summer, and also along its migratory course, from which it does not seem to deviate much. On the Pacific coast it has not yet been observed, and on the Atlantic shores it appears only in limited numbers. The great highway of the species is through the States just east of the Rocky Mountains, where it is seen in immense flocks in spring and fall. Dr. Coues says it is extraordinarily abundant in some places during the migration, as in Labrador, where it fairly swarms in August. I once found myself, unexpectedly, in close proximity to a solitary individual on the shore of the beach, near Hamilton, and secured it, but that is the only record I have of its occurrence in Ontario.

Dr. Macallum's experience with this species is similar to mine. He secured two, which were feeding in an old pasture field along with some Golden Plovers; but these are all he has seen in twenty years.

119. Black-bellied Plover. (270)

Adult in breeding season:—Rarely seen in the United States; face and entire under parts, black; upper parts, variegated with black and white, or ashy; tail, barred with black and white; quills, dusky with large white patches. *Adults at other times and young:*—Below, white, more or less shaded with gray; the throat and breast, more or less speckled with dusky; above, blackish, speckled with white or yellowish; the rump, white, with dark bars; legs, dull bluish. Old birds changing show every grade, from a few isolated feathers on the under parts to numerous large black patches. Length, 11-12; wing, 7 or more; tail, 3; bill, 1-1¼; tarsus, 2; middle toe and claw, 1⅛; hind toe, hardly ¼.

Hab.—Nearly cosmopolitan, but chiefly in the northern hemisphere, breeding far north, and migrating south in winter, in America, to the West Indies, Brazil and New Grenada.

Nest, a hollow in the ground, slightly lined with grass.

Eggs, four, dark gray color, blotched or spotted with brownish-black.

11

Although of nearly cosmopolitan distribution, this large and handsome Plover is nowhere abundant. It has been found breeding on the Arctic coast east of the Anderson River, where its eggs were taken by Mr. Macfarlane.

In its migrations, it prefers the sea coast on either side to the interior, but a few are also observed inland.

At Hamilton, it visits the beach in spring and fall in limited numbers. I once got two out of three very handsome individuals which I saw there on the 3rd of June. In the "List of Birds of Western Ontario," it is mentioned as a "common migrant" at St. Clair Flats. It is also seen passing throughout the North-West, and Mr. White has found it at Ottawa.

Until quite recently, the only breeding place I had heard named for this species was within the Arctic Circle. In the "Birds of Minnesota," Dr. Hatch says: "In the summer of 1875, a clutch of four eggs was sent to me with the female, which proved to be a Black-bellied Plover. It was obtained in the vicinity of upper Lake Minnetonka. Since then several nests have been reported by persons competent to determine them, and I accept the conclusion that this species breeds to a limited extent in some portions of the State."

A few are said to breed at the mouth of the Yukon in Alaska, but at St. Michael's, where Mr. Nelson was stationed, he did not find it in summer.

SUBGENUS CHARADRIUS LINNÆUS.

CHARADRIUS DOMINICUS (MÜLL.).

120. American Golden Plover. (272)

Plumage, speckled above, and, in the breeding season, black below, as in the last species, but much of the speckling bright yellow; and the rump and upper tail coverts, like the back; forehead and a broad line over the eye to the nape, white; tail feathers, grayish-brown, with imperfect white or ashy bars; axillars, gray or ashy. At other times, the under parts nearly as in the last species. Length, 10-11; wing, 7 or less; tail, under 3; bill, 1 or less.

HAB.—Arctic America, migrating southward throughout North and South America to Patagonia.

Nest, composed of dry grass, in a natural hollow in the ground.

Eggs, four, similar to those of the preceding species but not quite so large.

Old sportsmen tell us that Golden Plovers used to follow the line of the Detroit River in immense flocks, passing quickly to the north in the spring, and lingering along the shores and in the pasture fields on their return in the fall.

According to the " List of Birds of Western Ontario," they are still regular visitors there, but only in small numbers. Near Hamilton they have never been common. Small flocks of immature birds are seen passing south in the fall occasionally, but not regularly.

The Golden Plover in full breeding plumage is a very handsome bird, but, like the Snow-bird and some others which breed in high latitudes, it does not assume the nuptial dress till it reaches its northern home, and by the time it gets back within the bounds of civilization it has donned the sober garb of winter.

Mr. White reports the Golden Plover as a regular visitant at Ottawa during October. It is an abundant migrant in the North-West, but none remain during the summer. It breeds abundantly on the coast and islands of the Arctic Sea, and on the barren lands, and again puts in an appearance at the usual stopping places on its return trip in the fall.

Genus ÆGIALITIS Boie.

Subgenus OXYECHUS Reichenbach.

ÆGIALITIS VOCIFERA (Linn.).

121. Killdeer. (273)

Above, quaker-brown, with a greenish tinge, sometimes most of the feathers tipped and edged with orange-brown; rump and upper tail coverts, orange-brown; most of tail feathers, white at base and tip, suffused with orange-brown in part of their length, and with one to three black bars; secondaries, mostly white; and primaries, with a white space; a black bar across the crown; and two black bands on the neck and breast; forehead and entire under parts, except as stated, white; bill, black; feet, pale; eyelids, scarlet. Length, 9-10 inches; wing, 6 or more; tail, 3½, much rounded; tarsus, about 1½.

Hab.—Temperate North America, migrating in winter to the West Indies, Central America and northern South America.

Nest, in the grass or shingle, in the vicinity of water.

Eggs, four, clay color, marked with blackish-brown.

This is a noisy, well-known bird, generally distributed throughout Ontario, and abundant in the North-West. In April, even before the snow is quite gone, the shrill cry of the Killdeer is heard in the upper air, as it circles around, surveying its old haunts, and selecting a bare spot on which to settle.

11*

Its favorite resorts are pasture fields or waste places near water, where it spends much of its time on the ground, sometimes running with great speed, or sitting quietly as if aware that it is more likely to escape observation in that way than by moving. It can scarcely be called gregarious, yet, in the fall, when the young birds are getting strong on the wing, they may be seen in companies of ten or a dozen, visiting the muddy shores of streams and inlets, till about the end of September, when they all move off south.

It is common during summer in Manitoba and the North-West, but is not named among the " Birds of Alaska."

SUBGENUS ÆGIALITIS Boie.

ÆGIALITIS SEMIPALMATA Bonap.

122. Semipalmated Plover. (274)

Above, dark ashy-brown with an olivaceous shade; below, white; very broad coronal and pectoral black bars in the adult in spring; in fall and in the young the coronal bar hardly evident; the pectoral, grayish-brown; edges of eyelids, bright orange; bill, moderately short and stout, orange or yellow, black tipped; legs, yellowish; toes, conspicuously semipalmate. Length, about 7 inches; wing, 4¾; tail, about 2¼ rounded.

HAB.—Arctic and subarctic America, migrating south throughout tropical America, as far as Brazil and Peru.

Nest, a depression in the ground, lined with dry grass.

Eggs, four, clay color, marked with blackish-brown.

The Semipalmated is a solid, plump little bird of very pleasing plumage, particularly in spring, when the colors are clear and decided. In company with other beach birds, it is found along the shores of the lakes in Ontario from the middle till the end of May. In the fall it is again seen in increased numbers in similar places till about the end of September, when it disappears for the season. Dr. Coues found the Ring-necks breeding abundantly in Labrador, and mostly remaining there till the beginning of September. The distance between their summer and winter home is very great, but their flight is rapid, and as they seem to know the way, the journey is quickly made.

They have been found breeding in the latter part of June on the Arctic coast and in the Anderson River regions, as well as on the islands off the Alaskan coast and at the mouth of the Yukon.

It is probable that a few stop and raise their brood by the way, for in Manitoba, where the species is abundant as a migrant, Macoun says : " Abundant in company with the Killdeer, and evidently breeding, as I saw young with them at the Salt Springs on Red Deer River, July 22nd, 1881, at Lake Manitoba."

ÆGIALITIS MELODA (Ord.).

123. Piping Plover. (277)

Above, very pale ashy-brown; the black bands narrow, often imperfect; bill, colored as in the last, but shorter and stumpy; edges of eyelids, colored; no evident web between inner and middle toes, and only a slight one between middle and outer. Length, about 7 inches; wing, $4\frac{1}{2}$; tail, 2.

HAB.—Eastern Province of North America, breeding from the coast of New Jersey (at least formerly) northward; in winter, West Indies.

Eggs, four, deposited among the shingle of the beach, clay color, marked with spots of brownish-black, not exceeding a pin's head in size.

The Piping Plover is a more southern bird than the Ring-neck, and evidently does not penetrate far into Ontario. I have met with it at Hamilton Beach, but only on two occasions. It has also been found on the island at Toronto, but is more common along the north shore of Lake Erie, and Mr. Saunders reports it as breeding at Point Pelee, at the western end of that lake. When sitting quietly among the shingle of the beach, the colors of this little bird harmonize so well with its surroundings that quite a number may be close at hand without being observed. The birds seem aware of this, and if suspicious of danger, sit perfectly still till it is time to fly, when they rise simultaneously and move off with a soft, plaintive, piping note.

In looking over Mr. Thompson's "Birds of Manitoba," I was surprised to find this species reported as occurring there, Macoun having "shot it in company with some sanderlings on the shore of Lake Manitoba on the 12th June, 1881." Its centre of abundance during the summer is along the Atlantic coast, from the Carolinas north to the Gulf of St. Lawrence.

ÆGIALITIS NIVOSA Cass.

124. **Snowy Plover,** (278)

Male in breeding dress:—Above, pale ashy-gray, little darker than in *meloda*; top of head with a fulvous tinge: a broad black coronal bar from eye to eye; a narrower black post-ocular stripe, tending to meet its fellow on the nape, and thus encircle the fulvous area; a broad black patch on each side of the breast: no sign of its completion above or below; no complete black loral stripe, but indication of such in a small dark patch on either side of base of upper mandible; forehead, continuous with line over the eye, sides of head, excepting the black post-ocular stripe, and whole under parts, excepting the black lateral breast patches, snowy white; no white ring complete around back of neck; primaries, blackish, especially at bases and ends, the intermediate extent fuscous; shaft of first, white, of others, white for a space; nearly all the primaries bleaching toward bases of inner webs, but only some of the inner ones with a white area on outer webs; primary coverts like the primaries, but white-tipped; greater coverts like the back, but white-tipped; secondaries, dark brown, bleaching internally and basally in increasing extent from without inwards, their shafts white along their respective white portions; tertiaries, like back; several intermediate tail feathers like back, darkening toward ends; two or three lateral pairs entirely white; all the feathers more pointed than usual; bill, slender and acute, black; legs, black. Length, 6.50 to 7.00; extent, 13.50 to 14.00; wing, 4.00 to 4.25; tail, 2.00 or less.

HAB.—Western Province of North America; in winter, both coasts of Central America and western South America to Chili.

Eggs, three, placed among the shingle on the beach, pale buff or clay color, finely marked with blackish-brown spots.

The Snowy Plover is a western bird very seldom seen east of the Rocky Mountains, and would not have been mentioned here, but for the following notice of it which appears in the *Auk* for October, 1885. It is contributed by E. E. Thompson, of Toronto. "A specimen of this bird was shot here by Mr. J. Forman, May, 1880, and is now in the rooms of the Toronto Gun Club. It was at the time in company with some Piping Plovers. This specimen answers in general to the description in 'Coues' Key' and fully in regard to the bill; it differs in being much lighter in plumage. I had no opportunity to make measurements, but in the same case were *meloda* and *semipalmata*, and comparison with these makes me almost certain that it is *nivosa*. The bill is noticeably long, black and slender. I never met the bird before, and have no material to aid me in settling the point."

If Mr. Thompson has correctly identified the specimen described, it can only be regarded as a casual straggler from the far west which may not be seen here again.

FAMILY APHRIZIDÆ. SURF BIRDS AND TURNSTONES.

SUBFAMILY ARENARIINÆ. TURNSTONES.

GENUS ARENARIA BRISSON.

ARENARIA INTERPRES (LINN.).

125. Turnstone. (283)

Adult, in summer:—Pied above, with black, white, brown and chestnut-red, the latter color wanting in winter and in young birds; below, from the breast (which is more or less completely black); throat, most of the secondaries, bases of the primaries, and bases and tips of the tail feathers, white; bill, black; feet, orange. Length, 8-9 inches; wing, 5½-6; tail, 2½; bill, ⅞, almost recurved; tarsus, 1; tibiæ, bare but a little way.

HAB.—Nearly cosmopolitan. In America, from Greenland and Alaska, to the Straits of Magellan; more or less common in the interior of North America, on the shores of the Great Lakes and the larger rivers. Breeds in high latitudes.

Nest, a hollow scratched in the earth, lined with bits of grass.

Eggs, two to four; greenish-ash, spotted, blotched and dotted irregularly with yellowish and umber-brown.

In the "Birds of Ohio," Dr. Wheaton says that Mr. Sinnett observed this species on the coast of Texas in the breeding season, and he believes that they breed there.

The beautifully marked Turnstone is a bird of nearly cosmopolitan distribution. It is found in America on both coasts, and also in the interior. At Hamilton Beach it is a regular visitor in spring and fall, but there are seldom more than two or three found together.

They are very sociable in their habits, mixing freely with any other waders they chance to meet, and are seen here till the end of the first week in June.

They are observed again, young and old together, early in September, and linger around the shores of the bay till the end of that month, when they move farther south to spend the winter.

They are migrants throughout Ontario and the North-West, but breed abundantly on the barren lands of the Arctic coast and the Anderson River district, and sparingly in Alaska.

Order GALLINÆ. Gallinaceous Birds.

Suborder PHASIANI. Pheasants, Grouse, Partridges, Quails, etc.

Family TETRAONIDÆ. Grouse, Partridges, etc.

Subfamily PERDICINÆ. Partridges.

Genus COLINUS Lesson.

COLINUS VIRGINIANUS (Linn.).

126. **Bob-white.** (289)

Coronal feathers erectile, but not forming a true crest. Forehead, superciliary line and throat, white, bordered with black; crown, neck all round and upper part of breast, brownish-red; other under parts, tawny-whitish, all with more or fewer doubly crescentic black bars; sides, broadly streaked with brownish-red; upper parts, variegated with chestnut, black, gray and tawny, the latter edging the inner quills. *Female*:—Known by having the throat buff instead of the white, less black about the foreparts, and general colors less intense, rather smaller than the male. Length, 9-10; wing, 4½-5; tail, 2½-3.

Hab.—Eastern United States and southern Canada, from southern Maine to the South Atlantic and Gulf States, west to Dakota, eastern Kansas and eastern Texas.

Nest, on the ground in a natural or excavated hollow, lined with grass or leaves, usually sheltered by tall grass, weeds, bushes or brush.

Eggs, pure white, said to range in numbers from ten to forty, the larger lots supposed to include contributions from several females; fifteen being considered the usual set.

Bob-white may be claimed as a permanent resident in Southern Ontario, which is the northern limit of his range, but he has hard work to hold his own against the many influences that are continually operating against him. Birds of prey, crows, jays, weasels, dogs, cats, mowing machines, and sportsmen of all classes tend to thin the ranks; but worst of all are the vicissitudes of winter. The spells of cold weather, during which the mercury gets down below zero, and the occasional long-continued deep snow, tell so severely against this little bird, that were it not for his wonderful capacity for increase he would soon be exterminated.

The Quail follows in the wake of cultivation, and under ordinary circumstances thrives best near the abode of man. It is a good friend to the farmer, and is well entitled to his protection in return for the service it renders, not only in the consumption of large quantities of the seeds of noxious weeds, but also in the destruction of many sorts of insects whose ravages among the crops are often very severe and difficult to prevent. A recent writer mentions having examined the crop of one which was killed as it rose from a potato patch, and having found that it contained seventy-five potato bugs. This is only one of the many instances illustrating the value of this bird to the farmer.

Were I a farmer, I should hang on the end of my barn the motto, inscribed in goodly characters, "Spare the Quail."

Many interesting articles have from time to time appeared in sporting magazines concerning the query—Has the Quail the power to withhold its scent?

No one acquainted with the habits of the birds can deny that at times the best of dogs fails to find them where they have been marked down, but how this happens is a subject regarding which sportsmen still hold different opinions.

From among many instances given in illustration of the fact, we select the following by Dr. H. E. Jones, an enthusiastic sportsman and naturalist: "A few years ago I was out with a friend, and we flushed a very large bevy, and marked them down accurately on an elevated piece of ground in a woodland pasture. The grass was short and there was not even a weed or briar, but here and there a large tree. We moved forward with three dogs, expecting to bring on an engagement at once. We made the dogs approach cautiously, giving them warning that game was in the immediate vicinity, but they arrived at the identical spot where we had seen as many as thirty birds alight, without making the least demonstration whatever that there was anything unusual about the place. We knew better, and made them go over and over, crossing and recrossing, until it seemed every foot, every inch of ground had been most thoroughly examined. We did this until two sportsmen and three dogs gave up the pursuit. It was now past noon, and we sat down on the grass, uncorked our canteens and opened out our lunch. We were eating, talking and laughing, occasionally rewarding the dogs with a cracker, when my friend, by way of sport, said, 'Look at old Tom, he is on a point.' The dog was standing half up, half down, with his nose thrown under his chest between his front legs. Sure enough he was on a point, for there was the bird, with its bright black eyes, only partially concealed by a leaf, almost under the dog's body. My friend put his hat over it and caught it without moving from the dinner table. At that instant another dog made a point within six inches of my feet. I saw the bird at once, and tried to capture it with my hand, but it made its escape. This was the signal for a general move, and the whole covey now arose from all around and about us. The concert of action in the manner of going down, retaining their scent, remaining still under the most trying circumstances, and the mode of leaving —all indicated an understanding and education by command how to act in time of danger."

Some time ago the Government of Ontario passed an Act prohibiting the killing of quail under any circumstances for a period of three years, and this law, coincident with mild winters, had the effect for a time of increasing the numbers, but again they are greatly reduced and in need of the protection which they well deserve.

Subfamily TETRAONINÆ. Grouse.

Genus DENDRAGAPUS Elliot.

Subgenus DENDRAGAPUS.

DENDRAGAPUS OBSCURUS RICHARDSONII (Sab.).

127. Richardson's Grouse. (297b)

Adult male: Back and wings, blackish-brown, crossed with wavy lines of slaty-gray, mixed with yellowish-brown on the scapulars; long feathers of the sides, tipped with white; under parts, light slate color, mixed with white on the lower parts; cheeks, black; chin and throat, speckled with black; and white feathers on the sides of the neck slightly enlarged, covering a rudimentary air sack; tail, brownish-black, veined and marbled with gray, and having a broad terminal band of the same color. *Female:*—Smaller, more varied and generally lighter in color, but having the under parts and bar at the end of the tail, slaty-gray, as in the male. Length, 20-22 inches; wing, 9-10; tail, 7.

Hab. Rocky Mountains, from central Montana northward into British America.

Nest, on the ground, in shelter of a rock or log; it consists of only a few pine needles scratched together, on which the eggs are laid.

Eggs, eight to fifteen, creamy-buff, freckled all over with chocolate-brown.

For a notice of the occurrence of this species in Ontario I am indebted to C. J. Bampton, of Sault Ste. Marie, who has frequently seen it brought into market at that place.

It bears a strong resemblance to the Dusky Grouse (*Dendragapus obscurus* [Say.]), of which it is regarded as the northern form. The Dusky Grouse is found chiefly on the west coast, as far south as New Mexico and the White Mountains of Arizona. In the Rocky Mountains, toward the north, it gradually assumes the peculiarities of the present species, but many intermediate individuals are found which cannot positively be said to belong more to the one than to the other.

In *richardsonii*, the tail feathers are longer and broader than in *obscurus;* the slate-colored bar at the end is smaller, or wanting, and the general colors darker, specially so on the throat.

SUBGENUS CANACHITES STEJNEGER.

DENDRAGAPUS CANADENSIS (LINN.).

128. **Canada Grouse.** (298)

Adult male:—Tail, of sixteen feathers, rounded, black, with an orange-brown bar at the end; prevailing color, black, barred and spotted with white on the lower parts, and above, crossed with wavy lines of tawny and gray. *Female:*—Smaller, variegated all over with black, brown, white and tawny; tail bar, as in the male, but less decided. Length, 16; wing, 7; tail, 5.50.

HAB.—British America, east of the Rocky Mountains, from Alaska south to northern Michigan, northern New York and northern New England.

Nest, on the ground in secluded places, well concealed, built of twigs, leaves, moss and grass.

Eggs, twelve to fourteen, creamy-brown, sometimes dotted or blotched with a darker shade.

When young birds of different species are cast loose from parental oversight, and go out into the world on their own account, they are often very erratic in their movements, are frequently found in places where they have no business to be, and sometimes thereby come to grief.

It was from some such cause as this that I once got a specimen of the Canada Grouse in the Hamilton market. It was in the month of October that a farmer had seen this small dark-colored bird in company with some Ruffed Grouse, and, following them up, had singled it out as something new. They are not known to breed anywhere near Hamilton, but are common in the picturesque district of Muskoka, between Georgian Bay and the Ottawa River, where they breed and are resident.

They breed also in suitable places throughout the North-West, and with regard to Alaska, Mr. Nelson says : "This handsome Grouse is found throughout the wooded portion of Alaska, extending to the shores of Behring Sea, at the points where the spruce forests reach the vicinity of tide-water. It is more numerous, however, in the interior and along the upper portion of the Yukon. It is permanently resident wherever found."

They are plump, handsome little birds, but for the table are not equal to the Quail or the Ruffed Grouse.

GENUS BONASA STEPHENS.

BONASA UMBELLUS (LINN.).

129. Ruffed Grouse. (300)

Sexes nearly alike; variegated reddish or grayish brown; the back, with numerous oblong, pale, black-edged spots; neck-tufts, glossy-black; below, whitish, barred with brown; tail, with a broad subterminal black zone, and tipped with gray. Length, 16-18; wing, 7-8.

HAB.—Eastern United States, south to North Carolina, Georgia, Mississippi and Arkansas.

Nest, in a hollow in the ground, lined with grass or leaves, often placed by the side of a log or stump.

Eggs, eight to twelve, cream color, sometimes minutely spotted with chocolate-brown.

Notwithstanding the continual persecution to which the Ruffed Grouse is exposed, it is still a common species throughout Ontario, breeding in all suitable places from the shore of Lake Erie to the northern boundary of the Province, and even in Alaska.

It is a robust, hardy bird, well able to stand the rigors of our climate, and being exceedingly strong and active on the wing, gets away oftener from the sportsman than any other species he pursues. Occasionally, when the birds are found feeding among bushes of stunted growth, with a good dog a fair bag may be made, but following them through the tangled masses of foliage and fallen trees, where they are usually found, is attended with great fatigue, and, usually, very slim results. The birds get up with wonderful suddenness, and disappear as if by magic. They seem always to rise at the wrong time, from the wrong place, and to go off in the wrong direction to suit the sportsman.

Much has been written regarding the mode in which this bird produces the peculiar drumming sound so familiar to all who have had occasion to visit its haunts, but it is now generally believed to be caused by the rapid vibratory motion of the wings beating the air; a similar sound being produced in a similar way by the Humming-bird, and also by the Nighthawk. The Grouse, in the spring time, produces this music as a call to his lady fair, who, no doubt, delights to hear it, and responds accordingly. It is also heard, occasionally, late in the season, when he is possibly working off the exuberance of his spirits after some happy experience in his sylvan life.

At different points throughout its extensive habitat, this species is subject to considerable variation in plumage, and on this account

the American Ornithologists' Union Committee has separated from
the original *Bonasa umbellus* three sub-species, some of which had
been previously described, but were not generally recognized as
differing from the typical form. The three sub-species are thus
defined by Mr. Ridgway:

"300a. *Bonasa umbellus togata* (Canadian Ruffed Grouse). Darker, with
brown markings on lower parts very conspicuous, everywhere exposed, and
bordered by very distinct dusky bars; bars on flanks very dark brown or
brownish-black; upper parts, with more or less of gray, often mostly grayish;
the tail, usually gray (sometimes tinged with ochraceous). HAB.—Washington
Territory, east to Moose Factory, Nova Scotia, Maine, etc.

"300b. *Bonasa umbellus umbelloides* (Gray Ruffed Grouse). Upper parts,
mostly or entirely grayish; the tail, always gray. HAB. Rocky Mountains
and north to Alaska (Yukon Valley), east to Manitoba.

"300c. *Bonasa umbellus sabini* (Oregon Ruffed Grouse). Upper parts, dark
rusty, with little, if any, admixture of gray; the tail, usually deep rusty (very
rarely grayish). HAB. North-west coast, from northern California to British
Columbia."

At present we have no large public collection of specimens in
Ontario to refer to, and the number of private ones is so small that
it is difficult to tell just how these groups are represented among us,
or whether the forms are observed to be distinct.

While in Manitoba, Ernest E. Thompson found both 300a and
300b, the identification of which was subsequently confirmed by Mr.
Ridgway. We may, therefore, expect to find these two in suitable
places along our northern frontier. In Southern Ontario the species
varies considerably in plumage, some being grayish-white, others reddish, and a good many intermediate. A few years ago, some Indians
from the Grand River brought to the Hamilton market specimens of
a race decidedly red, their tails being fox-colored, but these were
seen during one fall only.

From all I have observed, I think that we have in Ontario individuals of both *Bonasa umbellus* and *Bonasa umbellus togata*, that
these two intergrade, and produce a mixed race, which is found
throughout Southern Ontario, but cannot properly be classed with
either of the varieties.

GENUS LAGOPUS Brisson.

LAGOPUS LAGOPUS (Linn.).

130. **Willow Ptarmigan.** (301)

Bill, stout, as high as the distance from the nasal groove to its tip; in summer, rufous or orange-chestnut on the head and neck; the feathers of the back, black, barred rather closely with yellowish-brown and chestnut. In winter, white; the tail, black, tipped with white. Length, 15-17; wing, about 8; tail, 5.50.

Hab.—Arctic America, south to Sitka and Labrador.

Nest, on the ground, a slight depression, lined with grass, leaves and a few feathers.

Eggs, ten to fifteen, fawn color, spotted with reddish-brown.

Ptarmigans are found both in the Old and New Worlds, as far north as vegetation extends, and so thoroughly boreal are they in their habits, that they seldom come within the bounds of civilization. C. J. Bampton, Registrar of the District of Algoma, who has furnished me with many interesting notes regarding the birds of that remote district, mentions the Willow Ptarmigan as a rare winter visitor at Sault Ste. Marie.

Their southern migrations depend, to some extent, on the peculiarities of the season, but usually they are common winter visitors throughout Manitoba and the North-West, and Dr. Bell reports them as common every winter near Fort Cumberland, in the Hudson's Bay region.

In Alaska, the species is a summer resident, frequenting the extensive open country, being most abundant along the barren sea coast region of Behring Sea and the Arctic. The birds assemble there in immense flocks, and we might suppose that in those forlorn lands, so seldom visited by their greatest enemy, man, the birds enjoy a time of peace and security, but it appears from Mr. Nelson's account that such is not always the case. He says: "Among the Alaskan natives, both Eskimo and Indian, especially those in the northern two-thirds of the Territory, this bird is one of the most important sources of food supply, and through the entire winter it is snared and shot in great abundance, and many times it is the only defence the people possess against the ever-recurring periods of scarcity and famine."

In some districts the Eskimo have a way of catching the birds during their migration which is very destructive. Taking a long fishing net, they fasten poles to it at regular distances apart, and lay it on the ground across some open valley or swale running north and south, along which the birds are known to travel. Soon after sunset the poles are set upright, and the net is thus stretched across the passage. Ere long the men who are on the watch see the Ptarmigan advancing, skimming close to the snow-covered earth in the dim twilight. A moment later, and the first birds of the flock strike the obstacle, and the men then throw the net down, so as to cover the struggling captives, usually to the number of fifty or sixty. While the men hold the net, the women and children rush from their hiding, and kill the birds by wringing their necks or biting their heads. On some evenings this process is repeated several times, and the party return to their homes heavily laden with the spoils.

In the Old Country this species is common, also in Scandinavia, Finland, Russia, and in many parts of Siberia; while in the north of Scotland, they are found breeding near the tops of the highest heathery hills in that mountainous country.

Their presence there is one of the attractions for strangers, and many a corpulent southern sportsman has expended much wind, and reduced his corporeal dimensions, scaling those precipitous hill-sides in the vain hope of securing a brace of Ptarmigan. In another connection it is used with more delicacy by a young Highlander, who, in persuading his Lowland maid to go with him to the " Braes aboon Bonaw," sings :

> " We'll hunt the roe, the hart, the doe,
> The ptarmigan sae shy, lassie;
> For duck and drake we'll sail the lake,
> Nae want shall e'er come nigh, lassie."

LAGOPUS RUPESTRIS (Gmel.).

131. **Rock Ptarmigan.** (302)

Bill, slender, distance from the nasal groove to the tip greater than height at base; in summer the feathers of back black, banded distinctly with yellowish-brown and tipped with white; in winter white, the tail black, tipped with white. *Male :-* With a black bar from the bill through the eye. Length, 14 to 15; wing, 7 to 7.50; tail, 4.50.

Hab.—Arctic America, from Alaska to Labrador.

Nest, on the ground, a hollow lined with grass and a few feathers.

Eggs, ten to fifteen, reddish-brown, spotted with darker brown.

This is another northern species reported by Mr. Bampton as being occasionally exposed in the winter time in the market at Sault

12

Ste. Marie. It resembles the preceding in general appearance, but is rather less in size, and in winter plumage the black band through the eye of the male serves at once to decide his identity.

The Ptarmigans have a most interesting history. Their small feet, covered densely with hair-like feathers, the wonderful change which their plumage undergoes to match their surroundings, and their life amid the rigors of an Arctic winter, are matters which invest the story of the group with peculiar charm.

The Rock is a more northern species than the Willow Ptarmigan. It is not mentioned among the birds of Manitoba, so we presume that it has not been seen there. Dr. Bell has observed it migrating at Hudson's Bay, and Mr. Macfarlane found it breeding in the vicinity of Fort Anderson. Mr. Nelson, speaking of its presence in Alaska, says: "This beautiful Ptarmigan is a common resident of the Alaskan mainland, from Behring Straits to the British border on the east, including the entire north and south extent of the main-land. Unlike the common White Ptarmigan, it frequents the sum-mits of the low hills and mountains during the summer season, where it remains until the severe weather of early winter forces it down to the lower elevations and under the shelter of the bush-bordered ravines and furrows marking the slopes. During the entire year these birds are resident north at least to Behring Straits, as I obtained specimens from that vicinity on one of my winter expeditions."

GENUS TYMPANUCHUS GLOGER.

TYMPANUCHUS AMERICANUS (REICH.).

132. Prairie Hen. (305)

Above, variegated with black, brown, tawny or ochrey and white, the latter especially on the wings; below, pretty regularly barred with dark brown, white and tawny; throat tawny, a little speckled, or not; vent and crissum, mostly white; quills, fuscous, with white spots on the outer web; tail, fuscous, with narrow or imperfect white or tawny bars and tips; sexes, alike in color, but the female smaller with shorter neck tufts. Length, 16-18; wing, 8-9; tail, about 5.

HAB.—Prairies of the Mississippi Valley, south to Louisiana, east to Kentucky and Indiana.

Nest, on the ground, in a tuft of grass or small shrub.

Eggs, eight to twelve, pale greenish-gray, sometimes minutely dotted with brown.

Southern Ontario has no prairie to meet the requirements of the Prairie Chicken, and therefore the bird is rarely here. From various sources I have heard of its being still found along the south-western frontier, but the numbers are on the decrease. In the " List of Birds of Western Ontario," it is stated that a few still breed at St. Clair. From W. E. Wagstaff, one of the oldest and most respected settlers in the county of Essex, I have a most interesting letter regarding the birds he has observed during his long residence there. Of this species he says : " I have never seen Prairie Chickens alive, but have heard of their being seen in bands about Sandwich. When I first came to Amherstburg, about 1840, I heard the old sports tell of having killed them in the gardens of the town."

From the foregoing, it would appear that the days of the Prairie Chicken in Ontario are numbered. It affords excellent sport to the gunner, and the facilities for reaching it in its remote haunts are now so much increased, that year by year, even in the United States, it is being driven to regions still more remote.

In the first week in May, 1886, some young men were practising flight shooting at any water-fowl that happened to be passing between the bay and the lake, near the canal at the Beach. Presently a bird of different flight and shape came buzzing along, and was brought down by one of the gunners, who was greatly astonished to find he had killed a male Prairie Chicken in fine spring plumage. I passed shortly afterwards and saw the bird just as it had been picked up. It had been going at a very rapid rate, but whence it came, or whither bound, was not apparent.

GENUS PEDIOCÆTES BAIRD.

PEDIOCÆTES PHASIANELLUS (LINN.).

133. Sharp-tailed Grouse. (308)

Adult male :—A decurved crest of narrow feathers, a bare space on each side of the neck, capable of being inflated; tail, short, much graduated, of sixteen feathers, all of which are more or less concave, excepting the two middle ones along the inner edge, obliquely and abruptly terminated, the two middle projecting an inch beyond the rest. Upper parts variegated with light yellowish-red, brownish-black and white, the latter in terminal triangular or guttiform spots on the scapulars and wing coverts; quills, grayish-brown, primaries with white spots on the outer webs; secondaries, tipped and barred with white; tail, white, variegated at the base, the two middle feathers like the back; loral space and a band behind the eye, yellowish-white, a dusky streak under the eye; throat, reddish-white, with dusky spots; fore parts and sides of the neck, barred with reddish-white; on the breast the dusky spots become first curved, then arrow-shaped, and so continue narrowing on the hind part of the breast and part of the sides of which the upper portion is barred; abdomen, lower tail coverts and axillars, white; tarsal feathers, light brownish-gray, faintly barred with whitish. *Female:*— Smaller, the tints of colors less bright. Length, 18-20; wing, 8-9; middle feathers of the tail, 4-6; outer feathers, 1½.

HAB.— British America, from the northern shore of Lake Superior and British Columbia to Hudson's Bay territory and Alaska.

Nest, in a tuft of grass on the prairie.

Eggs, five to twelve, grayish-olive or drab color, minutely dotted with brown spots the size of a pin's head.

Writing from the North-West, Prof. Macoun says of this species: "This is the Prairie Chicken of our western plains, the true Prairie Chicken not being observed here."

Dr. Coues, writing in the same strain, says: "This is the Prairie Chicken of the whole North-West, usually occurring where the Pinnated Grouse does not, although the habitats of the two species overlap to some extent." From the foregoing it appears that while the present species occupied the North-West, the Prairie Chicken flourished more in the south-east, but that now both are being driven farther to the north-west, as the prairies come under cultivation.

The Sharp-tail is abundant near Winnipeg, from which point it has reached the Hamilton market. It is also reported by Mr. Bampton as being found at Sault Ste. Marie.

FAMILY PHASIANIDÆ.   PHEASANTS, ETC.

SUBFAMILY MELEAGRINÆ.   TURKEYS.

GENUS MELEAGRIS LINNÆUS.

MELEAGRIS GALLOPAVO LINN.

134. Wild Turkey. (310)

Naked skin of head and neck, livid blue; general color, copper-bronze with copper and green reflection, each feather with a narrow black border; all the quills, brown, closely barred with white; tail, chestnut, barred with black and a broad subterminal black bar. Tip of tail feathers and upper tail coverts, lighter chestnut.   Length, 3-4 feet.

HAB.—United States, from southern Canada to the Gulf coast, and west to the Plains, along the timbered river valleys; formerly along the Atlantic coast to southern Maine.

Nest, on the ground.

Eggs, ten to fifteen, dark buff or cream color, thickly sprinkled with dark umber-brown.

Within the recollection of people still living, Wild Turkeys were comparatively common along our south-western frontier. Mr. Wagstaff, in his letter already referred to, says: "Wild Turkeys are getting scarce. They were once numerous in Kent and Essex, going about in flocks, but the severe winter of 1842 almost exterminated them. About 1856 they had again become numerous, but are gradually getting fewer in number, as the settler's axe clears away the timber." In the "List of Birds of Western Ontario," it is stated that a nest was found in the county of Middlesex in 1878.

That veteran sportsman and naturalist, Dr. Garnier, of Lucknow, writing under date of December 11th, 1884, says: "I have killed several Wild Turkeys in the county of Kent, and saw one there this season which I did not obtain.

"On the 21st of last October, I had a female of this species in my hands at Chatham station, which had just been killed near by. About four years ago, at Leguis farm, near Mitchell's Bay, I saw three gobblers, two of which I killed right and left, the third was shot the same day by a boy, from whom I bought it for a dollar.

"Most of the domestic turkeys in that section are either the wild species tamed or half-breeds, and are far superior in flavor to the ordinary stock. In 1856, I killed two out of a large flock within half a mile of Hagersville, which at that time consisted of a wagon shop, a toll gate, post-office, and a small shop called a 'store.' I also got a set of nine eggs, and found the female killed by a fox, lying close by, still warm but quite dead."

The Wild Turkey has never advanced into Ontario much beyond the southern boundary, the climate being evidently too severe, and the locality from other causes perhaps not very attractive. The few which still remain are more hunted as they become more rare, and to all appearances the day is not far distant when this valuable game bird will be sought for in vain in the Province of Ontario.

A second species is found, which is believed to be the parent of the domestic stock. It is more of a southern bird, being found chiefly in Texas, New Mexico, Arizona and southward.

ORDER COLUMBÆ. PIGEONS.

FAMILY COLUMBIDÆ. PIGEONS.

GENUS ECTOPISTES SWAINSON.

ECTOPISTES MIGRATORIUS (LINN.).

135. **Passenger Pigeon.** (315)

*Adult male :—*Dull blue above with olivaceous tinge on back; below, dull purplish-red, whitening on vent and crissum; sides of neck, golden and ruby; some wing coverts, black spotted: quills, blackish, with slaty, whitish and rufous edging; middle tail feathers, bluish-black; the others, white or ashy; the inner webs, basally black with chestnut patch; bill, black; feet, coral red. *Female and young:—*Duller and more brownish or olivaceous above; below, dull grayish, with a tawny tinge anteriorly, or quite gray; very young have the feathers skirted with whitish. Length, 15-17; wing, 7-8; tail, about the same.

HAB.—Eastern North America, from Hudson's Bay southward, and west to the Great Plains; straggling westward to Nevada and Washington Territory.

Nest, on bushes or small trees, loosely built of twigs.

Eggs, one or two, pure white.

As its name implies, this is a migratory species, but it has not, like many others, a regular migratory course which it instinctively follows year after year in the same direction. On the contrary, the movements of the Wild Pigeon are quite irregular, and guided only by the instinct which directs the birds in their search for food. A few straggling pairs are still found in the backwoods in Southern Ontario, where they probably breed, but the rising generation of sportsmen can have but inadequate conceptions of the vast flocks of pigeons which used in former years to pass over Hamilton.

They were annually looked for in April. The first who observed them circulated the news, "The pigeons are flying," and early in the morning a regular fusilade would be heard all along the edge of the "Mountain," where at daylight the gunners had taken up their stand at points where the flocks were likely to pass. These annual migrations seemed to attain their maximum in 1854, "the year of the cholera." During that season, from the middle of April till the end of June, flocks could be seen in every hour of every day passing to the west. The summer was unusually warm, and as the heat increased, the birds seemed weak and languid, with scarcely enough energy left to rise above the houses. Vast numbers were killed, till, fortunately for the birds, a rumor got abroad that eating too many pigeons caused the cholera, and then they were allowed to pass on their way unmolested.

After that year the flocks rapidly decreased in number, till at present the annual migrations have entirely ceased.

The food of the species consists chiefly of beech nuts, wild berries, and seeds of different kinds. These disappear as the country comes under cultivation, and the pigeons seek the less settled districts in search of their favorite fare. At present we hear of them being exceedingly abundant in the valley of the upper Mississippi, and as they are quite hardy, they probably extend up north into the "Great Lone Land."

Those who wish to see such flocks of pigeons as used to pass over Ontario will have to follow them there, for in all probability they will never be seen here again.

Such is the record for the species written in 1885, and since that date there has been little to add to their history in Ontario, for with the exception of one or two stragglers occasionally met with in the beech woods, they have left the Province.

I find, however, that two enthusiastic ornithologists have acted on the closing suggestion of the preceding notice, and have followed the birds to the far west. They did not succeed in finding any large roost, but gained much information regarding the habits of the birds, some of which I shall here repeat, for the questions are often asked, "Has the Wild Pigeon become extinct?" "Where has it gone?" It is to Wm. Brewster, Esq., of Cambridge, Mass., that I am indebted for the information which enables me to reply to these questions. In the interest of science he followed the birds to their haunts, and in the *Auk* for October, 1889, he gives a most interesting account of the trip. I would fain give the article in full, but I can only make a few extracts. He says: "In the spring of 1888, my friend Capt. Bendire wrote me that he had received news from a correspondent in central Michigan that Wild Pigeons had arrived there in great numbers and were preparing to nest. Acting on this information, I started at once in company with Jonathan Dwight, jun., to visit the expected 'nesting' and learn as much as possible about the habits of the breeding birds, as well as to secure specimens of their skins and eggs." On reaching Cadillac, on the 8th of May, they found that large flocks had passed north late in April, and the professional pigeon netters expected to hear of their settling not far off, and were busy getting their nets in order. From them much information was gained regarding the habits of the birds. "Our principal

informant was Mr. Stevens, of Cardillac, a veteran pigeon-netter of
large experience, and, as we were assured by everyone we asked
concerning him, a man of high reputation for veracity and careful-
ness of statement."

"'Small colonies,' said Mr. Stevens, 'still breed throughout Michigan, but the largest roost of late years was near Grand Traverse in
1881.' It was about eight miles long. The largest nesting place he
ever visited was in 1876 or 1877. It was near Petosky and extended
north-east past Crooked Lake for twenty-eight miles, averaging three
or four miles wide. For the entire distance every tree of any size
had one or more nests in it, and some were full of them, placed
generally not less than fifteen feet from the ground. The usual
number of eggs is two, but many nests have only one. Both birds
incubate and change regularly. The old birds never feed near the
'nesting,' leaving all the beech mast, etc., there for their young,
many of them going a hundred miles daily for food."

Pigeon-netting, as a business, assumed large proportions at Petosky
at the time referred to. At least five hundred men were engaged in
netting pigeons and sending them to market.

Mr. Stevens thought that each man captured 20,000 birds during
the season, for at one time as many as two car loads were shipped
south on the railroad each day, yet he believed that not one bird out
of a thousand of those present was taken.

The first birds sent to the market yield the netter about a dollar
per dozen, but at the height of the season the price sometimes falls
as low as twelve cents per dozen.

"All the netters with whom we talked believe firmly that there
are just as many pigeons in the west as there ever were. They say
the birds have been driven from Michigan and the adjoining States
partly by persecution and partly by the destruction of the forests,
and have retreated to uninhabited regions, perhaps north of the
Great Lakes in British North America."

In the *Auk*, Vol. VIII., page 310, appear some extracts from a
letter written by Mr. Caleb S. Cope, of West Chester, Pa., who is
well acquainted with the habits and appearance of the birds, having
trapped them many years ago. During the spring of 1887, Mr. Cope,
in company with his son, travelled extensively through the west,
straggling beyond the plains into California, Oregon, western Washington and Vancouver Island. Of the last-named place he says : "I
saw and heard more Wild Pigeons (*Ectopistes*) than I remember to
have ever met with in any other place. The locality where most

of the pigeons were observed was on an extensive plain in Pierce County, Washington, fifteen miles east of Puget Sound, between the Sound and the Cascade Mountains. This fertile plain was dotted over with clumps of pine and fir trees, in many instances bent down by flocks of Wild Pigeons, that feasted on the strawberries which, in some places, were so abundant as to give the sward a scarlet tinge. These flocks numbered several hundreds in each, and during the short time spent there (a few days) plainly showed they were but 'transitory visitants' passing northward, and unlikely to breed in that vicinity."

In Ontario we still see and hear of single birds or pairs observe l in the woods. In the report of the Ornithological Sub-section of the Canadian Institute for 1890-91, are two records of specimens being observed near Toronto, one of which, a young female, was secured.

In Dr. Bergtold's "Birds of Buffalo and Vicinity," the author says that the Wild Pigeon (*Ectopistes migratorias*) is tolerably common, and breeds there.

I have similar reports from many points in Ontario, where I have made inquiry, but on the whole this part of the subject is left much as it has been so long as I can remember. The migratory flocks are now seen no more, and if these vast roosts exist anywhere, it must be beyond the reach of railroads, or assuredly the birds would be slaughtered, as already described, and carried to market. If, on the other hand, they have been greatly reduced in number, the probability is that they will abandon their gregarious habit, as many pairs have already done, and breed throughout the country generally, each pair selecting a nesting place to suit its own ideas, as most other birds do.

Since writing the above, I have received a letter from Vernon Bailey, of the Department of Agriculture, Washington, dated October 14th, 1893, in which the following passage occurs: "I stopped at Elk River, Minn., and while there asked my brothers and several others about Wild Pigeons. My brother had seen two or three flocks of about four to six birds during the past summer, and had killed two pigeons, but had seen no nests, neither had he heard of any of the birds breeding there."

GENUS ZENAIDURA BONAPARTE.

ZENAIDURA MACROURA (LINN.).

136. Mourning Dove. (316)

Brownish-olive, glossed with blue on the crown and nape; below, purplish-red, becoming tawny white on the vent and crissum; neck, metallic-golden; a velvety-black spot on the auriculars and others on the wing coverts and scapulars; middle tail feathers, like back; the rest, ashy-blue at the base, then crossed by a black bar, then white or ashy-white; bill, very slender, black; feet, carmine; the *female* and *young* differ as in the wild pigeon. Length, 11-13; wing, 5-6; tail, 6-7.

HAB.—North America, from southern Maine, southern Canada and Oregon, south to Panama and the West Indies.

Nest, usually in a tree or bush, sometimes on a log or on the ground, composed mostly of twigs.

Eggs, two, pure white.

The Mourning Dove breeds sparingly throughout Southern Ontario, but is more common farther south. It feeds in the open fields on berries, buckwheat and the seeds of certain weeds, but on being disturbed, seeks shelter in the nearest woods.

It is a gentle, timid species, and as it does not occur with us in sufficient numbers to make it worth following, it is seldom disturbed. It is one of the most difficult birds the collector undertakes to handle, the skin being so tender that should the bird be brought down even from a moderate height the fall is almost sure to burst the skin and destroy the specimen. For this reason the greatest care is necessary when preparing the skin for the cabinet.

Ontario is about the northern boundary of this species, and it is, therefore, not so numerous as it is in many places farther south. Dr. Coues says that it is "the most widely and equally diffused of its tribe, abundant in most localities, in some, swarming. 'Millions' in Arizona, for example. Irregularly migratory, imperfectly gregarious; great numbers may be together, but scarcely in compact flocks." They leave Southern Ontario about the end of September, and are not seen again until April. They are recorded as rare summer residents in Manitoba.

OUR BIRDS OF PREY.

WE ARE now about to enter on the consideration of the Birds of Prey, which, as a class, have in all ages borne an evil reputation. Within the past few years, however, their lives have been subjected to a most searching scrutiny, which has placed them in a more favorable light, and I think it very desirable that all lovers of birds should be made aware of this fact, and give their friends justice.

In Scotland, I can well remember, they were regarded by sportsmen as the thieves and robbers of the bird creation, ready at all times to swoop down upon the grouse on the moors, the ducks on the lake or river, or to hang around the farm-house, waiting for a chance to steal the domestic poultry. Nothing suitable was supposed to be safe from their attack. Lambs were carried off from the hillsides by the eagles, and even the human species was not exempt, for instances are on record of young children having been carried away from the cottage door in remote parts of the country.

As the hawks and eagles were believed to live entirely by plunder during the day, so were the owls supposed to take up and carry on the work during the hours of darkness, reluctantly giving way to the other class as daylight appeared.

With such a record, it is not surprising that every man and boy who carried a gun believed it to be his duty to kill a hawk or an owl whenever opportunity offered. The keepers who were entrusted with the care of the game had instructions to this effect, and their industry in carrying out their orders was testified to by the numbers of eagles, hawks and owls which were to be seen nailed upon the outside of their houses.

But even in these old lands an occasional voice was raised in remonstrance against the indiscriminate slaughter of the Birds of Prey, some of which were believed to be totally innocent of the charges brought against them; while others, it was stated, if their habits were properly understood, might be found to be really beneficial. For instance, I have heard at least one observant keeper remark, in

regard to the common Buzzard of Britain, a bird which was believed
to be one of the most destructive among the game, that he took only
the wounded birds, or those of a weak and sickly constitution, which,
if allowed to come to maturity, would evolve offspring of a weak and
degenerate type, little valued by the sportsman.   If this hawk, indo-
lent and sluggish in its habits, never heeding the strong-winged,
active birds, was permitted to carry out the work Nature intended it
to do, in picking off the weaklings, the hardy, healthy character of
the birds would be kept up, and there would be fewer instances of
disease among the grouse to report.

It has been noticed, too, that some of the offenders have noble
qualities, which could not but excite the admiration of the sportsman
who might observe them, but still the cry is kept up, "Kill off the
hawks, and save the game and poultry," and they are still being
killed off to an extent which, to the student of bird-life, is a matter
for deep regret.

In this country the same prejudice, inherited, perhaps, has existed
against the hawks and owls among the people generally, and the
opinion is frequently expressed that they should all be killed off.   In
some parts of the country this opinion became general, and in order
that the work might be done more effectually and expeditiously than
it had been in older lands, the governments of different States in the
Union were petitioned to offer a bounty for the heads of all hawks
and owls, on account of the destruction they caused amongst the
game and poultry.

In several States a liberal bounty was offered, the slaughter
began, and for a time was carried on with pleasure and profit by
those who were working for the money.   In one county alone, in
the State of Pennsylvania, over five thousand dollars was paid in
one year for the heads of hawks and owls; and in Colorado, with
the best intentions, many thousands of hawks were killed.   But this
arrangement, though it satisfied the farmers for a time, was not
satisfactory to the ornithologists, many of whom expressed doubts as
to the correctness of the statements made regarding the extent of the
destruction of poultry.   Others felt sure that *all* the hawks and owls
were not *equally* guilty of the charges brought against them, and
some even went so far as to say that some of the species which
were being destroyed were positively beneficial.

In this shape the subject was brought before the Board of the
Department of Agriculture, at Washington, who, after due consideration, placed the matter in the hands of their indefatigable ornitholo-

gist, Dr. Merriam, and then began to dawn a brighter day for our beautiful and interesting Birds of Prey.

Besides being himself an enthusiastic ornithologist, Dr. Merriam has a large staff of assistants, both in the field and at Washington, and he delegated the work of preparing a "Report on the Hawks and Owls of the United States in their relation to Agriculture," to his first assistant, Dr. A. K. Fisher. The work has now been published, and is a credit to all connected with it. It is neatly bound, has twenty-six beautiful, colored illustrations, and about 200 pages of reading matter, every line of which is interesting to the lover of birds, and specially valuable to those interested in any branch of agriculture. It is seldom that one meets with a report on any subject which is at once so full and so concise; nothing seems wanting which should be there, and nothing is there which could be dispensed with. It is thoroughly scientific, and yet presented in such a form that all can read it intelligently. It is a work which should be spread broadcast over the land, and I feel sure that if the Government takes that course with it, the results will be for the good of generations yet to come.

On taking up the task assigned to him, Dr. Fisher at once found that nothing short of the examination of stomachs of each of the species named in the list would enable him to make a correct report on the nature of their food; and to meet all the requirements of the case, these would have to be obtained at far distant points and at different seasons of the year. This work has engaged much of the time and attention of Dr. Fisher for several years past, while collectors and occasional contributors have kept on sending in specimens from all parts of the continent. No fewer than 2,700 stomachs have been examined, and the results placed on record in the most careful and impartial manner. Week after week, as the various entries were made, it became apparent to the operators that the general result would be a surprise to nearly everyone, for some of the birds for whose destruction a bounty was being paid on account of the amount of game and poultry used by them, were found rarely, if ever, to touch such things at all. They had really been doing a vast amount of good, in the destruction of the innumerable field mice and other small mammals which formed the principal part of their food, for these are the greatest pests with which the farmer has to contend. Others, it was found, had been making themselves useful by devouring countless locusts and grasshoppers, which are often a serious scourge in many districts in the west.

In this way, forty-eight species (with their sub-species) of hawks
and owls have been considered in their relation to agriculture. The
food and habits of each species is given in detail, and, for conve-
nience, the whole have been classified and placed in four different
groups, as follows:

(*a*) Those wholly beneficial or wholly harmless   .   6
(*b*) Those chiefly beneficial   29
(*c*) Those in which the beneficial and harmful qualities seem
    to balance each other   .   7
(*d*) Those positively harmful   6

For our present purpose it will not be necessary to go over the
above list in detail, as many of the species enumerated are not found
in Ontario; but we have all the bad ones and a good many of the
others, and I shall now give these in detail. The total number is
twenty-eight.

(*a*) Those wholly beneficial or wholly harmless.   In this class
    we have only the Rough-legged Buzzard and the Swal-
    low-tailed Kite   2

(*b*) Those *chiefly* beneficial, that is, those which take game
    and poultry occasionally, but kill enough field mice
    and other vermin to *more* than pay for it, leaving a
    small balance in their favor. In this class we have
    the Marsh Hawk, Red-tailed Hawk, Red-shouldered
    Hawk, Swainson's Hawk, Broad-winged Hawk, Spar-
    row Hawk, Barn Owl, Long-eared Owl, Short-eared
    Owl, Great Gray Owl, Barred Owl, Richardson's Owl,
    Saw-whet Owl, Screech Owl, Snowy Owl, Hawk Owl .   16

(*c*) Of those whose good and bad deeds balance each other,
    we have the Great-horned Owl, Golden Eagle, Bald
    Eagle, Pigeon Hawk   4

(*d*) Of those which are positively injurious, we have the Gyr-
    falcon, Goshawk, Cooper's Hawk, Sharp-shinned Hawk,
    Peregrine Falcon, Fish Hawk   6

It is very gratifying to find the report so favorable to our Birds
of Prey, and even of Class *d* a few words in favor might be said.
Of course, Dr. Fisher, in the position of judge interpreting the law,
could not but condemn where the evidence was so conclusive; but
from our standpoint in Ontario, we cannot complain very much, for
the two largest and most destructive, the Gyrfalcon and Goshawk,
are peculiar to the far north, the former, especially, rarely coming
within our boundary; and even the Peregrine, though ranked as an

Ontario species, is by no means common. Occasionally, we hear of him striking down a duck at the shooting stations, but we also know that his food at such places consists to a great extent of the different species of waders known as "Mud Hens."

The Fish Hawk has been put upon the black list, because he kills fish. That he does this cannot be denied, but he is a magnificent bird, whose fine presence adds a point of beauty to many a lovely landscape by lake and river, and surely it would be poor policy to kill off this interesting feathered fisher because he takes a few suckers, with now and then a trout or bass by way of a change.

This leaves only two of the six to be considered, and I fear that not much can be said in their favor. Cooper's Hawk, the larger of the two, is not so common throughout Ontario as the Sharp-shinned, but he is the one who will come back day after day to rob the roost or the pigeon loft, and who watches to catch the grouse and quail wherever they appear. This is he, in short, whose misdeeds have brought disgrace on the whole class, and given occasion for the bad name which still clings to them all.

The Sharp-shinned is a smaller bird, content with humbler fare, but his eyes and claws are sharp as needles. He is always on the alert, and few birds escape which he tries to capture. The record is bad for these two, and Dr. Fisher says of them: "Unquestionably both species should be killed wherever and whenever possible."

I can hardly follow the Doctor to this extent, because we have been a long time in discovering the true standing of other species, and it may be that the two now under consideration have some redeeming points with which we are not yet acquainted. At all events we must give them both credit for reducing the number of *Passer domesticus*, which they do to a considerable extent. Every sparrow they take relieves us of a nuisance, and the greater the number of sparrows they use the less of anything else will be required for their support. Let us then be patient with them, give them another chance, and enjoy the fine exhibitions they give us of their dexterity while capturing their prey in the near neighborhood of our dwelling houses.

ORDER RAPTORES. BIRDS OF PREY.

SUBORDER SARCORHAMPHI. AMERICAN VULTURES.

FAMILY CATHARTIDÆ. AMERICAN VULTURES.

GENUS CATHARTES ILLIGER.

CATHARTES AURA (LINN.).

137. Turkey Vulture. (325)

Blackish-brown; quills, ashy-gray on their under surface; head, red; feet, flesh-colored; bill, white; skin of the head corrugated, sparsely beset with bristle-like feathers; plumage, commencing in a circle on the neck; tail, rounded. Length, about 2½ feet; extent, 6; wing, 2; tail, 1.

HAB. Temperate North America, from New Jersey, Ohio Valley, Saskatchewan region, and Washington Territory southward to Patagonia. Casual northward on the Atlantic coast to Maine. Breeds generally in communities.

Nest, on the ground, or in a hollow log or stump.

Eggs, usually two, creamy-white, spotted and blotched with different shades of brown.

So far as I am aware, the Turkey Buzzard has been observed in Ontario only in the south-western portion of the Province.

Mr. Wagstaff, in the letter already quoted, says: "Turkey Buzzards are frequently seen in Essex sailing around in search of carrion." I once saw it at Baptiste Creek some years ago, but have not heard of its having been seen farther east. Dr. Coues says: "This species has a curious habit of 'playing possum' by simulating death when wounded and captured, the feint being admirably executed and often long protracted."

The Turkey Buzzard is more frequently seen to the west of Ontario than to the east of it. A nest was found by Mr. Arnott at Kerwood, Middlesex County, in 1891.

It is reported as being generally distributed throughout Michigan and Minnesota, and has also been frequently seen in Manitoba.

Dr. Bell says: "I have shot the Turkey Buzzard on the upper Assiniboine, but have never heard of it near Hudson's Bay. The locality referred to is in about latitude 52°. It had not before been noticed north of Minnesota, while on the eastern part of the continent it is rarely found north of New York, or about latitude 41°."

It is a rare visitor to the south-west of Ontario, and to the east I have not heard of its having been observed.

SUBORDER FALCONES. VULTURES, FALCONS, HAWKS,
BUZZARDS, EAGLES, KITES, HARRIERS, ETC.

FAMILY FALCONIDÆ. VULTURES, FALCONS, HAWKS,
EAGLES, ETC.

SUBFAMILY ACCIPITRINÆ. KITES, BUZZARDS, HAWKS,
GOSHAWKS, EAGLES, ETC.

GENUS ELANOIDES VIEILLOT.

ELANOIDES FORFICATUS (LINN.).

138. Swallow-tailed Kite. (327)

Head, neck and under parts, white; back, wings and tail, lustrous black; feet, greenish-blue; claws, pale. Length, *female*, 23-25; wing, 16-17½; tail, 14; *male*, a little smaller.

HAB.—Southern United States, especially in the interior, from Pennsylvania and Minnesota southward, throughout Central and South America; westward to the Great Plains. Casual eastward to southern New England. Accidental in England.

Nest, on a tree, constructed of sticks, hay, moss, etc.

Eggs, two, rarely three, whitish, blotched and spotted with chestnut-brown.

In the course of its extensive wanderings, this bold, dashing Kite has been known to visit Ontario. In the "List of Birds of Western Ontario," mention is made of a pair having spent a summer about eight miles north-west of London, and there is also a record of one having alighted on the top of a flagstaff at Ottawa, where it was closely examined through a glass and satisfactorily identified.

The food of this species consists chiefly of snakes, lizards, grass-hoppers, locusts, etc., and the fact of these not being abundant in Ontario readily accounts for the absence of the birds. According to Audubon, the Swallow-tailed Hawk feeds chiefly on the wing, pounces on his prey upon the ground, rises with it and devours it while flying. "In calm weather," he further observes, "they soar to an immense height, pursuing the large insects called musquito hawks, and performing the most singular evolutions that can be conceived, using their tail with an elegance peculiar to themselves."

In Dr. Fisher's report it is classed as harmless in its relation to agriculture.

13

GENUS CIRCUS Lacépède.

CIRCUS HUDSONIUS (Linn.).

139. Marsh Hawk. (331)

Adult male:—Pale bluish-ash, nearly unvaried, whitening below and on upper tail coverts; quills, blackish towards the end. Length, 16-18; wing, 14-15; tail, 8-9. *Female:*—Larger; above, dark brown, streaked with reddish-brown; below, the reverse of this; tail, banded with these colors. *Immature male* is like the female, though redder, but in any plumage the bird is known by its white upper tail coverts and generic character.

Hab.—North America in general, south to Panama.

Nest, on the ground, composed of twigs and dried grass.

Eggs, four or five, pale greenish-white, sometimes spotted faintly with light brown or lilac.

In Southern Ontario the Marsh Hawk in the red plumage is a well-known bird, but in the blue phase it is seldom seen. It arrives from the south in April, as soon as the ice is gone, and from that time till November it may usually be seen coursing over the marshes and moist meadows in search of its food, which consists of mice, small birds, snakes, frogs, worms, etc. It breeds sparingly at the St. Clair Flats, becoming quite numerous in the fall on the arrival of those which have bred farther north.

It is one of our most abundant and widely dispersed birds, being found throughout the whole of North America. In Ontario it is found breeding in all suitable places. Dr. Bell mentions its occurrence at Hudson's Bay; it is common among the marshes in the North-West, and also appears in Alaska. One of the most remarkable characteristics of this species is its habit of turning somersaults in the air, which is thus described by Mr. Nelson: "While I was at the Yukon mouth, on May 19th, 1879, a pair of hawks was seen repeatedly crossing the river on different days at a certain point, the leader always performing, as he went, a succession of curious antics. He would turn over and over half a dozen times in succession, like a Tumbler Pigeon, and after descending nearly to the ground he would mount to his former height and repeat the performance, so that his progress became a perfect series of these evolutions. The other bird always flew slowly and smoothly along, as if enjoying the performance of its companion." Nearly all of those seen in Southern Ontario are in the brown plumage, and I have never observed one of them indulge in the eccentricities above described. It may be that they are peculiar to the old male in the blue dress.

This is one of those birds regarding which the idea prevailed that they took game and poultry, and should therefore be killed whenever they came within reach. But the evidence shows that the Marsh Hawk does so only on very rare occasions. It does so much good by the destruction of injurious rodents, that I hope in future it will be allowed to follow its course in peace.

GENUS ACCIPITER BRISSON.

SUBGENUS ACCIPITER.

ACCIPITER VELOX (WILS.).

140. Sharp-shinned Hawk. (332)

Feet, extremely slender; bare portion of tarsus longer than middle toe; scutellæ, frequently fused; tail, square. Above, dark brown (deepest on the head, the occipital feathers showing white when disturbed), with an ashy or plumbeous shade, which increases with age till the general cast is quite bluish-ash; below, white or whitish, variously streaked with dark brown and rusty, finally changing to brownish-red (palest behind and slightly ashy across the breast), with the white then only showing in narrow cross-bars; chin, throat and crissum mostly white with blackish penciling; wings and tail, barred with ashy and brown or blackish; the quills, white, barred basally; the tail, whitish tipped; bill, dark; claws, black; cere and feet, yellow. *Male:*—10-12; wings, 6-7; tail, 5-6. *Female:*—12-14; wings, 7-8; whole foot, 3½ or less.

HAB.- North America in general, south to Panama.

Nest, in trees.

Eggs, four or five, grayish-white, shaded with purple and splashed with brown, in endless shapes and shades.

This is a rather common summer resident in Southern Ontario, smaller in size than Cooper's Hawk, but similar in markings. It lives chiefly on small birds, and nothing can exceed the impetuosity with which it dashes down and captures them by sheer power of flight.

"Many have been the times," says Audubon, "when watching this vigilant, active and industrious bird, have I seen it plunge headlong into a patch of briers, in defiance of all thorny obstacles, and passing through, emerge on the other side, bearing off with exultation in its sharp claws a finch or a sparrow which it had surprised at rest."

This species is much given to variation in size and markings, making it difficult at times to distinguish between a large Sharp-shinned and a small Cooper's Hawk. In the present species, the legs and feet are relatively longer and more slender than in the other, the term "sharp-shinned" being no misnomer. They all seem to retire from Ontario in the fall, for none are observed during winter.

In the spring they appear about the middle of April, and are soon common throughout the country, making their way north through British America to Alaska, where, Mr. Nelson says, they are much prized by the natives, who use their feathers for shafting their arrows and for ornamenting their dancing costumes.

This species, Dr. Fisher says, lives entirely on flesh, and is so destructive to small birds, young grouse and chickens that "it should be killed whenever possible." If it could only be compelled to confine its attention to the European House Sparrow, it would soon repay us for all the other damage it has done.

ACCIPITER COOPERI (Bonap.).

141. Cooper's Hawk. (333)

Feet, moderately stout; bare portion of tarsus shorter than middle toe; scutellæ, remaining distinct; tail, a little rounded. Colors and their changes as in *A. velox*; larger. *Male:*—16-18; wing, 9-10; tail, 7-8. *Female:*— 18-20; wing, 10-11; tail, 8-9. Whole foot, 4 or more.

Hab.—North America in general, south to southern Mexico.

Nest, in trees, mostly in evergreens. The deserted nest of some other species is often used.

Eggs, four or five, white, tinged with green; sometimes faintly spotted with brown.

This is one of the Chicken Hawks, and it well deserves the name from the havoc it makes among the poultry. It is most common in spring and fall, but sometimes appears suddenly in winter, and shortens the days of *Passer domesticus* when nothing better is available.

Cooper's Hawk breeds sparingly throughout Southern Ontario, apparently preferring the vicinity of large marshes, where Blackbirds, Rails, etc., are easily obtained.

Extraordinary migrations of hawks are sometimes seen in the fall, when for two or three days in succession, along a certain section of country, individuals of this and the preceding species are continually in sight. Flocks of this description have often been observed at Point Pelee, near the western extremity of Lake Erie, where the birds probably gather when working their way round the west end of the lake, in preference to going across. Although a few remain during the winter, this species is mostly migratory, arriving in April and leaving in October.

This, and its near relative, the Sharp-shinned Hawk, are the two which Dr. Fisher says are so destructive to small birds, and young grouse and chickens, that they should be destroyed whenever it is possible.

Cooper's Hawk is the larger and stronger of the two. It is particularly fond of pigeons, and, if permitted, will return to the dovecot day after day till the last pigeon is taken.

It is a most active bird on the wing, the long tail, short rounded wings, and quick sight enabling it to capture at will anything suitable it may surprise while skimming over the meadows or through the open woods.

It is generally distributed throughout Ontario, but is more common farther south. In Alaska it has not been observed.

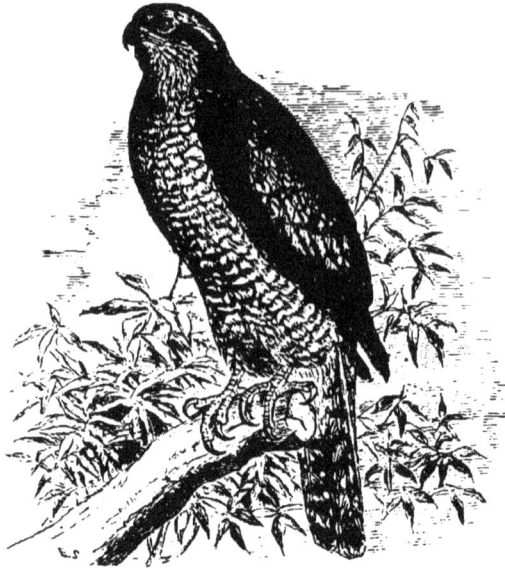

ACCIPITER ATRICAPILLUS (Wils.).

142. **American Goshawk.** (331)

Adult:—Dark bluish-slate blackening on the head, with a white superciliary stripe; tail, with four broad dark bars; below, closely barred with white and pale slate, and sharply streaked with blackish. *Young:*—Dark brown above, the feathers with pale edges, streaked with tawny-brown on the head and cervix; below, fulvous white, with oblong brown markings. *Female:*—2 feet long; wing, 14 inches; tail, 11. *Male:*—Smaller.

Hab.—Northern and eastern North America, breeding mostly north of the United States, south in winter to the Middle States. Accidental in England.

Nest, in trees, composed of sticks, twigs and weeds, lined with grass and strips of bark.

Eggs, two or three, soiled white, sometimes faintly blotched with brown.

The Goshawk and the Peregrine Falcon were both much prized in the olden time when hawking was a princely amusement in Europe, and the same spirit and courage which was the admiration of lords and ladies fair in those ancient days still characterize the birds in their native haunts. They never fail to attract the attention of the sportsman, as, unencumbered by hood or bell, they carry terror and dismay among the ranks of the water-fowl.

In Ontario, the Goshawk is an irregular winter visitor, sometimes appearing in considerable numbers, and again being altogether wanting for several years in succession. In the young plumage, it bears some resemblance to Cooper's Hawk, but is always much larger in size, and is more bold and daring in proportion, frequently carrying off poultry from the very doors of houses in the suburbs of the city.

It is one of the handsomest species of the family. A small-sized adult male in my collection is the finest I have ever seen, a perfect model in symmetry, the colors clear and bright, and the whole plumage smooth and compact, admirably suited for passing rapidly through the air with the least possible resistance.

In the "Birds of Alaska," Mr. Nelson says: "This is the handsomest, as well as one of the most abundant, of the birds of prey resident in Northern Alaska. It is present everywhere throughout all the wooded region, and in spring and autumn, especially during the latter season, it is a common visitor to the open country bordering on the shore of Behring Sea and the Arctic coast, and is a characteristic bird in the fur countries, breeding nearly to the Arctic coast.

"The Goshawk has a bad reputation among the natives, from its habit of stealing birds from their snares, as well as for hunting the Ptarmigan, upon which, at certain seasons, the Eskimo depend largely for a food supply. Although many of these birds remain in the north during the winter, I believe that a considerable number of them pass farther to the south."

This is one of the six *injurious* species named in Dr. Fisher's report, but we see it so seldom it does not annoy us much.

GENUS BUTEO CUVIER.

BUTEO BOREALIS (GMEL.).

143. **Red-tailed Hawk.** (337)

Four outer quills emarginate on inner web. *Adult:*—Dark brown above, many feathers with pale or tawny margins, and upper tail coverts showing much whitish; below, white or reddish-white, with various spots and streaks of different shades of brown, generally forming an irregular zone on the abdomen; tail above, bright chestnut-red, with subterminal black zone and narrow whitish tip, below pearly-gray; wing coverts, dark. *Young:*—With the tail grayish-brown barred with darker, the upper parts with tawny streaking. A large, stoutly-built hawk. *Female:*—23; wing, 15½; tail, 8½. *Male:*—20; wing, 14; tail, 7.

HAB.—Eastern North America, west to the Great Plains.

Nest, placed on a high tree, composed of sticks, twigs, grass, moss, etc.

Eggs, two to four, dull white, sometimes blotched with rich brown of different shades.

This is a large and powerful bird, strong of wing and stout of limb, but incapable of performing the feats of dexterity common to the hawks and falcons. It is most frequently seen sitting bolt upright on a stub in a field, or by the edge of the woods, carefully scrutinizing the ground below in search of the small quadrupeds on which it feeds. It is resident in Ontario, having been seen both in summer and winter, but is most frequently observed during the period of migration in spring and fall, from which it may be inferred that many individuals spend the winter farther south. Occasionally, in spring, this species may be seen singly, or in pairs, soaring to a vast height, sailing round in wide circles, apparently enjoying the warm sunshine and the return of life to the landscape below.

The Red-tail breeds in Southern Ontario, is generally distributed throughout the Province, and is included in the list of birds observed by Prof. Macoun in the North-West.

It is one of the " hen-hawks " which have generally been supposed to live on game and poultry, but recent investigations show that it does not touch either, save when sorely pressed with hunger, and then it will eat carrion sooner than starve. Its principal food consists of the smaller mammals and reptiles, and this fact entitles it to the protection of the farmer. It has been placed in Class b, for the good it does is supposed to be in excess of the evil.

BUTEO BOREALIS CALURUS (Cass.).

144. Western Red-tail. (337b)

The extreme form is chocolate-brown or even darker, quite unicolor, with rich red tail crossed by several black bars, from which the erythro-melanism grades insensibly into ordinary borealis. The usual case is increase over borealis of dark rufous and dusky shades in bars and spots underneath, particularly on the flanks and crissum, and presence of other than the subterminal black bar on the tail.

HAB.—Western North America, from the Rocky Mountains to the Pacific, south into Mexico, casually to Illinois, Minnesota, Michigan and Canada West.

Nesting habits and eggs identical with those of the common Red-tail.

For the privilege of including this sub-species among the " Birds of Ontario," I am indebted to the following notice which appears in the *Auk*, Vol. V., page 205 :

"I have obtained from Mr. J. Dodds, of St. Thomas, Ontario, a fine adult male of the Western Red-tail (*Buteo borealis calurus*) which was killed near that place in the fall of 1885, by John Oxford, This seems to be the first recorded occurrence of this species in Ontario."—W. E. SAUNDERS, London, Ontario.

Dr. Coues, who is always averse to making too many sub-species, after describing the ordinary Red-tail, says of this subject : "Such is the ordinary Hen-hawk, so abundant in eastern North America, where it is subject to comparatively little variation. In the west, however, where it is equally numerous, it sports almost interminably in color, and not always conformably with geographical distribution. Several of these phases have received special names as given beyond. I am willing to spread them upon my page, but too much of my life is behind me for me to spend time in such trivial mutabilities."

As the subject may still be interesting to younger students, I place the record here for their consideration.

BUTEO LINEATUS (GMEL.).

145. Red-shouldered Hawk. (339)

Four outer primaries emarginate on inner web ; general plumage of the adult of a rich fulvous cast ; above, reddish-brown, the feathers with dark brown centres ; below, a lighter shade of the same, with narrow dark streaks and white bars ; quills and tail, blackish, conspicuously banded with pure white ; the bend of the wing, orange-brown. *Young:* Plain dark brown above ; below, white with dark streaks ; quills and tail, barred with whitish. Nearly as long as *B. borealis,* but not nearly so heavy ; tarsi, more naked. *Female :*—22 ; wing, 14 ; tail, 9. *Male :*—19 ; wing, 13 ; tail, 8 (average).

HAB. Eastern North America, west to Texas and the Plains, south to the Gulf coast and Mexico.

Nest, in trees, composed of sticks and twigs, lined with grass and a few feathers.

Eggs, two to four, variable in color, usually dull white, blotched with rich brown.

In Southern Ontario this species is a common summer resident, breeding freely in the less settled parts of the country, where it is more frequently seen than any other of the "chicken hawks."

In the fall it becomes quite numerous on the return of those which have bred farther north, accompanied by their families.

Southern Ontario is probably its centre of abundance in summer, for though it occurs in Manitoba and has been found by Dr. Bell at York Factory, it is not common so far north. During the winter it has not been observed.

Like others of the family, this species varies greatly in plumage according to circumstances. The young birds do not show any of the rich reddish-orange of the adult, and were at one time described as a separate species under the name of Winter Falcon. From western Texas to California, and south into Mexico, the colors become much brighter and more decided, which has led to this western form being described as a sub-species under the name of *Buteo lineatus elegans* (Cass.). Occasionally we meet here with an adult in full plumage which might well be included in this group, but generally all are much brighter in the west.

This is another of the chicken hawks which has borne the reputation of robbing the poultry yard, but it is pleasant to observe that the accusation is not verified by the recent examination of stomachs. Dr. Fisher says: "The diet of the Red-shouldered Hawk is probably more varied than that of most other birds of prey. For example, the writer has found in the stomachs of the different individuals which have come under his notice the remains of mammals, birds, snakes, frogs, fish, insects, centipedes, spiders, crawfish, earthworms and snails, which represent eleven classes of animal life. This Hawk is very fond of frogs, and, although these batrachians are mentioned by Audubon and other writers as forming a very considerable portion of their sustenance, yet mice furnish fully sixty-five per cent. of their food.

"Besides this very injurious group of rodents, other small mammals, such as squirrels, young rabbits, shrews and moles are taken. Some authors insist that the Red-shouldered Hawk is destructive to poultry, but the writer in all his field experience has never seen one attack a fowl, nor has he found the remains of one in the stomachs of those examined."

With this record, he is fairly entitled to a place in the class whose good acts are in excess of the evil.

BUTEO SWAINSONI Bonap.

146. Swainson's Hawk. (342)

It is hardly possible, within the limited space at my disposal, to give anything like a detailed description of the various phases of plumage which this interesting Buzzard assumes, according to age, sex, or the season of the year. Suffice it to say, that individuals differ so much from one another as to have led to the description of about a dozen different individuals as new species, all of which are now attributable to *Buteo swainsoni*.

In measurement, this species is about the same as its nearest relative, the Red-tail, averaging about 20 inches in length by about 50 in extent, but it is less stoutly built, has the wings longer and more pointed, and has only three of the primaries emarginate, whereas the Red-tail has four. The entire upper parts are dark brown, many of the feathers with tawny edgings, those on the head showing white when disturbed. Tail feathers, ashy-gray, crossed with numerous dark bars, and tipped with yellowish white; upper tail coverts, chestnut and white, with blackish bars; under parts, white, more or less shaded with chestnut. A broad pectoral area of bright chestnut, usually with a glaucous shade, and displaying sharp black shaft lines; this area contrasting strongly with the pure white throat.

In the younger birds, the upper parts are much as already described; the lower parts, including the lining of the wings, are nearly uniform fawn color, thickly spotted with blackish-brown. These large dark spots, for the most part circular or guttiform, crowd across the fore breast, scatter on the middle belly and tibiæ, and are wanting on the throat. In all stages of plumage the iris of the eye is brown.

Hab.—Western North America, from Wisconsin, Illinois, Arkansas and Texas to the Pacific coast; north to the Arctic Regions, and south to Buenos Ayres. Casual east to Massachusetts.

Nest, in a tree, at a height varying from ten to forty feet from the ground.

• Eggs, two to four, greenish-white or buffy-white, often stained or blotched with rusty-brown.

Dr. Coues gives an admirable history of this species in his " Birds of the North-West " (page 356), from which I shall here make a few extracts:

"This large Hawk is very abundant in Northern Dakota, where it came under my almost daily observation during the summer of 1873." "Swainson's Buzzard may be seen anywhere in the region mentioned—even far out on the prairie, miles away from the timber, circling overhead, or perched on the bare ground. In alighting, it generally takes advantage of some little knoll commanding a view around, though it often has no more prominent place than the heap of dirt from a badger's hole, from which to cast about for some imprudent gopher espied too far from home, or still more ignoble game."

"The quarry of Swainson's Buzzard is of a very humble nature. I never saw one swoop upon wild fowl or Grouse, and though they strike rabbits, like the Red-tails, their prey is ordinarily nothing larger than gophers. Though really strong and sufficiently fierce birds, they lack the 'snap' of the Falcons and Asturs; and 1 scarcely think they are smart enough to catch birds very often. I saw one make the attempt on a Lark Bunting. The Hawk poised in the air, at a height of about twenty yards, for fully a minute, fell heavily, with an awkward thrust of the talons—and missed. The little bird slipped off, badly scared, no doubt, but unhurt, while the enemy flapped away sulkily, very likely to prowl around a gopher-hole for his dinner, or take potluck at grasshoppers."

From the foregoing it will be seen that the home of Swainson's Buzzard is on the prairies of the North-West, while in Ontario it is only a casual visitor. I first met with it at an agricultural fair in Hamilton in 1865, where a young specimen was observed in a collection which was competing for a prize. Being called upon to name the species to which it belonged, I turned to the works of reference then available, and made it out to be *Buteo bairdi* (Hoy.), which is now known to be the young of *Buteo swainsoni*. Since that time, I have occasionally seen birds in similar plumage flying overhead, but did not again meet with it close enough for examination till the present summer (1886), when I saw one in the hands of a local taxidermist, where it had been left to be "stuffed." It, too, was a young bird, but in fine plumage, with the characteristic markings fully displayed.

When we have more naturalists among our sportsmen, such a bird as this will be more frequently brought to light. At present, should a hawk come along, when there is nothing better in sight, it is killed in the interest of the game, but is seldom picked up.

In *Forest and Stream* for December, 1888, Dr. Merriam gives a most interesting account of a flock of this species which he saw feeding on grasshoppers in Oregon. I have room only for a short extract, to show the influence these birds exercise in the interest of the farmer:

"One hundred and fifty hawks were counted, which did not quite include all those present. We counted as many as thirteen in one tree. Two of the three stomachs we examined contained grasshoppers and nothing else; the third contained, in addition to grasshoppers, the head of a meadow mouse. One contained eighty-eight grasshoppers, another ninety-six, and the third 106. Assuming that each hawk captured 200 grasshoppers a day, and that there were 200 hawks, the daily catch would thus be 40,000 grasshoppers. At this rate these hawks would destroy 280,000 grasshoppers in a week, or 1,200,000 in a month." And yet this is one of the species for whose destruction a bounty has recently been offered !

BUTEO LATISSIMUS (WILS.).

147. Broad-winged Hawk. (343)

Three outer primaries emarginate on inner web; above, umber-brown, the feathers with paler, or even with fulvous or ashy-white edging, those of the hind head and nape cottony-white at base; quills, blackish, most of the inner webs white, barred with dusky; tail, with three broad dark zones alternating with narrow white ones, and white tipped; conspicuous dark maxillary patches: under parts, white or tawny, variously streaked, spotted or barred with rusty or rufous, this color usually predominating in adult birds, when the white chiefly appears as oval or circular spots on each feather; throat, generally whiter than elsewhere, narrowly dark-lined. In the *young* the upper parts are duller brown, varied with white; the under parts, tawny-whitish, with linear and oblong dark spots, the tail, grayish-brown, with numerous dark bars. *Female*, 18; wing, 11; tail, 7. *Male*, less.

Hab.—Eastern North America, from New Brunswick and the Saskatchewan region to Texas and Mexico, and thence southward to Central America, northern South America and the West Indies.

Nest, in a tree, built of sticks and twigs, lined with grass and leaves.

Eggs, two or three, grayish-white, marked with spots and blotches of umber-brown.

This species was first described by Wilson, who met with two individuals in the woods near the Schuykill, but does not appear to have seen it again.

In Southern Ontario the Broad-winged Hawk is often very common in the spring. Toward the end of April, or early in May, should the weather be clear, great numbers are seen soaring at a considerable height, and moving in circles toward the North-West.

About the same time, singly or in pairs, it may be met with in the woods, usually sitting quietly on the lower branch of a tree near some wet place watching for frogs. A few pairs remain during summer, but the greater number pass on to the North-West, and in winter none have been observed.

Late in April, or early in May, the Broad-winged Hawk has been reported at Ottawa, Toronto, Hamilton and London. It does not go much farther north, for it is rare in Manitoba, and is not found in Alaska.

Its food consists chiefly of small mammals, insects, reptiles and frogs, and it is not known ever to have touched poultry, so that on this account it is placed in Class *b*, the good work being in excess of the evil.

GENUS ARCHIBUTEO BREHM.

ARCHIBUTEO LAGOPUS SANCTI–JOHANNIS (GMEL.).

148. American Rough-legged Hawk. (347*a*)

Below, white, variously dark colored, and often with a broad black abdominal zone, but generally no ferruginous; above, brown, varying from dark chocolate in the adult to light umber in the young; the back, scapulars and shorter quills, strongly cinereous; the head above, more or less white, dark streaked; upper tail coverts and tail at base, white, the former tipped with blackish; the latter barred near the tip with one, and sometimes with several bands of black or dark brown. In this plumage the bird has been known as *A. lagopus*, the Rough-legged Buzzard, while to a melanotic variety of the same, found in this country only, the name *sancti-johannis* has been given. This variety is entirely glossy-black, except the occiput, forehead, throat, inner webs of quills, base of tail and broad tail-bars, white. As it is now generally conceded that these are varieties of the same species, the original name, *lagopus*, is retained, and the American form considered a geographical variety of the European characterized as variety *sancti-johannis*. Length, about 2 feet; wing, 16-17; tail, 8-10.

HAB.--Whole of North America north to Mexico, breeding chiefly north of the United States.

Nest, on trees or rocks, composed of sticks, grass, weeds and other material matted together.

Eggs, two or three, soiled white, blotched with reddish-brown.

Another large and powerful bird which, from some cause, seems contented with very humble fare, living chiefly on mice, lizards, frogs, etc, while its appearance would lead us to suppose it capable of capturing much larger game. It is sometimes found in a melanotic state, the plumage being nearly black, and in this garb it was formerly described as a distinct species, but this idea has now been abandoned.

It can always be recognized by the legs, which are feathered down to the toes, the latter being short.

In Southern Ontario this is only a visitor during the season of migration, being most plentiful in the fall, when it is often seen frequenting the marshy shores of Hamilton Bay. It has not been observed during the breeding season, neither does it occur in winter.

This species in appearance and habits bears some resemblance to the owls, its full, soft plumage, feathered legs, large head and flat face all pointing in that direction. It is also observed to be fond of hunting in the dim twilight, after all the other hawks have retired and only the owls are abroad. Its manner is quite respectable. It is neither given to whining nor to ferocity, but is one of the "hen-hawks" for whose head a bounty has been offered. We can readily imagine the dignified look of injured innocence he would assume, if he could be placed on the perch and made to understand that he was charged with robbing the hen roost, because the evidence now goes to show that he never touched a hen in his life.

The Rough-legged Buzzard has the highest record of any of our birds of prey, for its food consists almost entirely of the small mammals which are the farmers' worst enemies. It is rather irregularly distributed, but has been found breeding in Labrador, and north even within the Arctic circle. On the sea coast and along the Yukon River it is replaced by the Old World form.

GENUS AQUILA BRISSON.

AQUILA CHRYSAËTOS (LINN.).

149. Golden Eagle. (349)

Dark brown, with a purplish gloss; lanceolate feathers of head and neck, golden-brown; quills, blackish. In the young, tail, white, with a broad terminal black zone. About 3 feet long; wing, upwards of 2 feet; tail, a foot or more.

HAB.—North America, south to Mexico. Northern parts of the Old World.

Nest, an accumulation of sticks, usually placed on an inaccessible rocky crag, more rarely in a tree.

Eggs, two, rarely three, soiled white, marked with brown or chestnut.

This fierce and daring Eagle has its home among the rugged and inaccessible cliffs of Canada East; but in the fall it is seen following the flocks of water-fowl, which, at this season, visit the lakes to rest and recruit themselves as they travel southward. Some years ago I asked a boy, whose home I considered a favorable point for getting birds of prey, to shoot any hawks or owls he saw and bring them to me. A few days afterwards I saw him approaching my house with a

sack over his shoulder, which, judging from the bulk, might contain
a dozen hawks, but great was my surprise when he shook out a fine
large female Golden Eagle, he had shot that morning as it flew over
the place where he happened to be standing.

Shortly afterwards I got a young male which had been caught
near Stony Creek, and I have seen several that were procured near
Toronto. Dark brown Eagles are often observed hovering along the
shores of Lake Ontario during the fall, but at a distance it is impossible to distinguish between this and the young of the Bald Eagle,
which is also uniformly brown throughout. The quickest way of
identifying the species, on close inspection, is by referring to the legs,
which, in the Golden Eagle, are feathered down to the toes, differing
as much in this respect from the Bald Eagle as the Rough-legged
Buzzard does from any of the other hawks.

Except as a wanderer during the season of migration, the Golden
Eagle is seldom seen in Ontario.

In the far North-West it is more common. In the "Birds of
Alaska," Mr. Nelson says regarding it: "It nests rather commonly
on the Lower McKenzie and Anderson Rivers, and extends its range
to the Arctic shores of the mainland, and perhaps reaches some
of the adjacent islands north of British America. In spite of the
courageous and even fierce character of this fine bird, it sometimes
descends to feed upon carrion. On one occasion a pair was disturbed by a friend of mine while they were feeding upon the remains
of a hog in Northern Illinois. As my friend approached, the birds
arose and swooped fiercely at him. Both birds were shot almost at
the muzzle of the gun. The first fell dead almost at his feet, but this
apparently served only to increase the rage of the survivor, which
renewed the attack until it, too, was disabled."

In sections of the country where prairie dogs, rabbits or gophers
are abundant, the Golden Eagle is very beneficial, confining its
attention mainly to these noxious animals; but in places where wild
game is scarce, it is often very destructive to the young of domestic
animals.

From its record, where best known, it has been placed in Class c
its good and bad deeds being about equal in quantity.

GENUS HALIÆETUS Savigny.

HALIÆETUS LEUCOCEPHALUS (Linn.).

150. Bald Eagle. (352)

Dark brown; head and tail, white after the third year; before this, these parts like the rest of the plumage. About the size of the last species. Immature birds average larger than adults.

Hab.—North America at large, south to Mexico.

Nest, of huge dimensions, built of sticks, placed on a tree.

Eggs, two, white, often soiled by their surroundings.

This is more frequently seen than the preceding species, and may be considered resident, for it is often observed during winter, and breeds in suitable places throughout the country, usually on or near the shore of a lake. In a letter from Dr. McCormick, dated Breeze Place, Pelee Island, June 12th, 1884, the writer says: "I chanced to observe an interesting incident a few days since, showing what looked very much like reasoning powers in a Bald-headed Eagle. The wind was blowing quite strong from the west, and the eagle had caught a large fish. Rising in the air with his dying prey in his talons, he tried to fly directly to windward, towards his nest, but the wind was too strong, and after several unsuccessful attempts, he dropped the fish (now dead) into the water. Then flying off toward the north for some distance, apparently to try the wind in that direction, and finding he could progress more easily, he turned round, went back to the fish, took it up again in his claws, and flying north with a beam wind, made the shore. Then in shelter of a friendly grove of trees, he flew away toward the west and his nest, with his scaly treasure, thus exercising what appeared to be a reasoning process of cause and effect."

A favorite haunt of this species used to be along the Niagara River below the Falls, where they would sit on the dead trees by the river bank and watch for any dead or dying animals that came down the stream. This habit becoming known to collectors, a constant watch was kept for the appearance of the birds. Many were picked off with the rifle, and although a few still visit the old haunts, their numbers are greatly reduced.

Twenty years ago, I knew a youth who shot one of these birds as it flew over him while he lay concealed among the rushes on the shore of Hamilton Bay watching for ducks. On taking it up, he found an unusual appendage dangling from the neck, which proved,

14

on examination, to be the bleached skull of a weasel. The teeth had
the "death grip" of the skin of the bird's throat, and the feathers
near this place were much confused and broken.

The Eagle had probably caught the weasel on the ground, and
rising with his prize, a struggle had ensued in the air, during which
the weasel had caught the bird by the throat and hung there till he
was squeezed and clawed to pieces.

Bald Eagles are, during some winters, common at Hamilton Beach,
where they pick up any dead fish and "cowheens" that are shaken
out of the fishermen's nets. Knowing the habits of the birds, the
fishermen often capture them by placing a poisoned carcase near the
edge of the ice. The bait is sure to be taken by the first Eagle
that comes along, and usually the bird dies before leaving the spot.

They still rear their young in suitable places throughout Ontario,
but as the country becomes more thickly settled, the birds seek for
greater retirement elsewhere. I learn from Dr. Macallum that a
few pairs still breed along the north shore of Lake Erie. "Quite
recently," he says, "I had a set of eggs taken from a nest which was
placed in an oak 100 feet up. It is known to have been there for
sixteen years, and from annual patching had got to be eight feet
deep and five feet across, and so firm that a sailor named Fox, who
took the eggs, was able to dance on the top, and kneel down on the
edge to lower the eggs to those below. The eggs were placed in the
middle of the platform, which was composed of sticks and clay, they
were three in number, pure white, and were quite fresh when taken,
April 3rd. I once saw an Eagle swoop down on a Herring Gull,
which it carried off in its claws to a large tree, where it was
devoured. It had done the same thing in the same place the day
before."

The food of this species consists entirely of fish, where they can
be obtained, and the bird is then harmless, and should be protected,
but where fish are scarce, and the birds begin to interfere with the
domestic animals, they should be kept within proper limits. They
have thus been placed in Class c, along with others whose good
deeds are supposed to balance the evil they do.

They are very abundant in Alaska and throughout the Aleutian
Chain, where they are resident. In summer they feed upon fish and
the numerous wild fowl that breed among these islands. In winter
they feed on Ptarmigan and the sea-fowl which reside there during
that season. When at Salmon Neck, in Sanborn Harbor, Mr. Dall
saw seventeen eagles all within 100 yards. There, as elsewhere, they

were accused of carrying off young chickens, but no authenticated instance was known.

Two years ago, I saw one that had been killed at Hamilton Beach under circumstances which made it appear that his reasoning powers were not equal to those of the hunter.

It was in January, the bay was firmly frozen over, and a keen, strong wind was blowing from the west. Half a mile out from the shore sat a Bald Eagle, tearing at the carcase of a small quadruped, which was frozen into the ice. There was no cover or chance of approach, for his view was open all round, and he seemed to enjoy his feast in safety.

Willie Smith, one of the Beach boys, looked on him with longing eyes, and his ingenuity was equal to the occasion. Carefully examining his gun, he set sail with a comrade in his iceboat, and by short tacks worked his way up in shelter of the shore to a point about a mile from where the Eagle sat, but directly to windward. Then the tackling was examined and set, and away went the boat at a rate at which only iceboats before the wind can go.

The Eagle must have seen the boat, but probably had little idea of the rate at which it was approaching. When it was still a good way off he got up to fly, but it is necessary for such birds to rise against the wind, and in doing this he went directly to meet his adversary. In vain he tried to sheer off to the right, still flying low. The boat was on his track at once, and for some seconds it looked as if they would pass each other about forty yards apart, but there was a flash, a crack, and the Eagle fell on the ice. The boat went tearing on, taking a long sweep, and then came round to the spot where lay the bird, which was picked up and brought off in triumph.

It takes at any time a quick eye and a steady hand to bring down a bird on the wing, but the peculiarity in this case was that both the hunter and his game were moving rapidly, and the hunter perhaps the more quickly of the two.

151. **White Gyrfalcon.** (353)

One of the largest and most powerful of the *Falconinæ*. Feet, very stout;
tarsus, rather longer than the middle toe without claw, feathered fully half way
down in front and on sides, with narrow bare strip behind; elsewhere, seticulate.
Wing, pointed by second quill, supported nearly to the end of the third, only
the first decidedly emarginate on inner web. Tail, rounded; sexes, alike.
Young:– Little different. Wing of male, 13.50-14.50; tail, 8.50-9.50. Wing
of female, 15-16; tail, 9-10. *Adults:*– General plumage of upper parts, barred
with dark brown and pale ash, the former predominating, especially on the
head and neck; tail, closely barred with light and dark in about equal amounts;

lower parts, white, immaculate on throat; elsewhere, streaked and variously spotted with dusky. *Young:*– Darker than the adults at an early stage, some of the lighter markings tinged with ochraceous.

HAB. -Arctic Regions, including Arctic America and Greenland.

Nest, placed on a shelf of a cliff.

Eggs, two to four, light brownish-red, faintly speckled with umber-brown.

This is *the* Gyrfalcon of America, and perhaps also of Europe, for the difference between the two, if any, is very trifling. There are now two different species and two sub-species of Gyrfalcon described as American, and we bow to the decision of the revision committee who have so decided, but still hold our own opinion, that when this fever of subdivision has cooled off a little and we become better acquainted with this group, one species will be sufficient to include the whole.

At all events I am much pleased at being able to place this one among the "Birds of Ontario," and for this privilege I am indebted to Mr. G. R. White, of Ottawa, who secured, on the 23rd December, 1890, a fine adult male that was bestowing unsolicited attention upon his domestic pigeons.

This is so decidedly a northern bird, that it is rare to find it even as far south as Ottawa. It is of circumpolar distribution, and has been found breeding in Greenland. In Alaska this form is rather rare, though some of the others are quite common. Speaking of *Falco rusticolus gyrfalco*, Mr. Nelson says : "Throughout all Alaska, from the Aleutian Islands north, both along the coast and through the interior, extending from Behring Straits across the northern portion of British America, the present falcon is the commonest resident bird of prey. It frequents the vicinity of cliffs and rocky points about the sea coast, or the rocky ravines of the interior, during the breeding season, and the remainder of the year, especially in the fall, it is found wandering over the country wherever food can be obtained. In a series of skins of this species from various parts of its range, there is found an interminable gradation from the whitest *islandus* to the darkest *gyrfalco* and *rusticolus*.

"Specimens in the National Museum collection from Greenland show the widest extremes, which are bridged by connecting specimens, so that it is impossible to definitely separate them. Newton's separation of *gyrfalco* from *islandus* on the assumption that the head is lighter than the back in one, and uniform with the back in the other, rests upon a purely individual character, as shown by my Alaskan series of skins."

In the Hudson's Bay country, Dr. Bell says: "This species is known as the Partridge, or Winter Hawk, although it remains also during the summer and breeds in the country."

This is one of the destructive class (*d*), which is said to live on game without compensation, and should therefore be destroyed, but throughout Ontario it is not likely to be troublesome.

In a letter from Mr. W. E. Brooks, of Mount Forest, dated January 17th, 1894, the writer says: "A few days ago, a fine Gyrfalcon passed over our fowls as they were being fed. It was a little too high to be reached with an ordinary charge, and, as often happens, there was no gun at hand, so that the chance of securing a rare specimen was lost."

Mr. Brooks is well acquainted with these birds, for he had shot them during his residence in British Columbia.

These are the only positive records of the species having been observed in Ontario, but as they are powerful birds on the wing, and much given to wandering during the winter, I have no doubt that all of the forms described will yet be found along our northern frontier. To assist in the identification of any that may be secured, I shall here give Mr. Ridgway's description of the different varieties:

354. F. RUSTICOLUS (LINN.). GRAY GYRFALCON.

Adult: —With upper parts banded with dusky and bluish-gray (sometimes uniform dusky anteriorly); the flanks and thighs, barred, banded or transversely spotted with dusky. *Young:* Without transverse bars on upper parts (except sometimes on tail); and lower parts, with all the markings longitudinal.

HAB. Extreme northern portions of Europe (except Scandinavia), Asia and North America (including Iceland and southern Greenland); south in winter to northern border of United States.

354a. F. RUSTICOLUS GYRFALCO.

Adult: With anterior upper parts (back, scapulars and wing coverts), rather indistinctly barred with bluish-gray, often nearly plain dusky; flanks, heavily banded or spotted with dusky; and thighs heavily barred with slaty (the white ground color tinged with bluish-gray posteriorly).

HAB.—Northern Europe and Arctic America, from northern Labrador and coasts of Hudson's Bay to Alaska.

354b. F. RUSTICOLUS OBSOLETUS. BLACK GYRFALCON.

Lower parts, with dusky prevailing, sometimes entirely dusky, except on lower tail coverts, which are always spotted with white.

HAB.—Coast of Labrador, south in winter to Maine, Canada and New York.

SUBGENUS RHYNCHODON NITZSCH.

FALCO PEREGRINUS ANATUM (BONAP.).

152. Duck Hawk. (356)

Tarsus feathered but little way down in front, elsewhere irregularly reticulated in small pattern, not longer than middle toe; first quill, alone decidedly emarginate on inner web, not shorter than the third. Above, blackish-ash, with more or less evident paler waves; below and the forehead, white, more or less fulvous tinge, and transverse bars of blackish; conspicuous black ear patches. *Young:*—With the colors not so intense and tending to brown; the tawny shade below stronger; the lower parts longitudinally striped. Length, about 18; wing, 13-14; tail, 7-8.

HAB.—North America at large.

Nest, in a tree, or on a rock, or on the ground.

Eggs, three to five, dull white, blotched with different shades of reddish-brown.

This is the Bullet Hawk, the terror of the ducks and the admiration of sportsmen at the shooting stations, where he is often seen, either capturing game on his own account, or appropriating what has been killed by the gunner before he has time to pick it up. As this species is known to breed in Massachusetts, on the coast of Labrador, and in Alaska, it will also be found most likely to do so in suitable places in Ontario, but at present we have no satisfactory record of the fact. The steep rocky ledges which overhang the blue waters of Lake Superior offer inducements which the birds will hardly overlook, and we expect yet to hear of their being found breeding there.

While here, the Peregrine is no loiterer, but follows the migratory course of the water-fowl, and fares sumptuously every day. Ducks are his favorite game, and he need never be at a loss, yet (by way of a relish perhaps) we see him sometimes scoop up a sandpiper or a mudhen and pick its bones on an elevation which commands a clear view for some distance around. In Southern Ontario, the Peregrine is seldom seen except in the fall.

The Peregrine was held in great esteem in the olden time when "hawking" was a princely amusement in Europe, and a very high price was often paid for a well-trained pair. Here he is under no restraint, but, handicapped neither by bell nor hood, he follows the bent of his own wild nature, exceeding even the Gyrfalcon in the skill and activity he exhibits when capturing his prey. He is placed in Dr. Fisher's destructive class (d), but he is so seldom seen in Ontario that his habits are not often the subject of complaint.

SUBGENUS ÆSALON KAUP.

FALCO COLUMBARIUS LINN.

153. **Pigeon Hawk.** (357)

Tarsus, scarcely feathered above, with the plates in front enlarged, appearing like a double row of alternating scutellæ (and often with a few true scutellæ at base); first and second quill emarginated on inner web. *Adult male:—* Above, ashy-blue, sometimes almost blackish, sometimes much paler; below, pale fulvous or ochreous, whitish on the throat, the breast and sides with large oblong dark brown spots with black shaft lines; the tibiæ, reddish, streaked with brown; inner webs of primaries with about eight transverse white or whitish spots; tail, tipped with white, and with the outer feather whitening; with a broad subterminal black zone and three to four black bands alternating with whitish; cere, greenish-yellow. *Female:* - With the upper parts ashy-brown; the tail, with four to five indistinct whitish bands. Length, about 13; wing, 8; tail, 5. *Male:* - Smaller.

HAB.—The whole of North America, south to the West Indies and northern South America.

Nest, in a hole in a tree, or on a branch, or on rocks.

Eggs, four, yellowish-brown, blotched with brown of a darker shade.

This handsome little falcon is a miniature of the Peregrine, and is quite its equal in courage and spirit, often attacking birds of much greater weight than itself. It is not a common species anywhere, and in Southern Ontario can only be regarded as a migratory visitor in spring and fall. It is at all times a difficult matter to define the precise breeding range of birds that are rare everywhere, and regarding the summer haunts of the Pigeon Hawk we have yet much to learn. As it has been known to breed in Maine and in Alaska, it is quite likely to breed also in Ontario, where there is plenty of room for it to do so without being observed. In the fall when the black-birds get together in flocks, they are frequently followed by the "little corporal," who takes his tribute without much ceremony. I once saw him "stoop" on a flock as they hurried toward the marsh for shelter. How closely they huddled together, as if seeking mutual protection, but he went right through the flock and came out on the other side with one in each fist.

This species has a wide distribution, going south in winter as far as the West Indies and northern South America. In summer it has been found breeding in Nova Scotia and in several of the New England States. Mr. Macfarlane found it common in the Anderson River regions. And Mr. Nelson says : "It is one of the most numerous and familiar birds of prey throughout the wooded portion of Northern Alaska, ranging during migrations to the barren coasts of Behring Sea and the Arctic.

"On the 6th of October, 1878, I saw one of these birds dart down and strike its talons into the back of a Burgomaster Gull as the latter was flying over the sea. After holding on for a moment, the gull continuing its flight unimpeded, the falcon let go, and rising almost directly up for thirty or forty yards went clear off, apparently not desirous of renewing the attack." Although the Pigeon Hawk is a true falcon and consumes a great many small birds of different sorts, Dr. Fisher believes it kills sufficient injurious insects and mammals to balance the account, and has accordingly placed it in Class *c*. In Ontario it is never sufficiently numerous to do serious injury.

SUBGENUS TINNUNCULUS VIEILLOT.

FALCO SPARVERIUS LINN.

154. American Sparrow Hawk. (360)

Tarsus and quills, as in *columbarius*: crown, ashy-blue, with a chestnut patch, sometimes small or altogether wanting, sometimes occupying nearly all the crown; conspicuous black maxillary and auricular patches, which, with three others around the nape, make seven black places in all, but a part of them often obscure or wanting; back, cinnamon-brown, in the *male*, with a few black spots or none; in the *female*, with numerous black bars; wing coverts, in the *male*, ashy-blue, with or without black spots; in the *female*, like the back; quills, in both sexes, blackish, with numerous pale or white bars on inner webs; tail, chestnut, in the *male*, with one broad black subterminal bar, white tip, and outer feather mostly white, with several black bars; in the *female*, the whole tail with numerous imperfect black bars; below, white, variously tinged with buff or tawny, in the *male*, with a few small black spots or none; in the *female*, with many brown streaks; throat and vent, nearly white and immaculate in both sexes; bill, dark horn; cere and feet, yellow to bright orange; 10-11; wing, 7; tail, 5, more or less.

HAB.   Whole of North America, south to northern South America.

Eggs, four or five, deposited in the hollow of a decayed limb, or deserted woodpecker's hole.   In color, variable, usually yellowish-brown, blotched all over with brown of a darker shade.

The peculiar and handsome markings of this little falcon serve, even at a distance, to prevent its being mistaken for any other species.   Though sometimes seen near the farm-house, it does not bear the stigma of having felonious intentions towards the occupants of the poultry yards, but is credited with the destruction of large numbers of mice, and is, therefore, regarded with favor by the farmer.   It also feeds freely on snakes, lizards, grasshoppers, etc., but has the true falcon etiquette of taking only what is newly killed.   It is generally distributed throughout Ontario, arriving on the southern frontier about the end of April, and leaving for the south in September.

In the "Birds of Minnesota," page 203, is given an account of an experiment made by M. de Lantrie, to test the value of this species to the farmer.   He says: "I took five little Sparrow Hawks and put them in a cage.   The parents immediately brought them food, and I was not surprised to see that it consisted of twelve mice, four large lizards, and six mole crickets.   A meal of like size was brought every day for a month.   At one time there were fifteen field mice, two little birds, and a young rabbit.   Last year I made the same experiment, with similar results : one meal consisting of twelve small birds, one lark, three moles and one hedgehog.   In one month, the five baby-hawks rid the world, by actual count, of 420 rats and mice, 200 mole crickets, and 158 lizards."   In view of the foregoing record, this species is well entitled to the place which Dr. Fisher has given it in Class *b*, as one of those whose *good* deeds are in excess of the *evil*, and it is, therefore, entitled to the protection of all interested.

GENUS POLYBORUS VIEILLOT.

POLYBORUS CHERIWAY (JACQ.).

155. Audubon's Caracara. (362)

General color, blackish; the throat all around, and more or less of the fore back and breast, whitish, spotted and barred with blackish; upper and lower tail coverts and most of the tail, white, the latter crossed with numerous bars of black, of which color is the broad terminal zone; the shaft, white along the white portion of each feather; basal portion of primaries, also barred with whitish; iris, brown; feet, yellow; claws, black. *Young:*—Similar, but rather brownish, the markings of the body running lengthwise; the tail, barred. Length, 21-23; wing, 14½-16½; tail, 8-10.

HAB.—Southern border of the United States (Florida, Texas, Arizona), and Lower California, south to Ecuador and Guiana.

Nest, on trees, bushes, or cliffs, coarse and bulky, composed of sticks and twigs, lined with leaves and grass.

Eggs, two or three, beautiful in color, varying from light cinnamon to umber-brown, with markings of yellowish, umber-brown, chestnut, claret-brown, or reddish-brown.

A description of this southern bird may seem very much out of place among the "Birds of Ontario," but as the bird is known to have once paid us a visit, it is only right to place it on record.

The specimen referred to was captured on the north shore of Lake Superior, near Port Arthur, on the 18th July, 1892, and was so reported to the Natural History Society of Ontario, by Mr. Geo. E. Atkinson.

As this species is not a migrant, but considered resident wherever it is found, it is difficult to account for the occurrence of this individual so far north of its usual habitat. Whatever the cause may have been, the journey cost the bird its life—a penalty very often paid by such wanderers.

The centre of abundance of the Caracara seems to be on both sides of the southern boundary of the United States. It is also common in Florida, Texas and Oregon. In its habits it resembles both the eagles and vultures, catching some of its prey alive, but also feeding greedily on dead animals. In its flight it is sometimes slow and sluggish, like a vulture, but again, when occasion requires, it can be as quick and active as a falcon.

Captain B. F. Goss says regarding this species: "Brown Pelicans breed in great numbers off the coast of Texas. When these birds were returning to their breeding ground with pouches filled with fish, the Caracaras would attack them until they disgorged, and then alight and devour the stolen prey.   These attacks were made from above, by suddenly darting down on the Pelicans with shrill screams, and striking at them with their talons.   I saw this manœuvre repeated a number of times by a pair of these birds that nested on the island, and by others that came from the shore.   They did not attack outgoing birds, but invariably waited for the incoming ones, and as soon as these were over land, so that the contents of the pouches would not fall in the water, they pounced on them, and kept it up until their object was attained."

They are very fond of fish and frogs ; they also hunt prairie dogs and other small mammals, but are not accused of disturbing poultry.

SUBFAMILY PANDIONINÆ.   OSPREYS.

GENUS PANDION SAVIGNY.

PANDION HALIAËTUS CAROLINENSIS (GMEL.).

156. American Osprey. (364)

Plumage lacking after-shafts, compact, imbricated, oily to resist water ; that of the legs short and close, not forming the flowing tufts seen in most other genera, that of the head lengthened, acuminate ; primary coverts stiff and acuminate.   Feet, immensely large and strong, the tarsus, entirely naked, granular-reticulate, the toes all of the same length, unwebbed at base, very scabrous underneath, the outer versatile ; claws, very large, rounded underneath.   Hook of the bill, long, nostrils touching edge of cere.   Above, dark brown ; most of the head and neck and the under parts, white, latter sometimes with a tawny shade and streaked with brown.   Length, 2 feet ; wing, 16-18 inches : tail, 8-10.

HAB.—North America, from Hudson's Bay and Alaska, south to the West Indies and northern South America.

Nest, in a tree, composed of sticks, often very bulky, from annual additions.

Eggs, two to four, variable in color, usually creamy-brown, blotched with various darker shades of brown.

The Fish Hawk is generally distributed throughout Ontario, breeding by the lakes and rivers in the less thickly settled parts of the country.   Along the sea coast it is more abundant, frequently breeding in communities of several hundreds.   In such cases the nests are placed indifferently on rocks or trees, and sometimes the eggs have been deposited on the sand.   Near these breeding places,

the Bald Eagle has every opportunity of tyrannizing over the Fish
Hawks, and compelling them to drop the fish they have just caught.
On the inland waters of Ontario, the Bald Eagle is of less frequent
occurrence, and the Osprey is allowed to enjoy the results of his
industry in peace.

The Fish Hawks arrive in Ontario as soon as the ice breaks up in
the spring, and are soon distributed throughout the country. Here
and there in Ontario and elsewhere, a pair will settle and remain
for the summer, but many of them do not slack in their northern
flight till they are within the Arctic circle, where they rear their
young on the banks of the clear streams of the interior, and along
the Yukon River in Alaska. Only one brood is raised in the
season, and in the fall they again work their way south, calling at
many intermediate stations. In Southern Ontario they are seen
during October, but continue their southern route by easy stages, till
many of them reach the West Indies and northern South America,
where they soon again prepare for the spring trip. Fish are protected by a law which ordinary fishers are bound to respect, but this
Hawk lives on fish and nothing else, and he takes the sort he can get
without regard to the season of the year. On this account he has
been placed on Dr. Fisher's black list, in Class d, of which the evil
deeds exceed the good; but in Ontario I feel sure that the vote would
be to let him take all he requires, in consideration of the additional
attraction his splendid presence gives to the scenery of many a lake
and river throughout the Province.

SUBORDER STRIGES. OWLS.

FAMILY STRIGIDÆ. BARN OWLS.

GENUS STRIX LINNÆUS.

STRIX PRATINCOLA BONAP.

157. American Barn Owl. (365)

Tawny or fulvous brown, delicately clouded or marbled with ashy or white, and speckled with brownish-black ; below, a varying shade from nearly a pure white to fulvous, with sparse sharp blackish speckling ; face, white to purplish-brown, darker or black about the eyes ; the disk, bordered with dark brown ; wings and tail barred with brown, and finely mottled like the back ; bill, whitish ; toes, yellowish. *Female* : Length, 17 inches ; wing, 13 ; tail, 5½. *Male* :—Rather less.

HAB.—Warmer parts of North America, from the Middle States. Ohio Valley and California, southward through Mexico.

Breeds, in hollow trees, frequently in the tower of a church or other high buildings.

Eggs, five to eleven ; soiled white.

Although this species, so much like the Barn Owl of Britain, has long been known as an American bird, coming as far north as Massachusetts, it is only within the last few years that it has been observed in Canada. In May, 1882, a specimen was killed by young Mr. Reid, gardener, York Street, Hamilton, and in the fall of the same year another was found in an empty outhouse near the canal leading to Dundas. On calling the attention of Dr. Garnier, of Lucknow, to these facts, he mentioned having seen one several years before, near where he lives, and from Mr. C. J. Bampton comes a report of his having seen two individuals near Sault Ste. Marie. Compared with the British Barn Owl, the American species is a little larger, but by many they are regarded as identical. The British bird is noted for its partiality to ruinous church towers and other lonely places. Strange to say, Mr. Reid's specimen was killed in the cemetery, while one of those seen by Mr. Bampton was perched on the cross on the spire of the Catholic church.

The Barn Owl has a sharp, inquisitive visage, and is said to be an expert mouser. In Ontario it can only be regarded as an accidental visitor from the south.

No recent instance of the occurrence of this species in Ontario has come to my knowledge. We are farther to the north than its usual habitat. In the south, it is quite common across the continent, from the Atlantic to the Pacific coast, its centre of abundance being, apparently, in southern California. Its food consists almost entirely of rats, mice, gophers, and other destructive little mammals, which leads one to expect that it would be placed in Class *a*, as entirely beneficial, but for some reason, not apparent, it is included in Class *b*, among those whose good work is greater than the reverse. This rank entitles our friend to exemption from the persecution which is accorded to all our birds of prey, a practice we hope to see entirely changed in the near future. This is the bird we find so often alluded to in the superstitions of the old land poets and historians, ancient and modern, all associating his name with evil.

Shakespeare frequently refers to the Owl as a bird of evil repute. For instance, in speaking of the omens which preceded the death of Cæsar, Casca says :

> " And yesterday, the bird of night did sit
> Even at noonday upon the market-place,
> Hooting and shrieking."

In Burns' account of that memorable midnight ride, when Thomas Graham, of the farm of Shanter, was privileged to get a glimpse of the proceedings of a social science meeting of the moving spirits of the time, the farmer's progress homeward is thus described :

> " Kirk Allowa was drawin' nigh,
> Whaur ghaists and hoolets nightly cry."

In the rural districts of Scotland, where superstition still lingers, the "hoolet" continues to be regarded with aversion, and its visits to the farm-house are looked upon as the forerunners of disaster to the family. Its cry, when heard at night, is described as most penetrating and depressing, and it is often referred to in this way in the literature of the country.

In a song by Tannahill, the fellow-townsman and brother poet of Wilson, the Ornithologist, the hero is entreating admission to the chamber of his lady-love, and in describing his uncomfortable position outside, mentions among other causes, that the "cry o' hoolets mak's me eerie." Such sounds, when heard unexpectedly at night, in a lonely place, are not calculated to inspire courage in a breast already depressed with superstitious fear, but the effect produced must, to a great extent, depend on the train of thought passing through the mind of the hearer at the time. Many a stalwart Scot may have quailed at the cry of the hoolet when heard under certain conditions, but it is a matter of history that the sons of that romantic land, when roused to enthusiasm by similar sounds extorted from the national instrument, have performed deeds of personal valor which will live in song and story so long as poets and historians seek such themes.

FAMILY BUBONIDÆ. HORNED OWLS, ETC.

GENUS ASIO BRISSON.

ASIO WILSONIANUS (LESS.).

158. American Long-eared Owl. (366)

General plumage above, a variegation of dark brown, fulvous and whitish, in small pattern; breast, more fulvous; belly, whiter; the former, sharply striped; the latter, striped and elaborately barred with blackish; quills and tail mottled and closely barred with fulvous and dark brown; face, pale, with black touches and eye patches; bill and claws, blackish. Ear-tufts, of 8 to 12 feathers. Length, 14-15; wing, 11-12; tail, 5-6.

HAB. Temperate North America.

Nest, of sticks, loosely put together, lined with a few feathers, variable as to situation, frequently in a thick evergreen.

Eggs, four to six, oval, white.

The Long-eared Owl is strictly nocturnal in its habits, and is seldom seen abroad by day, except when disturbed in its retirement among the evergreens. So far as I have observed, it is not a common species in Ontario, but, from its retiring habits, it may be more so than we are aware. Those observed near Hamilton have been found in the fall, the season when birds of all kinds wander away from their summer resort, before retiring south to spend the winter. Along the sea coast it is more common, and in New England resides throughout the year. That it breeds in Ontario is vouched for by Mr. Robert Elliot, who found a nest near his home at Bryanston during the summer of 1886.

It is reported a tolerably common summer resident in Manitoba, and is also plentiful in the Saskatchewan and Hudson's Bay districts. It is one of the best of the farmers' feathered friends, consuming immense numbers of rats, mice, moles, beetles, etc.

We find it placed in Class *b*, with a balance of good work at its credit, which entitles it to our protection.

ASIO ACCIPITRINUS (Pall.).

159. **Short-eared Owl.** (367)

Fulvous or buffy-brown, paler or whitey-brown below ; breast and upper parts, broadly and thickly streaked with dark brown ; belly, usually sparsely streaked with the same, but not barred crosswise ; quills and tail, buff, with few dark bands and mottling ; facial area, legs, and crissum, pale, unmarked ; eye-patch, blackish ; ear-tufts, of from 3 to 6 feathers. Size of *wilsonianus*.

Hab.— Throughout North America : nearly cosmopolitan.

Nest, on the ground, consisting of a few sticks, blades of grass and feathers, loosely thrown together.

Eggs, four to six, white, nearly round.

This is a much more common species than the preceding, and probably more northern in its range. I have reports of its occurrence at different points throughout Ontario, and it was observed in the North-West by Prof. Macoun. It is less nocturnal in its habits than the preceding, and is somewhat gregarious, being occasionally seen during the day in the fall, in flocks of ten or twelve, hunting in company. It has not been my fortune to fall in with any of these migratory groups, but I have observed the species skimming noiselessly over the inlets and moist meadows along the shores of Hamilton Bay.

It is a most expert mouser, destroying large numbers of the farmers' foes, and is, therefore, entitled to his protection ; but all

15

birds of prey are regarded as enemies by the sportsman, who allows none to pass that come within his reach.

A few remain in summer, and raise their young in Southern Ontario, but the greater number pass on farther north. They are tolerably common in Manitoba, and are also reported from Hudson's Bay and Alaska. The examinations of the stomachs of this species made by Dr. Fisher, show that its food consists mainly of field-mice and shrews. It has been placed in Class c, among those whose good and evil habits are about equal.

Genus SYRNIUM Savigny.

SYRNIUM NEBULOSUM (Forst.).

160. Barred Owl. (368)

Above, cinereous-brown, barred with white, often tinged with fulvous; below, similar, paler, the markings in bars on the breast, in streaks elsewhere; quills and tail feathers, barred with brown and white, with an ashy or fulvous tinge. Length, about 18; wing, 13-14; tail, 9.

Hab.—Eastern United States, west to Minnesota and Texas, north to Nova Scotia and Quebec.

Nest, in a hollow tree, or in the deserted nest of a hawk or crow.

Eggs, two to four, round, white.

Along the southern boundary of Ontario the Barred Owl is by no means rare, but farther north I have not heard of it being observed. It does not occur west of the Rocky Mountains, but is very abundant along the south Atlantic and Gulf States. It is occasionally seen abroad by day, but at such times its sight seems to be rather uncertain, so that the capture of the small animals on which it feeds is accomplished during the hours of darkness.

Regarding its uncertain vision by day, Mr. Giraud, in his "Birds of Long Island," says: "My friend, Mr. J. G. Bell, informs me that when on a collecting tour in South Carolina, and while looking for the Blue-winged Yellow Warbler whose note he had a moment before heard, he was startled by feeling a sudden pressure on his gun. Judge of his surprise when he perceived perched on the barrels a Barred Owl, which, at the same moment, discovered its mistake, but too late to correct the fatal error, as it was shot down by the astonished gunner."

Audubon mentions seeing one alight on the back of a cow, which it left so suddenly, when the cow moved, as to show that it had mistaken the object on which it perched for something else.

In former years, I used to find the Barred Owl regularly every fall in the ravines along the south shore of the Dundas marsh, but now many of the pines and hemlocks which formed an inviting retreat are cut down, and the bird has sought for greater seclusion elsewhere. Its black eyes are at all times a ready mark to distinguish it from any other member of its family.

This is one of the noisiest of the Owls, and his weird, uncanny cries are often a subject of interest to the lonely wayfarer while dozing by his camp-fire after dark. Nearly all the earlier writers give the Barred Owl credit for being one of the most destructive to poultry and game-birds. Dr. Fisher, after giving a number of extracts to that effect, says : " To all this testimony, which could be increased by the addition of many other notes attesting the destructiveness of the species to poultry and game, the investigations of the writer are in direct variance. Of the 109 stomachs examined, only four contained the remains of poultry, and in one, traces of a game-bird were found."

One peculiarity of this species, which, we are glad to say, is not very common among birds, is given by Dr. Fisher, as follows : " This Owl seems to be more given to cannibalistic habits than any of the other species. In seven stomachs examined, the writer found the remains of smaller owls among the contents, and from two different sources comes the record of the remains of Screech Owls being found under similar circumstances. Now and then small birds are killed by this species, but mammals furnish a large proportion of its food. The remains of mice, rabbits, squirrels (red, gray and flying), shrews, moles, and occasionally weasels, have been found in the stomach contents."

The species, owing to its large size, is capable of consuming numbers of mice at one meal. Dr. Merriam found the remains of at least a dozen red-backed mice in a single specimen killed near Moose River, in northern New York.

Dr. Fisher's evidence, which is indisputable, places this species in a much more favorable light, as a friend to the farmer, than it has hitherto occupied, and we hope it will get the benefit and be duly protected, as it deserves to be. It is placed in Class c, in which the good done balances the evil.

GENUS ULULA CUVIER.

ULULA CINEREA (GMEL.).

161. Great Gray Owl. (370)

Above, cinereous brown, mottled in waves with cinereous white ; below, these colors rather paler, disposed in *streaks* on the breast, in *bars* elsewhere : quills and tail, with five or six darker and lighter bars ; the great disk similarly marked in regular concentric rings. An immense owl, one of the largest of all, much exceeding any other of this country. Length. 2½ feet : wing, 1½ : tail, a foot or more.

HAB.—Arctic America, straggling southward, in winter, to the northern border of the United States.

Nest, in trees, composed of sticks and twigs, lined with moss and a few feathers.

Eggs, two or three, not quite round, white.

This beautifully marked and solemn-looking bird is usually described as the largest of North American Owls, but it can only be regarded so by measurement, for in weight, strength and ferocity it is inferior to both the Snowy and the Great Horned Owl. The lengthy tail, and the long, loose feathers with which its body is densely clothed, give it the appearance of a very large bird of prey, but when closely examined, the legs, claws and bill are smaller and weaker than those of either of the two species named.

The Great Gray Owl is said to be more northern in its range than even the Snowy Owl. In Southern Ontario, it is a casual visitor in the winter only. I have had two individuals brought to me which were got near Hamilton, and have seen several in the hands of other parties. During the present winter, I saw one which was sent down from Muskoka, where it was shot in the woods in the month of December.

It is truly a bird of the far north, those we see here in winter being only stragglers from the main body of the species, which is resident throughout all the wooded parts of Alaska, from Sitka north to the northern tree limit, and from the vicinity of Behring Straits east throughout the Territory, extending all over the fur countries.

Mr. Dall speaks of it being so exceedingly stupid that along the Yukon it can be caught by the hand in the daytime. Its food consists mainly of hares, mice, and other species of the smaller mammals, and also small birds. Mr. Dall took no less than thirteen skulls and other remains of Red-poll Linnets from the crop of a single bird.

It has been placed in Class *b*, as a friend of the farmer, but farmers are so rare in the regions it frequents that they are not likely to be much affected for good or ill.

GENUS NYCTALA BREHM.

NYCTALA TENGMALMI RICHARDSONI (BONAP.).

162. Richardson's Owl. (371)

Upper parts, grayish-brown, tinged with olive ; feathers of the head and neck, spotted with white ; scapulars, quills and tail also with white spots ; ruff and lower parts, yellowish-white ; throat, white. *Male :*—11 inches. *Female :*—12 inches.

HAB.—Arctic America, south occasionally in winter into the northern United States.

Nest, in trees.

Eggs, two to four, round, white.

This comparatively small and timid-looking owl is, perhaps, more hyperborean in its range than any of the others we have had under consideration, inasmuch as the records of its occurrence do not extend so far south as those of either the Great Gray or the Snowy Owl. It is warmly clad in a dense coat of soft, silky feathers, which, no doubt, enables it to withstand the severity of the winter. In the search for food, it evidently finds a supply, for the species is spoken of by Sir John Richardson as being abundant in the region of the Saskatchewan, but only a very few come as far south as Southern Ontario. The two in my collection were both found during winter in the neighborhood of Toronto, but besides these I have very few records of its having been observed anywhere throughout the country.

Proceeding farther north, this species is more frequently observed. In Manitoba, it is reported as a tolerably common winter visitor. In northern Alaska, it is found wherever trees or large bushes occur to afford it shelter. The Eskimo have a name for it, unpronounceable in our language, which means "the blind one," owing to its being frequently caught alive by hand, on account of its defective sight in the daylight. They are very gentle little birds, and are frequently kept as pets by the Eskimo children. They are placed in Class *b*, which entitles them to our hospitality whenever they elect to visit us.

NYCTALA ACADICA (Gmel.).

163. Saw-whet Owl. (372)

Size, small; bill, black ; the cere tumid, the circular nostrils presenting anteriorly; above, chocolate-brown, spotted with white, the tail with transverse white bars ; facial area and forehead variegated with white, the face and superciliary line grayish-white ; the lower parts, white, with streaks of the color of the back. Length, 7½-8 ; wing, 5½ ; tail, 2¾.

Hab.—Temperate North America, breeding from the northern States and southern Canada southward.

Nest, in a hole in a tree.

Eggs, four to six, round, white.

This is the smallest member of the family found east of the Rocky Mountains. For some reason all the owls are of irregular occurrence in the settled parts of the country. I have seen as many as six or eight of this species in one winter, and again for several years have not seen one. Without being migratory, in the ordinary sense of the word, I think it is highly probable that during the fall these birds associate in groups, and move from one section of the country to another in search of food. In this way a good many may be observed at one point, while for many miles around they may be altogether absent.

The Saw-whet is evidently partial to a medium temperature, for it is most common in the northern States, and does not penetrate far into British America. In the opposite direction, it has been found breeding as far south as Mexico, but mostly in the wooded mountain ranges. In Southern Ontario, these birds are most at home in the thick shelter of the evergreens in the depths of the woods, but when deep snow covers the ground, they are often found in the barn, or other outhouse near the farmer's dwelling, where they are forced to seek for food and shelter when their supply outside is cut off.

The food of this species consists almost entirely of mice, which renders it worthy of our friendship, although the small size of the bird limits the amount of its usefulness in that direction. It takes little else when mice can be had, and therefore the number destroyed in a season is considerable, so that we cordially endorse its position in Class b, where it has been placed.

Occasionally, there is a small owl found, the plumage of which is uniform chocolate-brown, with white eyebrows. For a time this was described as a distinct species, under the name of White-fronted Owl, also as Kirtland's Owl, but of late these names have been dropped from the lists, for it is by many believed to be the young of the Saw-whet. Ultimately, conclusive evidence that such was the case was given by Dr. J. W. Velie, of Chicago, who kept a live *albifrons* till it moulted, and became a fine specimen of *Nyctala acadica*. The transition is described in detail in Baird, Brewer and Ridgway's "Birds of North America," from which the following remarks are abridged :

The bird was placed in a large cage, where it had abundant room to fly about, and was well supplied with food. Through June and July there was absolutely no change in its plumage. On the 1st of August, a few medially spotted feathers were observed pushing their way through the uniformly brown ones on the fore part of the crown. Through the next two weeks they gradually increased and developed, until the full-face aspect of the head was that of an adult Saw-whet. At this stage there was no indication of any second plumage on the other parts, but about August 15th a few streaked feathers appeared along the central line of the breast and abdomen, while a little later the moulting began over the back and wings, and quickly became general. Through the last two weeks of the month the plumage gained daily, and by September 1st the final stage was perfected, and the bird had become a remarkably beautiful Saw-whet Owl.

From this it appears that the brown plumage is simply that which succeeds the downy stage, and it is even a little longer than in most other owls before the autumnal dress is assumed.

I am sorry to lose the little *albifrons* which I first heard described by the late Dr. Kirtland, but much pleased to see the truth so fully established regarding it.

GENUS MEGASCOPS KAUP.

MEGASCOPS ASIO (LINN.).

164. Screech Owl. (373)

One plumage: general aspect gray, paler or whitish below. Above, speckled with blackish: below, patched with the same; wings and tail, dark-barred, usually a lightish scapular area.

Another: general aspect brownish-red, with sharp black streaks: below, rufous white, variegated; quills and tail with rufous and dark bars. These plumages shade insensibly into each other, and it has been determined that they bear no definite relations to age, sex or season. Length, about 10; wing, 7; tail, 3½.

HAB.—Temperate eastern North America, south to Georgia, and west to the Plains. Accidental in England.

Nest, in a hole in a tree, lined with feathers.

Eggs, four to six, round, white.

This is the most abundant of the owls in this part of the country, yet, like the others, it is of very irregular occurrence. I have met with it once or twice in the woods in summer, but it is most frequently seen in winter, when the ground is covered with snow. It is then forced to approach the dwellings of man in search of food, and during some winters there is scarcely a farm in the country which has not its Screech Owl in the barn. There it sits on a rafter, snoozing away the hours of daylight, occasionally opening its round, yellow, cat-like eyes, and glowering at the farm hands as they move about like shadows below. After dark it is all alive ; not a mouse can stir without being observed, and so quick and noiseless is the flight of the bird, that few escape which expose themselves. It thus renders good service to the farmer, in consideration of which it is protected by the more intelligent of that class, but it is persecuted almost to extinction by the " boys."

As will be seen by the description of the markings given above, individuals of this species assume different phases of plumage, and are spoken of as the "red" and "gray." For many years great difference of opinion prevailed on this subject, some believing the red bird to be the male, and *vice versa*. It is now fully understood that the color is entirely independent of age, sex or season. It is one of those seeming irregularities which we find in nature, and all we can do is to bear witness to the fact, without being able to tell the reason for it.

During the long winter of 1883-84, I kept a record of the birds of this species I heard of, in or near Hamilton, and the total number reached forty. In 1884-85, they were less common, and during 1885-86, I am not aware of a single individual having been observed.

The Screech Owl seems partial to the south-west of Ontario, for it is common at Toronto, Hamilton and London, while Mr. White has not yet observed it at Ottawa, nor is it named among the birds of Manitoba. Of the two colors, the gray is the more common here, the red being rarely met with.

During summer it has a very varied bill of fare, but in the severe winters it lives chiefly on mice, though it sometimes takes small birds. It has been placed in Class *b*, as one of the beneficial class, worthy of protection.

The genus is widely distributed, chiefly east and west. It contains eight different species, only one of which occurs in Ontario. They all resemble each other, but are sufficiently different to warrant specific distinction.

Genus BUBO Cuvier.

BUBO VIRGINIANUS (Gmel.).

165. Great Horned Owl. (375)

Distinguished by its large size, in connection with the conspicuous ear-tufts; the other species of similar dimensions are tuftless. The plumage varies interminably, and no concise description will meet all its phases; it is a variegation of blackish, with dark and light brown, and fulvous. A white collar is the most constant color mark. Length, about 2 feet; wing, 14-16 inches; tail, 9-10.

Hab. —Eastern North America, west to the Mississippi Valley, and from Labrador south to Costa Rica.

Nest, sometimes in a hollow tree, or cleft of a rock, or among the branches of a high tree in the thickest part of the woods, very often the nest of the Red-tailed Hawk is appropriated, or that of other species where available.

Eggs, two or three, round, white.

The Great Horned Owl is well known in Ontario, being generally distributed throughout the Province. During the day it hides away in the deep impenetrable parts of the woods, but at night sallies forth in quest of prey, and does not hesitate to rob the hen roost, returning for that purpose night after night, unless stopped by a snap shot in the dark, or caught in a trap baited for the purpose. Individuals vary greatly in plumage, so much so that they have been described as distinct species. Near Hamilton I have found them varying from light silvery gray to deep fulvous brown. I once obtained a very handsome specimen in the latter dress which I was unable to utilize from its having been recently in contact with a skunk. It is strictly nocturnal in its habits, but when obliged by the attention of crows, or other disturbing causes, to move during the day, it makes good use of its eyes, and gets quickly away to the nearest thicket for shelter.

This is not the most numerous of the owls which occur with us, but no other member of the family receives so much attention. This is owing to its large size, its loud call, which is heard at a great distance on still evenings in the fall, but chiefly to the raids it makes on the hen roosts. In older countries, where, by many of the inhabitants, all the owls are supposed to be to some extent in communication with the supernatural agents of evil, this would be regarded as a most "unlucky" visitor; but in this new land we have no birds of evil omen, and the owl receives his proper place in science and in literature. Longfellow speaks of him as "a grave bird, a monk who chants midnight mass in the great temple of nature."

His visits to the farm-house are well understood, and if followed by disaster, it is usually to the poultry, or to the bird himself, if the farmer's boys can so arrange matters. He is of a most savage and untamable nature, excelling any of the owls in strength and ferocity. If brought to trial in Ontario, he would undoubtedly be condemned to be killed off at once, on account of his depredations in the hen roost and dove cot ; but we are told that in some parts of the west, where rabbits are so numerous that it is next to impossible to bring to maturity any large proportion of the crops, the Great Horned Owl feeds on this destructive rodent, almost to the exclusion of other food, and thus becomes a most valuable ally to the farmer. On this account he has been placed in Class *c*, among those birds whose good and evil deeds are about equal.

On account of the variation in size and plumage, according to the region they inhabit, the American Ornithologists' Union Committee records three sub-species of the Great Horned Owl emanating from the original *Bubo virginianus*. These Mr. Ridgway describes as follows :

375*a*. *Bubo virginianus subarcticus*—Western Horned Owl. General aspect of plumage : above, grayish, with more or less of buffy admixture, dark markings of lower parts distinct.

H.ab. Western United States (except north-west coast), eastward across great plains, straggling to northern Illinois, Wisconsin and Western Canada, north to Manitoba, south over table-lands of Mexico.

375*b*. *Bubo virginianus arcticus*—Arctic Horned Owl. General aspect of plumage : above, white through fading of the ground color and restriction of dark markings ; beneath, pure white with dark markings, usually much restricted.

H.ab. Arctic America, chiefly in the interior, south in winter to northern Rocky Mountains and Great Plains, Dakota, Montana, etc.

375*c*. *Bubo virginianus saturatus*—Dusky Horned Owl. Extremely dark-colored ; the face, usually sooty brownish, mixed with grayish-white ; the plumage, usually without excess of ochraceous or tawny, sometimes with none.

H.ab.—North-west coast from Oregon north to Alaska, Labrador.

The common form in Southern Ontario is the original *Bubo virginianus*, but I have also had one or two which answer the description given of the western variety, and several apparently intermediate. From Manitoba I have one as white as a female Snowy Owl, and from the dense forests of British Columbia I have several which are very dark. All of these varieties are smaller than the typical species, the best developed specimens of which are found farther south. A full plumaged female of *Bubo virginianus* is a very handsome bird, rich in markings, and the strongest and fiercest of all the American owls, a perfect flying tiger when loose among the game or poultry.

GENUS NYCTEA STEPHENS.

NYCTEA NYCTEA (LINN.).

166. Snowy Owl. (376)

Pure white with more or fewer blackish markings. Length, nearly 2 feet : wing, 17 inches ; tail, 10.

HAB.—Northern portions of the northern hemisphere. In North America breeding north of the United States, and extending beyond any point yet reached by navigators; in winter migrating south to the Middle States, straggling to South Carolina, Texas and the Bermudas.

Nest, on the ground, or on rocks, made up of sticks, grass, and lined with feathers.

Eggs, five to ten, laid at intervals, so that the nest may contain young birds and fresh eggs at the same time. (*Coues Key.*)

The Snowy Owl is an irregular winter visitor to Ontario, sometimes appearing in considerable numbers, and again entirely absent for several years in succession. Near Hamilton its favorite resort is on the beach, or along the shore of the bay, where it may be seen sitting watchful on the top of a muskrat heap, or pile of driftwood, frequently turning its head right round to look out for approaching danger. It hunts by day as well as by night, but is most active in the morning and evening. I once saw a large female make several attempts to capture a wounded duck which was swimming in a patch of open water among the ice on the bay near the canal. The owl skimmed along close to the ice and tried in passing to grasp the duck, which quickly went under water and appeared again cautiously at a different place. The owl passed several times over the pond in this way, resting alternately on the pier of the canal and on the shore, till, getting into a favorable position, I shot it on one of the return trips, and, subsequently, I also shot the duck, on which I had a first claim.

The number of these birds which occasionally descend from the north in the early part of the winter must be very great, for their migrations extend over a wide extent of country, and at Hamilton, which is only one of the points they pass, I have known as many as thirty to be captured in a single season. During the winter they are seen as far south as Texas and the Carolinas. How interesting it would be to know how many of these individuals which travel so far south are permitted to return.

The Snowy Owl is not migratory in the ordinary sense of the term. It is distributed over a very wide extent of country, and those we see in Ontario have merely come from some particular locality where at certain seasons their food supply has given out, and they have to seek it elsewhere. The bulk of the species pass the winter near the northern line of trees.

While at home in the north in summer, its food consists almost exclusively of lemmings and mice, but in winter, when these can not be had, it takes readily to fish, hares, rats, ducks, and even offal. I once killed a large female which was so gorged with tallow that she could scarcely fly.

It has been classed in *b* as a friend of the farmer, but so little of its time is spent in the agricultural districts that its influence for good or ill is very small.

GENUS SURNIA DUMÉRIL.

SURNIA ULULA CAPAROCH (MÜLL.).

167. **American Hawk Owl.** (377a)

Dark brown; above, more or less thickly speckled with white; below, closely barred with brown and whitish, the throat alone streaked; quills and tail, with numerous white bars; face, ashy, margined with black. Length, about 16 inches; wing, 9; tail, 7, graduated, the lateral feathers 2 inches shorter than the central.

HAB. Arctic America, migrating in winter to the northern border of the United States. Occasional in England.

Nest, of sticks, grass, moss and feathers, in trees or on rocks.

Eggs, four to seven, soiled white.

In Southern Ontario the Hawk Owl can only be regarded as a rare winter visitor. Farther north it seems more common, as I have heard of it being frequently seen in the district of Muskoka. While here in winter it has no particular haunt, but takes the country as it comes, like a hawk, and is evidently as sharp in the sight as it is active on the wing. The two in my collection were obtained in the neighborhood of Hamilton.

The Hawk Owl, like some other boreal birds of prey, occasionally comes south in the winter in large numbers, and is welcomed by collectors wherever it appears. These extensive migrations occur most frequently in the east. In Quebec, some years ago, in the month of March, I saw them exposed in the market day after day, and when coming west I noticed many perched on trees near the railroad track.

In Manitoba it is reported as an irregular winter visitor, but in Alaska Mr. Nelson says it is perhaps the most abundant resident bird of prey throughout the entire wooded region in the north. On May 5th, 1868, Mr. Dall secured six eggs of this bird from the top of an old birch stub about fifteen feet high, in the vicinity of Nulato, on the lower Yukon. The eggs were laid directly on the wood, and the male was sitting. When he climbed to the nest the sitting bird dashed at him and knocked off his cap, showing the fierce and bold character of the species.

The food of the Hawk Owl varies with the season of the year. In summer it feeds on the smaller mammals, such as lemmings, mice and ground squirrels, with a few grasshoppers by way of change, but in winter, when these are not available, ptarmigan constitute its chief article of diet. When in search of these birds, it is said to follow the hunter, often pouncing upon his game and carrying it off ere he can reach it.

It has been placed in Dr. Fisher's beneficial list, but it lives too far north to have much influence on agriculture.

Order COCCYGES. Cuckoos, etc.

Suborder CUCULI. Cuckoos, etc.

Family CUCULIDÆ. Cuckoos, Anis, etc.

Subfamily COCCYGINÆ. American Cuckoos.

Genus COCCYZUS Vieillot.

COCCYZUS AMERICANUS (Linn.).

168. Yellow-billed Cuckoo. (387)

Above, uniform satiny olive gray or "quaker color," with bronzy reflections; below, pure white; wings, extensively cinnamon-rufous on inner webs of the quills; central tail feathers, like the back; the rest black with large white tips, the outermost usually edged with white; bill extensively yellow below and on the sides. Length, 11-12 inches; wing, 5-5½; tail, 6-6½; bill, under 1 inch.

HAB.—Temperate North America, from New Brunswick, Canada, Minnesota, Nevada and Oregon, south to Costa Rica and the West Indies. Less common from the eastern border of the Plains westward.

Nest, on a bough, or in the fork of a low tree, composed of twigs, leaves and soft vegetable material.

Eggs, four to eight, pale glaucous green.

It is a well-known fact that the British Cuckoo entirely ignores family responsibilities by depositing its eggs in the nest of a bird of a different species, and with a pleasant "cuckoo" bids good-bye to the whole connection.

The two kinds we have in Canada are not so totally depraved. They usually build a nest and bring up a family, but even to them the duty does not seem to be a congenial one, and they are sometimes known to slip an egg into each other's nests or into that of a different species. The nest they build is of the most temporary description, and the eggs are deposited in such a desultory manner that it is no uncommon occurrence to find fresh eggs and young birds therein at the same time.

Of the two Cuckoos we have in Ontario, the Yellow-billed seems the more southern, apparently finding its northern limit along our southern border, where it is rather scarce and not generally distributed.

This species seems partial to orchards and cultivated grounds along the banks of rivers. It is very common about Niagara Falls, though not regularly distributed throughout the country. Last summer a pair had their nest and reared their young within fifty feet of my residence. They were very quiet and retiring, were seldom seen near the nest except while sitting on it, and only occasionally did we hear the emphatic *kow-kow-kow* which reminded us of their presence.

The nest was a flimsy affair, placed near the outer end of a horizontal branch of a maple, about eight feet from the ground. As soon as the young were able to fly, young and old disappeared, and were not again seen during the season.

Southern Ontario seems to be the northern limit of this species. It occurs at London, Hamilton, Toronto, but at Ottawa Mr. White has only found it once. To the north of Ottawa I have not heard of it being observed.

COCCYZUS ERYTHROPHTHALMUS (Wils.).

169. Black-billed Cuckoo. (388)

Above, uniform satiny olive-gray, or "quaker color," with bronzy reflections; below, pure white, sometimes with a faint tawny tinge on the fore parts; wings, with little or no rufous; lateral feathers not contrasting with the central, their tips for a short distance blackish, then obscurely white; bill, blackish, except occasionally a trace of yellowish below; eye-lids, red; bare circum-ocular space, purplish. Length, 11-12; wing, 5-5½; tail, 6-6½; bill, under 1.

Hab.—Eastern North America, from Labrador and Manitoba south to the West Indies and the valley of the Amazon; west to the Rocky Mountains. Accidental in the British Islands and Italy.

Nest, loosely constructed of twigs, grass, strips of bark, leaves, etc., placed in a bush.

Eggs, two to five, light greenish-blue.

The Black-billed Cuckoo is a regular summer resident in Ontario, where it arrives about the end of May, after which its peculiar note may often be heard, especially before rain, and its lithe, slim form be seen gliding noiselessly among the evergreens. Though not an abundant species, it is generally distributed throughout the Province, and well known to the country people as the " rain-crow."

The food of the cuckoos consists chiefly of caterpillars, with an occasional change to ripe fruit in the season. They also stand charged with sucking the eggs of other birds. They retire to the south early in September.

Of the two cuckoos the present species is the more common in Ontario, being more generally distributed. It also migrates a little farther north, having been reported as a tolerably common resident in Manitoba, where the other has not been observed.

We have heard that the Yellow-billed Cuckoo occasionally shirks the duties of incubation by leaving its eggs in the nest of another bird, and now come three separate charges of similar misconduct against the present species.

In the Report of the Ornithological Sub-section of the Canadian Institute for 1890, Dr. C. K. Clarke, of Kingston, Ontario, brings forward three cases of parasitism in the Black-billed Cuckoo, of the correctness of which there can be no doubt.

The first birds Dr. Clarke observed being imposed upon were a pair of chipping sparrows, who raised the young cuckoo at the expense of the family.

Next came a pair of yellow warblers, whose *protégé* soon crowded out the legitimate occupants of the nest. They were raised from the ground and placed within reach, but the big boy required all the attention of the foster-parents, and the others died. During the whole period, the old cuckoo was always to be found flitting about in a restless manner, as if she had some doubt in regard to the ability of the warblers to take care of her child.

The third case was another pair of chipping sparrows, in whose nest the cuckoo was observed sitting, and from which she did not move till the observers almost touched her. The result was the same as in the other cases. The young cuckoo threw the sparrows out as soon as he had strength to do so.

16

Suborder ALCYONES. Kingfishers.

Family ALCEDINIDÆ. Kingfishers.

Genus CERYLE Boie.

Subgenus STREPTOCERYLE Bonaparte.

CERYLE ALCYON (Linn.).

170. Belted Kingfisher. (390)

Upper parts, broad pectoral bar, and sides under wings, dull blue with fine black shaft lines ; lower eye-lid, spot before eye, a cervical collar and under parts, except as said, pure white ; the *female* with a chestnut belly band, and the sides of the same color ; quills and tail feathers, black, speckled, blotched and barred with white on the inner webs ; outer webs of the secondaries and tail feathers, like the back ; wing coverts, frequently sprinkled with white ; bill, black, pale at the base below ; feet, dark. Length, 12 or more ; wing, about 6 ; tail, 3½ ; whole foot, 1⅛ ; bill, about, 2¼.

Hab.—North America, south to Panama and the West Indies.

Nest, none.

Eggs, six to eight, white, deposited in an enlargement at the end of a tunnel, four to eight feet deep, dug by the bird into a sand bank or gravel pit.

The Kingfisher is generally distributed throughout Ontario. It arrives early in April, and soon makes its presence known by its loud, rattling cry, as it dashes along and perches on a horizontal bough overhanging the river. On some such point of observation it usually waits and watches for its scaly prey, but when passing over open water of greater extent, it is often observed to check its course, hover hawk-like at some distance above the surface, and then dash into the water after the manner of a Tern. If a fish be secured, it is carried in the bill to some convenient perch, on which it is hammered till dead, and then swallowed head downwards.

The Kingfisher is a strong flier, and is sometimes seen careering at a considerable height, as if for exercise.

Although many of them breed throughout Ontario, numbers proceed much farther north. In Manitoba and the North-West they occur in all suitable places, and in Alaska they are found along the entire course of the Yukon River, reaching the shores of Behring Sea. They have also been taken at Sitka, and frequent all the clear streams of the interior, nesting as they do elsewhere, in a deep burrow in a bank dug out by themselves.

They are not sensitive to cold, for in open seasons I have seen them remaining till January, but when the frost forces the fish to retire to deep water, the Kingfisher's supply of food is cut off, and he has to move to the south.

ORDER PICI. WOODPECKERS, WRYNECKS, ETC.

FAMILY PICIDÆ. WOODPECKERS.

GENUS DRYOBATES BOIE.

DRYOBATES VILLOSUS (LINN.).

171. **Hairy Woodpecker.** (393)

Back, black, with a long white stripe ; quills *and wing coverts*, with a profusion of white spots ; four middle tail feathers, black ; next pair, black and white ; next two pairs, white ; under parts, white ; crown and sides of head, black, with a white stripe over and behind the eye, another from the nasal feathers running below the eye to spread on the side of the neck, and a scarlet nuchal band in the *male*, wanting in the *female* ; *young*, with the crown mostly red or bronzy, or even yellowish. Length, 9-10 ; wing, nearly 5 ; tail, 3½.

HAB.—Middle portion of the Eastern United States, from the Atlantic coast to the Great Plains.

Nest, in a hole in a tree.

Eggs, four or five, pure white.

This is a resident, though not very abundant species, noticed more frequently in winter than in summer. It is generally distributed through Southern Ontario.

The Hairy Woodpecker is one of the most retiring of the family, spending much of its time in the solitudes of the woods, and when these are thinned out or cleared away, moving to regions still more remote. It is a strong, hardy, active bird, and the noise it makes while hammering on a tree, when heard in the stillness of the woods, might well be supposed to be produced by a bird of much greater size.

This is one of those birds which increase in size as they approach their northern limit, and decrease in proportion when found in the south. On this account the American Ornithologists' Union Committee has separated from the original *Dryobates villosus* two subspecies, the first of which, under the name of Northern Hairy Woodpecker, is said to be found in North America, south to about the northern border of the United States. The other has been named the Southern Hairy Woodpecker, and its habitat is said to extend over the southern portion of the United States east of the plains.

These subdivisions have been decided upon after comparing a large number of specimens from the different regions indicated, but the rule cannot be always correct, for I have some from British

Columbia which are no larger than those found in Southern Ontario. It is well to know that the species varies in size according to its habitat, north or south, but I hardly think the subdivision necessary, for we see among the members of the human family individuals who differ in size far more than the woodpeckers do, and yet we do not make subspecies of them. However, if the new arrangement suits the majority, so let it be.

DRYOBATES PUBESCENS (Linn.).

172. Downy Woodpecker. (394)

Coloration, exactly as in *P. villosus*, except that the outer tail feathers are barred with black and white. Length, 6-7; wing, under 4; tail, under 3.

Hab.—Northern and Eastern North America, from British Columbia and the eastern edge of the Plains northward and eastward.

Nest, a hole in a tree.

Eggs, four or five, pure white.

This is a miniature of the preceding species, which it resembles in habits as well as in appearance, although it is of a more sociable disposition, being often found in winter in company with the Chickadees and Brown Creepers. It is also an occasional visitor to the orchard, where it goes over the apple trees carefully, examining all injured or decayed parts in search of insects.

It is commonly known as the "little sapsucker," but the name is incorrectly applied, for any holes drilled by this species are made while it is in search of insects, those which allow the sap of the tree to exude being the work of the Yellow-bellied Woodpecker.

Like its big brother, the Downy Woodpecker is a resident species, but more plentiful in spring and fall than in summer, the numbers being increased at those seasons by passing migrants.

They do not, however, migrate in the ordinary sense of the word, but in the fall often shift about in search of better feeding ground. They are found in Manitoba, the North-West, and in Alaska, and are resident throughout their range. They have been taken along the entire course of the Yukon, as well as at various points on the coast of Behring Sea, and thence south at Kadiak and Sitka. Those found in the north do not differ from the southern individuals, except in being rather larger in size.

Genus PICOIDES Lacépède.

Picoides ARCTICUS (Swains.).

173. Arctic Three-toed Woodpecker. (400)

Crown, with a yellow patch in the *male*; back, uniform black; sides of head, striped; of body, barred with black and white; under parts, otherwise white; quills, with white spots; tail feathers unbarred, the outer white, the central black. Length, 8-9; wing, 4½-5; tail, 3½-4.

Hab.—Northern North America, from the Arctic Regions south to the northern border of the United States; much farther south in the western part of the United States (Nevada, California), along the mountain ranges.

Eggs, four to six, white, deposited in a hole in a tree.

This is a northern bird, seldom, even in winter, coming so far south as the southern border of Ontario. In November, 1859, I killed one on a pine tree on the south shore of Dundas marsh, which is the only time I have ever seen it alive. I have heard of one or two others having been obtained in Southern Ontario, but as the species is common farther north, these can only be regarded as wanderers.

In the district of Muskoka it is resident and quite common, frequenting certain tracts of country which the fire has gone through and left the trees standing dead and decaying. It belongs to a small group, the members of which have only three toes. Whether this is a special adaptation of the bird to its life among the pines is not apparent, but it seems quite as able to shift for itself with three toes as its near relatives are with four.

Although this species is a northern bird, so far as its not going south is concerned, it does not bear out the name by going as far north as its near relative, *americanus*. In some of these distant regions it seems almost unknown. Mr. Nelson speaks of it in Alaska: "Very little appears to be known concerning the movements and habitat of this bird, especially in the north. I secured but a single specimen, which was brought to me by Mr. McQuesten from Fort Reliance on the upper Yukon, and its rarity as compared with the other Three-toed Woodpecker appears from the fact that dozens of the latter were brought to me each winter. No one among the various naturalists who have visited this region before has secured it, and *arcticus* is totally unknown west of the point where my specimen was found."

PICOIDES AMERICANUS Brehm.

174. American Three-toed Woodpecker. (401)

Three-toed ; entire upper parts glossy, bluish-black with a few spots of
white on the wing quills ; below, white from the bill to the tail ; the sides,
flanks and lining of the wings, barred with black ; four middle tail feathers
black, the rest white. *Male :*—With a square patch of yellow on the crown,
wanting in the *female ;* bill and feet, dull blue. Length, 9-10 inches.

Hab.—Northern North America, from the Arctic Regions southward, in
winter, to the Northern States.

Nest, a hole in a tree at no great height above the ground.

Eggs, four or five, creamy white.

This is a more northern species than the preceding, and is nowhere
so abundant. The two are sometimes seen in company, and were
found by Dr. Merriam breeding in the same district in northern New
York, but, strange to say, the present species has not been found
breeding in Muskoka, where the other is common and resident.
During the past two years my friend Mr. Tisdall has been much in
the woods in that district, and though he has seen scores of the
Black-backed during that time, he has only once met the other.

Since the above was written, a hunter who was shooting grouse in
a different section of Muskoka from that referred to, promised to
send me any Three-toed Woodpeckers he came across during a day's
excursion. In due course he sent me sixteen, five of which were of
this species. It was late in October, and he most likely came upon a
group that had just arrived from the north. Both species are quite
rare at Ottawa, Mr. White having obtained only two of each. In
Manitoba the present species is said to be very rare. Richardson
says it is the most common woodpecker north of the Great Slave
Lake. In Alaska the northern form is very common and generally
distributed, but here again the work of subdivision has been going on
with vigor. Up to the time of writing, three species have been made
out of the one, and how many more will be produced time will show,
but we are likely to have only the original *americanus* in Ontario.

Genus SPHYRAPICUS Baird.

SPHYRAPICUS VARIUS (Linn.).

175. **Yellow-bellied Sapsucker.** (402)

Crown, crimson, bordered all around with black ; chin, throat and breast, black, enclosing a large crimson patch on the former in the *male* ; in the *female*, this patch white ; sides of head with a line starting from the nasal feathers and dividing the black of the throat from a trans-ocular black stripe, this separated from the black of crown by a white post-ocular stripe ; all these stripes frequently yellowish : under parts, dingy yellow, brownish and with sagittate dusky marks on the sides ; back, variegated with black and yellowish-brown : wings, black with large oblique white bar on the coverts, the quills, with numerous paired white spots on the edge of both webs ; tail, black, most of the feathers white-edged, the inner webs of the middle pair and the upper coverts mostly white. *Young* birds lack the definite black areas of the head and breast and the crimson throat patch, these parts being mottled-gray. About, 8½ ; wing, 4½-5 ; tail, 3¼.

Hab.—North America, north and east of the Great Plains, south to the West Indies, Mexico and Guatemala.

Eggs, four to six, white, deposited in a hole in a tree.

In Ontario this beautiful species is strictly migratory, not having been observed during winter, but from the fact of its being seen late in the fall and again early in spring, we infer that it does not go far south.

It is decidedly a sapsucker, the rows of holes pierced in the bark of sound, growing trees being mostly made by this species. It is not endowed with the long, extensile tongue peculiar to many of the woodpeckers, but feeds largely on insects, which it finds on the outer bark of the trees or catches on the wing. It has been accused of doing serious injury to growing trees, by girdling them to get at the inner bark, on which it is said to feed. Dr. King, of River Falls, in his "Economic Relations of our Birds," exonerates it from this charge, and says that in the stomachs of thirty specimens which he examined, he found in only six a small amount of material resembling the inner bark of trees, and further adds: "No instance in which the bark of trees has been stripped off has come under my observation, nor do I know of a single case in which their puncturings of the bark have been fatal, or even appreciably injurious to the tree." In Southern Ontario a few remain and raise their young, but the majority go farther north.

This species occurs in Manitoba, and Dr. Coues says of it: "Common summer resident of wooded sections, plentiful at Pembina, where it was breeding in June; again seen on the Moose River, not observed farther north."

GENUS CEOPHLŒUS Cabanis.

CEOPHLŒUS PILEATUS (Linn.).

176. Pileated Woodpecker. (405)

Black : the head, neck and wings much varied with white or pale yellowish ; bill, dark. *Male :*—Scarlet crested, scarlet moustached. *Female :*—With the crest half black, half scarlet, and no maxillary patches. Length, 15-19 ; wing, 8½-10 ; tail, 6-7.

HAB.—Formerly, whole wooded region of North America ; now rare or extirpated in the more thickly settled parts of the Eastern States.

Nest. a hole in the trunk or limb of a tall tree.

Eggs. four to six, oval, white.

This is one of the grand old aborigines who retire before the advance of civilization. It used (so we are told) to be common near Hamilton, but seclusion among heavy timber is necessary for its existence, and such must now be sought in regions more remote.

It is not strictly a northern species, being found resident in suitable localities both north and south, but varies considerably in size according to latitude, the northern individuals, as usual in such cases, being the larger. Many spend the winter in the burnt tracts in Muskoka, and in spring disperse over the country to breed in the solitude they seem to like.

They are wild, shy birds, difficult of approach, but their loud hammering is at all times a guide to those who wish to follow them in the woods. A nest was taken in the county of Middlesex, in May, 1885, by Mr. Robt. Elliot.

Dr. Macallum reports that they still breed in suitable places along the north shore of Lake Erie, near Dunnville. Their distribution seems to be more influenced by the size of the timber than by the temperature, as they are common in Manitoba and abundant in British Columbia, but are not named among the "Birds of Alaska." Though now rare, or absent, in the thickly settled parts of the Eastern States, they are still common in the timbered swamps and secluded woods in the south. The nest is usually made in a retired part of the woods, and at so great a height from the ground that it is seldom reached by the oölogist. This fact will help to prevent the species being reduced in numbers.

Genus MELANERPES Swainson.

Subgenus MELANERPES.

MELANERPES ERYTHROCEPHALUS (Linn.).

177.   Red-headed Woodpecker.   (406)

Glossy blue-black ; rump, secondaries and under parts from the breast, pure white ; primaries and tail feathers, black ; whole head, neck and breast crimson in both sexes, grayish-brown in the *young*.   About 9 ; wing, 5½ ; tail, 3½.

Hab. United States, west to the Rocky Mountains, straggling westward to Salt Lake Valley ; rare or local east of the Hudson River.

Nest, in a hole in a tree, varying greatly in height.

Eggs, four to six, white.

In Ontario the Red-headed Woodpecker is a summer resident only, arriving early in May and leaving again in September. It is quite common and perhaps the best known of any of the woodpeckers, both on account of its decided markings and from its habit of visiting the orchard during the season of ripe fruit. It is also an expert fly-catcher, frequently taking its position on the top of a dead pine, from which it darts out after the passing insect in true fly-catcher style. Though a very showy bird when seen in the woods, it does not look so well in collections, the red of the head evidently fading after death.

It is generally distributed throughout Ontario, but becomes rare in Manitoba, north of which I have not heard of its having been observed.

Besides the ordinary food of woodpeckers, it shows a taste for grasshoppers and beetles, in search of which it may often be observed on the ground. It is also very fond of ripe fruit, and incurs the serious displeasure of the gardeners by mutilating or carrying off the finest of the apples, pears, cherries and other fruits.

They are rather noisy and quarrelsome birds, but this trait may be partly assumed.

They all leave Ontario in October, and during the winter none are observed.

SUBGENUS CENTURUS Swainson.

MELANERPES CAROLINUS (Linn.).

178. Red-bellied Woodpecker. (409)

Back and wings, except larger quills, closely banded with black and white ; primaries with large white blotches near the base, and usually a few smaller spots ; whole crown and nape, scarlet in the *male*, partly so in the *female* ; sides of head and under parts, grayish-white, usually with a yellow shade, *reddening* on belly ; flanks and crissum, with sagittate-black marks ; tail, black, one or two outer feathers white, barred ; inner web of central feathers white with black spots ; outer web of same black, with a white space next the shaft for most of its length ; white predominating on the rump. Length, 9-10 ; wing, about 5 ; tail, about 3½.

HAB.—Eastern United States, to the Rocky Mountains ; rare or accidental east of the Hudson River.

Nest, a hole in a tree.

Eggs, four to six, white.

This handsome woodpecker is gradually becoming more common in Southern Ontario, and like some others, such as the Lark-finch, Orchard Oriole and Rough-winged Swallow, it evidently makes its entrance to the Province round the west end of Lake Erie.

It seems to find its northern limit in Southern Ontario, and even there confines itself chiefly to the south-west portion. Stragglers have been found near Toronto and Hamilton, while near London it breeds and is tolerably common, but Mr. White has not yet found it in Ottawa, and it is not included among the "Birds of Manitoba."

In the "Birds of Ohio," Dr. Wheaton mentions it as a common summer resident, but it is not named among the "Birds of Minnesota," and in the "Birds of New England" it is spoken of as the rarest of all the woodpeckers. Its centre of abundance is evidently farther south, where it is resident.

West of the Rocky Mountains it has not been observed. It is of shy, retiring habits, frequenting the gloomy depths of the forest, though its dress fits it for appearing in the best society. The feathers are of a peculiar soft and silky texture, and are so regularly barred across with black and white that it is known to many as the Zebra Bird, and is considered the handsomest of all the woodpeckers.

In Ontario it is migratory, leaving about the end of October.

GENUS COLAPTES SWAINSON.

COLAPTES AURATUS (LINN.).

179. Flicker. (412)

Back, wing coverts and innermost quills, olivaceous-brown, thickly barred with black ; rump, snowy-white ; quills and tail, golden yellow underneath, and shafts of this color. A scarlet nuchal crescent and large black pectoral crescent in both sexes. *Male :*--With black maxillary patches, wanting in the *female :* head and nape, ash ; chin, throat and breast, lilac-brown ; under parts, with numerous round black spots ; sides, tinged with creamy-brown ; belly, with yellowish. About 10 inches long ; wing, about 6 ; tail, 4½.

HAB. Northern and Eastern North America, west to the eastern slope of the Rocky Mountains and Alaska. Occasional on the Pacific slope, from California northward. Accidental in Europe.

Nest, a hole in a tree.

Eggs, five to seven, white.

Early in April, if the weather is mild, the loud cackling call of the "Higholder" may be heard from his perch at the top of a tall dead limb, where he watches to welcome his comrades as they hourly arrive from the south. For a week or two at this season they are very abundant, but many soon pass on farther north, and the others are distributed over the country, so that they are less frequently seen.

In habits this species differs considerably from all the other members of the family. It is more terrestrial, being often observed on the ground demolishing ant hills and devouring the inmates, for which achievement its curved bill and long slimy tongue are admirably adapted. It is also fond of fruit, and of corn, either green or ripe.

It is by no means confined to the forest, but is often seen peeping
from its hole in a stub by the roadside. When alighting upon a
tree, it perches on a bough in the ordinary manner, being seldom seen
clinging to the trunk like other members of the family, except when
entering its nest. In Southern Ontario it is seen till late in October,
but only on one or two occasions have stragglers been observed
during the winter.

This is at once the most abundant and widely distributed woodpecker in Canada. Crossing our southern border, it works its way
up north, leaving representatives in Manitoba, the North-West and
other territories through which it passes, till it finally reaches the
shores of the Arctic Sea. In Alaska, Mr. Nelson says of it : "This
handsome woodpecker breeds from one side of the Territory to the
other, wherever wooded country occurs. It has been sent to the
National Museum from the lower Anderson River, and is well
known to breed along the entire course of the Yukon, reaching to the
mouth of that river.

"It is a regular summer resident at the head of Norton Bay, and
reaches the Arctic on the shore of Kotzebue Sound."

It is also reported as an accidental visitor in Greenland.

Albinos of this species are of frequent occurrence. Once when
driving north in the township of Beverley, a cream-colored specimen
kept ahead of me for half a mile. How beautiful he looked in the
rich autumnal sunlight, as with long swoops he passed from tree to
tree by the roadside ! I could not but admire him, and that was all
I could do, for I was unarmed.

Order MACROCHIRES. Goatsuckers, Swifts, etc.

Suborder CAPRIMULGI. Goatsuckers, etc.

Family CAPRIMULGIDÆ. Goatsuckers, etc.

Genus ANTROSTOMUS Gould.

ANTROSTOMUS VOCIFERUS (Wils.).

180. **Whip-poor-will.** (417)

General color of the upper parts, dark brownish-gray, streaked and minutely sprinkled with brownish-black; quills and coverts, dark brown, spotted in bars with light brownish-red; four middle tail feathers like those of the back, the three lateral white in their terminal half; throat and breast, similar to the back with a transverse band of white on the foreneck; rest of the lower parts, paler than above and mottled. *Female:* Similar, but with the lateral tail feathers reddish-white toward the tip only, and the band across the forehead pale yellowish-brown. Length, 9½; wing, 5½; tail, 4½.

Hab. Eastern United States to the Plains, south to Guatemala.

Eggs, two, deposited in a hollow or a rotten log, or on the ground on a dry bank among leaves. They are elliptical, of moderate polish with a ground color of white or cream, handsomely marked with spots of yellowish-brown; deep shell marks are about as numerous as the surface marks and are of a lilac-gray or lavender tint.

This well-known bird crosses the southern frontier of Ontario about the 10th of May, and should the weather be mild its loud and well-known cry is soon heard at night at many different points throughout the country. It is seldom seen abroad by day, except when disturbed at its resting place in some shady part of the woods, when it glides off noiselessly like a great moth. Disliking the glare of the light, it avoids the city, but not unfrequently perches on the roof of a farm-house, startling the inmates with its cry, which they hear with great distinctness.

This is the only song of the Whip-poor-will, and it is kept up during the breeding season, after which it is seldom heard. We see so little of these birds that it is difficult to tell exactly at what time they leave us, but it is most likely early in September that they " fold their tents like the Arabs, and as silently steal away."

It is reported as a common summer resident in Manitoba, and Dr. Bell records its presence at Norway House, to the north of which I have not heard of it having been observed.

It is a delicately formed bird, though strong on the wing. Its legs and feet are very slender, but they are not often called into use, and are easily carried from the fact of their being light. On the middle toe is a curiously pectinated claw, which is supposed to be useful for ridding the bird of the insects with which it is troubled. When disturbed in the woods, if it alights on a branch it always sits lengthways, in which position it is very apt to be mistaken for a growth, and escape observation.

It is one of the few birds whose call can be intelligibly put into words. The experiment is often tried with other species, but in very few instances can they be printed so as to be recognized when heard out of doors.

GENUS CHORDEILES SWAINSON.

CHORDEILES VIRGINIANUS (GMEL.).

181. **Nighthawk.** (420)

Above, mottled with black, brown, gray and tawny, the former in excess : below from the breast, transversely barred with blackish and white or pale fulvous ; throat in the *male* with a large white, in the *female* tawny, cross-bar : tail, blackish, with distant pale marbled cross-bars and a large white spot (wanting in the *female*) on one or both webs of all the feathers toward the end : quills, dusky, unmarked except by one large white spot on five outer primaries about midway between their base and tip ; in the *female* this area is restricted or not pure white. Length, about 9 ; wing, 8 ; tail, 5.

HAB.—Northern and Eastern North America, east of the Great Plains, south through tropical America to Buenos Ayres.

Eggs, two, deposited on rocks or on the ground, or among the gravel of a flat-roofed house in the city. They vary from pale olive-buff to buffy and grayish-white thickly mottled and daubed with varied tints of darker gray slate, olive or even blackish mixed with a marbling of purplish-gray, both pattern and tints being very variable.

This is a well-known and abundant summer resident, arriving from the south early in May. Though a Nighthawk, it is often seen abroad by day during cloudy weather, and in the evening, just as the sun is sinking below the horizon, numbers of these birds are occasionally seen careering around high overhead, uttering their peculiar cry, so readily recognized, yet so difficult either to imitate or describe. While thus in the exercise of their most wonderful powers of flight, and performing many graceful aerial evolutions, they will suddenly change their course and plunge headlong downwards with great rapidity, producing at the same time a singular booming sound which can be heard for some distance. Again, as quickly, with a few bold strokes of their long, pointed wings, they will rise to the former height, and dash hither and thither as before.

Poets, in all ages, have sung the praises of their favorite birds, and even to-day, from the unromantic plains of Chatham come the following lines on the habit of the Nighthawk, just described :

> " With half closed wings and quivering boom,
> Descending through the deepening gloom,
> Like plummet falling from the sky,
> Where some poor moth may vainly try
>   A goal to win-
> He holds him with his glittering eye
>   And scoops him in."

Towards the end of August, when the first frosts begin to cut off their supply of insect food, large gatherings of Nighthawks may be seen in the evenings moving toward the south-west, not in regular order like ducks or pigeons, but skimming, darting and crossing each other in every imaginable direction, and still with a general tendency toward the south, till darkness hides them from our view.

SUBORDER CYPSELI. SWIFTS.

FAMILY MICROPODIDÆ. SWIFTS.

SUBFAMILY CHÆTURINÆ. SPINE-TAILED SWIFTS.

GENUS CHÆTURA STEPHENS.

CHÆTURA PELAGICA (LINN.).

182. Chimney Swift. (423)

Sooty brown with faint greenish gloss above ; below, paler, becoming gray on the throat ; wings, black. Length, about 5 ; wing, the same ; tail, 2 or less.

HAB.—Eastern North America, north to Labrador and the Fur Countries, west to the Plains, and passing south of the United States in winter.

Nest, a basket of twigs glued together, and to the side of the chimney or other support by the saliva of the bird.

Eggs, four or five, pure white.

The Swift is a late comer, and while here seems ever anxious to make up for lost time, being constantly on the wing, darting about with great rapidity, sometimes high overhead, sometimes skimming the surface of the pond, often so closely as to be able to sip from the water as it passes over it, or snap up the insects which hover on the surface.

The original nesting place of the Swifts was in a hollow tree, often of large diameter, and frequented year after year by a great many of the birds, but now they seem to prefer a city chimney. There they roost, and fasten their curious basket nests to the wall, inside the chimney, a few feet down, to be out of reach of the rays of the sun. A fine exhibition of bird-life it is to watch the Swifts, in the evening about sunset, circling a few times round the chimney, raising their wings above their backs and dropping like shuttlecocks down to their nest, near which they spend the night clinging to the wall with their claws. The sharp spines at the end of the tail feathers, pressed against the surface, form their chief support.

They arrive about the 10th of May, and leave for the south early in September.

SUBORDER TROCHILI. HUMMINGBIRDS.

FAMILY TROCHILIDÆ. HUMMINGBIRDS.

GENUS TROCHILUS LINNÆUS.

SUBGENUS TROCHILUS.

TROCHILUS COLUBRIS LINN.

183. Ruby-throated Hummingbird. (428)

Male :—With the tail forked, its feathers all narrow and pointed ; no scales on crown ; metallic gorget reflecting ruby-red, etc. ; above. golden green ; below. white ; the sides, green ; wings and tail, dusky purplish. *Female* :—Lacking the gorget ; the throat, white ; the tail. somewhat double-rounded, with black bars, and the outer feathers white-tipped. Length, 3¼ ; wing, 1⅝ ; bill, ⅔.

HAB.—Eastern North America to the Plains, north to the Fur Countries. and south, in winter, to Cuba and Veragua.

Nest, a beautiful specimen of bird architecture, usually placed on the horizontal branch of a tree in the orchard, composed of gray lichens, lined with the softest plant down.

Eggs, 2, pure white, blushed with pink while fresh.

The Hummingbirds begin to arrive towards the middle of May, and by the end of the month, when the lilacs are in bloom, they are quite numerous. About that time many pass on to breed farther north, while others engage in the same occupation here.

In September they again become common, showing a strong liking for the *impatiens fulva*, or wild balsam, which grows abundantly in moist places, and later they crowd about the *bignonia* or trumpet-

17

creeper. This is a late flowering plant, and the tiny birds, as if loath to leave it, are seen as late as the middle of September rifling it of its sweets.

There are about sixteen different species of Hummingbirds now known as North American, but this is the only one found east of the Mississippi River. Though small, it is very pugnacious, often attacking birds much larger than itself who may venture near its nest. On such occasions it produces an angry buzzing sound with its wings, but it has no voice save a weak chirp, like a cricket or grasshopper.

ORDER PASSERES. PERCHING BIRDS.
SUBORDER CLAMATORES. SONGLESS PERCHING BIRDS.
FAMILY TYRANNIDÆ. TYRANT FLYCATCHERS.
GENUS MILVULUS SWAINSON.
MILVULUS FORFICATUS (GMEL.).

184. Scissor-tailed Flycatcher. (443)

First primary alone emarginate: crown patch, orange or scarlet; general color, hoary ash, paler or white below, sides at the insertion of the wings scarlet or blood-red, and other parts of the body tinged with the same, a shade paler; wings, blackish, generally with whitish edgings; tail, black, several outer feathers extensively white or rosy. Wing, about 4½; tail, over 12 inches long.

HAB.—Texas and Indian Territory, casually north to Kansas and Missouri; south to Central America. Accidental in Virginia, New Jersey, New England, Manitoba, and at York Factory, Hudson Bay.

Nest, like the Kingbirds'.

Eggs, four or five, white, blotched with reddish and lilac shell-spots.

The home of this beautiful bird is in Texas, but it is evidently much given to wandering, appearing unexpectedly at points far distant from its usual habitat.

The first record I have of its occurrence in Ontario is furnished by Dr. Garnier, of Lucknow, Bruce County, who reports having seen one near his place some years ago. He had no means of securing the bird, but saw it by the roadside as he drove past, opening and closing its tail feathers with the usual scissor-like motion.

Dr. Bell, of the Geological Survey, speaking of birds which he occasionally found far from their usual habitat, says: "The most singular discovery in regard to geographical distribution is the finding of the Scissor-tailed Flycatcher (*Milvulus forficatus Sw.*) at York Factory.

"Hitherto its range has been considered to extend only from Mexico to Central Texas. Baird, Cassin and Lawrence say with regard to it: 'This exquisitely beautiful and graceful bird is quite abundant on the prairies of Southern Texas, and is everywhere conspicuous among its kindred species. It is usually known as the Scissor-tail from its habit of opening and closing the long tail feathers as if they were the blades of a pair of scissors.' The specimen in the Government Museum was shot at York Factory in the summer of 1880, and I have since learned that these remarkable birds were commonly seen at the posts of the Hudson's Bay Company all the way westward to the valley of the Mackenzie River."

The Scissor-tails are beautiful birds which we should gladly welcome to Ontario if they find the attractions sufficient to induce them to extend their habitat in this direction, but of these they are the best judges, and they will no doubt act accordingly.

Genus TYRANNUS Cuvier.

TYRANNUS TYRANNUS (Linn.).

185. Kingbird. (444)

Two outer primaries obviously attenuate; above, blackish, darker on the head; crown with a flame-colored patch; below, pure white; the breast shaded with plumbeous; wings, dusky, with much whitish edging; tail, black, broadly and rather sharply tipped with white, the outer feathers sometimes edged with the same; bill and feet, black. *Young*:—Without the patch; very young birds show rufous edging of the wings and tail. Length, about 8 inches; wing, $4\frac{1}{2}$; tail, $3\frac{1}{2}$; bill, under 1.

Hab. Eastern North America, from the British Provinces south to Central and South America. Rare west of the Rocky Mountains (Utah, Nevada, Washington Territory, etc.).

Nest, large for the size of the bird, placed on the horizontal bough of an isolated tree, composed of vegetable fibrous materials and sheep's wool compactly woven together.

Eggs, three to five, creamy or rosy-white, spotted and blotched with reddish, brown and lilac shell-spots.

The Kingbird arrives in Ontario from the south about the 10th of May, and from that time till it leaves again in September, it is one of the most familiar birds in the rural districts. It is generally

distributed, each pair taking possession of a certain "limit," which
is valiantly defended against all intruders, no bird, however large,
being permitted to come with impunity near where the Kingbird's
treasures are deposited. It is partial to pasture fields, a favorite
perch being the top of a dry mullein stalk. Here the male sits like
a sentinel, issuing his sharp note of warning, and occasionally darting
off to secure a passing insect. When the breeding is over and the
young are able to shift for themselves, he gets over his local attach-
ments and quietly takes his insect fare wherever he can find it,
allowing other birds to do the same.

The food of this species consists almost entirely of insects, which
it takes while on the wing. It may be considered a friend of the
farmer, for although it does take a few bees now and then, it more
than compensates for these by the large number of injurious insects
which it destroys.

The contents of the twelve stomachs examined by Dr. King, of
Wisconsin, were as follows:—Four had eaten seventeen beetles;
four, four dragon flies; one, a bee; one, six crane flies; one, a large
moth; one, a butterfly; and three, a few raspberries.

It is common throughout Manitoba and the North-West, retiring
south when its supply of food is cut off by the advance of the season.

GENUS MYIARCHUS CABANIS.

MYIARCHUS CRINITUS (LINN.).

186. Crested Flycatcher. (452)

Decidedly olivaceous above, a little browner on the head, where the
feathers have dark centres; throat and fore-breast, pure dark ash, rest of
under parts, bright yellow, the two colors meeting abruptly; primaries
margined on both edges with chestnut; secondaries and coverts, edged and
tipped with yellowish-white; tail, with all the feathers but the central pair,
chestnut on the whole of the inner web, excepting, perhaps, a very narrow
stripe next the shaft; outer web of outer feathers, edged with yellowish; the
middle feathers, outer webs of the rest, and wings, except as stated, dusky
brown. Very young birds have rufous skirting of many feathers, in addition
to the chestnut above described, but this soon disappears. Length, $8\frac{1}{2}$-$9\frac{1}{2}$;
wing and tail, about 4; bill and tarsus, each $\frac{3}{4}$.

HAB.—Eastern United States and Southern Canada, west to the Plains,
south through Eastern Mexico to Costa Rica.

Nest, in hollow of trees, sometimes in the deserted hole of a woodpecker, composed of straw, leaves, rootlets and other vegetable materials, lined with feathers; about the edge are always to be found pieces of the cast-off skins of snakes.

Eggs, four or five, light buffy-brown, streaked lengthwise by lines and markings of purplish and darker brown.

This species is a regular summer resident along the southern frontier, where it arrives early in May, and soon makes its presence known by the loud note of warning which is heard among the tree tops long before the bird is visible.

Dr. Wheaton, in his "Birds of Ohio," states that this species is very numerous near Columbus, where the country being well cleared and the usual breeding places difficult to find, the birds have taken to the use of boxes put up for bluebirds and martins, and have been observed to dispossess the legitimate owners. It has also been noticed that the snake skins are left out, when the nests are in boxes.

This Flycatcher is found sparingly in Manitoba and the North-West; farther south, it is more common. Its food consists of insects, while these are obtainable; but in the fall, before leaving, it readily takes grapes and other berries.

It is the largest of the flycatchers which visit us, and it shows to great advantage in the woods in spring, when its clear colors harmonize with the opening leaves.

Many spend the winter in Mexico.

Genus SAYORNIS Bonaparte.

SAYORNIS PHŒBE (Lath.).

187. Phœbe. (456)

Dull olivaceous-brown; the head much darker fuscous-brown, almost blackish, usually in marked contrast with the back; below, soiled whitish, or palest possible yellow, particularly on the belly; the sides and the breast, nearly or quite across shaded with grayish-brown; wings and tail, dusky; the outer tail feather, inner secondaries and usually the wing coverts, edged with whitish; a whitish ring around the eye; bill and feet, black, varies greatly in shade. The foregoing is the average spring condition. As the summer passes, the plumage becomes much duller and darker brown from wearing of the feathers, and then, after the moult, fall specimens are much brighter than in spring, the under parts being frequently decidedly yellow, at least on the belly. Very young birds have some feathers edged with rusty, particularly on the edges of the wing and tail feathers. Length, 6¾-7 ; wing and tail, 3-3½.

HAB.—Eastern North America, from the British Provinces south to Eastern Mexico and Cuba, wintering from the South Atlantic and Gulf States southward.

Nest, under bridges or projection about outhouses. When away from human habitation it is often found among the roots of an upturned tree or in a hollow tree, composed of vegetable material mixed with mud and frescoed with moss.

Eggs, four or five, usually pure white, sometimes faintly spotted.

This is one of the earliest harbingers of spring, and its quick, querulous notes are hailed with joy as a prelude to the grand concert of bird music which is soon to follow.

Early in April, the male Pee-wee appears in his former haunts, and is soon joined by his mate. They are partial to the society of man, and their habits, as shown in their nestings, have been somewhat changed by this taste. The original, typical nest of the Pee-wee, we are told, was placed on a ledge under a projecting rock, over which water trickled, the nest itself often being damp with the spray. We still see one, occasionally, in such a position, but more frequently it is placed on the beams of a bridge, beneath the eaves of a deserted house, or under a verandah or the projection of an outhouse. They raise two broods in the season, and retire to the south in September.

They are generally distributed throughout Ontario, but are most common in the south. A few have made their way to Manitoba, where they have been seen near Winnipeg and elsewhere. They are strongly attached to a chosen locality, and will return year after year to repair their old nest, or entirely rebuild it in the same spot should the old fabric be removed. They are often imposed upon by the Cow-bird, and accept the situation without remonstrance.

GENUS CONTOPUS CABANIS.

CONTOPUS BOREALIS (SWAINS.).

188. Olive-sided Flycatcher. (459)

Dusky olivaceous-brown, usually darker on the crown, where the feathers have black centres, and paler on the sides ; chin, throat, belly, crissum and middle line of the breast, white, more or less tinged with yellowish ; wings and tail, blackish, unmarked, excepting inconspicuous grayish-brown tips of the wing coverts, and some whitish edging of the inner quills ; feet and upper mandible, black ; lower mandible, mostly yellowish. The olive-brown below has a peculiar *streaky* appearance hardly seen in other species, and extends almost entirely across the breast ; a peculiar tuft of white fluffy feathers on the flanks. *Young :*—Birds have the feathers, especially of the wings and tail, skirted with rufous.   Length, 7·8 ; wing, 3½-4½, remarkably pointed ; second quill longest, supported nearly to the end by the first and third, the fourth abruptly shorter ; tail, about 3 ; tarsus, middle toe and claw together, about 1¼.

HAB.—North America, breeding from the northern and the higher mountainous parts of the United States northward, in winter, south to Central America and Columbia.

Nest, a shallow structure, composed of weeds, twigs, rootlets, strips of bark, etc., loosely put together, saddled on a bough or placed in a fork high up in a tree.

Eggs, three or four, creamy-white, speckled with reddish-brown.

So far as at present known, this species is rare in Ontario, and not very abundant anywhere.   Towards the end of May, 1884, when driving along the edge of a swamp, north of the village of Millgrove, I noticed a bird on the blasted top of a tall pine, and stopping the horse, at once recognized the Olive-sided Flycatcher by the loud *O-whee-o, O-whee-o*, so correctly described as the note of this species by Dr. Merriam in his "Birds of Connecticut." I tried to reach it with a charge of No. 8, and it went down perpendicularly into the brush, but whether dead, wounded or unhurt I never knew, for I did not see it again.   That is the only time I have ever seen the species alive.

It has a wide distribution, having been found breeding in New Jersey, Pennsylvania, and north on the Saskatchewan, near Cumberland House.   In the west it has been observed in Colorado and along the Columbia River.

It has occurred as an accidental visitor in Alaska and also in Greenland, but in all of these places it is reported as rare or accidental.   It is spoken of as *common* in Maine, New Hampshire and Vermont, where it seems to have been observed more frequently than elsewhere.   It is a species which is not likely to be overlooked when present, for its notes and habits readily attract notice.

CONTOPUS VIRENS (Linn.).

189. Wood Pewee. (461)

Olivaceous-brown, rather darker on the head ; below, with the sides washed with a paler shade of the same nearly or quite across the breast ; the throat and belly, whitish, more or less tinged with dull yellowish ; under tail coverts, the same, usually streaked with dusky : tail and wings, blackish, the former unmarked, the inner quills edged and the coverts tipped with whitish : feet and upper mandible, black ; under mandible, usually yellow, sometimes dusky. Spring specimens are purer olivaceous.   Early fall birds are brighter yellow below ; in summer, before the now worn feathers are renewed, quite brown and dingy-whitish.   Very young birds have the wing-bars and pale edging of quills tinged with rusty, the feathers of the upper parts skirted, and the lower plumage tinged with the same ; but in any plumage the species may be known from all the birds of the following genus by these dimensions.   Length, 6-6½ : wing, 3¼-3½ ; tail, 2⅞-3 ; tarsus, about ½, not longer than the *bill*.

Hab.   Eastern North America to the Plains, and from Southern Canada southward.

Nest, composed of bark fibre, rootlets and grass, finished with lichens ; on the outside it is compact and firm round the edge, but flat in form, and rather loose in the bottom.   It is sometimes saddled on a bough, more frequently placed on the fork of a twig ten or twelve feet or more from the ground.

Eggs, three or four, creamy-white, blotched and variegated at the larger end with reddish-brown and lilac-gray.

This species resembles the Phœbe in appearance, but is smaller, and has an erect, hawk-like attitude, when seen perched on a dead twig on the outer limb of a tree.   It is a late comer, being seldom seen before the middle of May, after which its prolonged, melancholy notes may be heard alike in the woods and orchards till the end of August, when the birds move south.   To human ears, the notes of the male appear to be the outpourings of settled sorrow, but to his mate the impressions conveyed may be very different.

In the breeding season, it is generally distributed throughout Ontario, and a few are found in Manitoba.

Its visit here is comparatively short, for it does not appear till the middle of May, and leaves again early in September.   Its food consists chiefly of insects, caught while on the wing.

GENUS EMPIDONAX CABANIS.

EMPIDONAX FLAVIVENTRIS BAIRD.

190. **Yellow-bellied Flycatcher.** (463)

Above, olive-green, clear continuous and uniform as in *acadicus*, or even brighter ; below, not merely *yellowish*, as in the succeeding, but emphatically *yellow*, bright and pure on the belly, shaded on the sides and anteriorly with a paler tint of the color of the back ; eye-rings and wing-markings, yellow ; under mandible, yellow ; feet, black. In respect of color, this species differs materially from all the rest : none of them, even in their autumnal yellowest, quite match it. Size of *traillii* or rather less ; feet, proportioned as in *acadicus*; bill, nearly as in *minimus*, but rather larger; first quill, usually equal to sixth.

HAB.— Eastern North America to the Plains, and from Southern Labrador south through Eastern Mexico to Panama, breeding from the Northern States northward.

Nest, in a mossy bank, composed mostly of moss, with a few twigs and withered leaves, and lined with black wiry rootlets and dry grass.

Eggs, four, creamy-white, spotted and blotched with reddish-brown and a few black markings chiefly near the larger end.

Several of the small Flycatchers resemble each other so closely that it is often difficult for the general observer to identify them correctly. The clear yellow of the under parts of the present species serves to distinguish it from the others, but it is everywhere scarce and little known except to collectors.

Near Hamilton, I have noticed one or two every spring, and sometimes the same number in the fall. During the summer it has not been observed.

It is only within the past five years that correct information has been obtained regarding the nest and eggs of this species, one of the first and best descriptions being given by Mr. Purdie in the Nuttall Bulletin for October, 1878. The nest in this case was placed among the roots of an upturned tree.

All the nests I have seen described have been found in Maine, but the species will no doubt yet be found breeding in Ontario and elsewhere in the interior.

In the " Birds of Manitoba," Mr. Thompson has the following : " Duck Mountain, June 11th, 1884.—Shot a Flycatcher, which was uttering continually a note like '*chee blie*.' It was all over of a greenish color, but yellow on the belly. It answers fairly well the

266

BIRDS OF ONTARIO.

description of *flaviventris*, but is very like an Acadian shot yester-
day. Evidently the species is breeding here."

Mr. Thompson's identification appears to be correct, for I find that
the bird gave utterance to the same note when found near its nest in
Maine.

EMPIDONAX ACADICUS (Gmel.).

191. Acadian Flycatcher. (465)

Above, olive-green, clear, continuous and uniform (though the crown may
show rather darker, owing to dusky centres of the slightly lengthened, erectile
feathers) ; below, whitish, olive-shaded on sides and nearly across breast,
yellow-washed on belly, flanks, crissum and axillars ; wings, dusky ; inner
quills, edged, and coverts tipped with tawny yellow ; all the quills whitish-edged
internally ; tail, dusky olive-glossed, unmarked : a yellowish eye-ring ; feet
and upper mandible, brown ; lower mandible, pale ; in midsummer, rather
darker ; in early fall, brighter and more yellowish below ; when very young,
the wing markings more fulvous, the general plumage slightly buffy-suffused.
Length, 5⅞-6¼ ; wing, 2¼-3 (rarely 3¼) ; tail, 2½-2⅞ ; bill, nearly or quite ½, about
¼ wide at the nostrils ; tarsus, ¾ ; middle toe and claw, ½ ; point of wing
reaching nearly an inch beyond the secondaries ; second, third and fourth quills
nearly equal and *much* (¼ inch or more) longer than first and fifth, which about
equal each other.

First plumage : Above, nearly pure olive, with indistinct narrow transverse
bands of darker ; wing bands, pale reddish-brown ; under parts, soiled yellowish
white, with an olivaceous cast on the sides and breast.

HAB. Eastern United States, chiefly southward, west to the Plains, south
to Cuba and Costa Rica.

Nest, composed of catkins, grasses, weed fibres, shreds of bark, rather
slovenly in appearance, from three to twenty feet from the ground, in a
horizontal fork, fastened by the brim, bottom without support.

Eggs, two to four, yellowish-buff, spotted round the larger end with rusty
brown.

I mention this little bird more as one to be looked for than as one
known to occur here, for I have no positive record of its having been
found in Ontario. I have always thought it would be discovered on
the north shore of Lake Erie, and the nearest approach I have to it
is in the description of a pair of small Flycatchers which nested in
Dr. Macallum's orchard near Dunnville. In this case the nest and
eggs were taken, but the birds were allowed to escape, and, unfortunately for identification, the eggs of this species are indistinguishable
from those of Traill's Flycatcher.

The position of the nest of *acadicus* is given on the opposite page, and that of Traill's Flycatcher is said to be always in an upright fork. Strange to say, the nest found by Dr. Macallum did not correspond with either, being "placed on the upper surface of a tolerably large limb." I still hope to hear of the species being found in the district indicated, and hope the above description may lead to its identification.

EMPIDONAX PUSILLUS TRAILLII (Aud.).

192. Traill's Flycatcher. (466a)

Above, olive-brown, lighter and duller brownish posteriorly, darker anteriorly, owing to obviously dusky centres of the coronal feathers; below, nearly as in *acadicus*, but darker, the olive-gray shading quite across the breast; wing markings, grayish-white, with slight yellowish or tawny shade; under mandible, pale; upper mandible and feet, black. Averaging a little less than *acadicus*, 5½-6; wing, 2⅜-2¾, more rounded, its tips only reaching about ⅔ of an inch beyond the secondaries, formed by second, third and fourth quills as before, but fifth not so much shorter (hardly or not ¼ of an inch), the first ranging between fifth and sixth; tail, 2½; tarsus, ¾ as before, but middle toe and claw three-fifths, the feet thus differently proportioned owing to length of the toes.

HAB.—Eastern North America, breeding from the Middle States (Southern Illinois and Missouri) northward, in winter, south to Central America.

Nest, in an upright fork, firmly secured in its place with the stringy fibres of bark, deeply cupped, composed chiefly of vegetable fibres, lined with dry grass and thistle down.

Eggs, three or four, variable, usually creamy-white, blotched, chiefly toward the larger end, with reddish-brown.

Traill's Flycatcher is not much known in Ontario, the number of collectors being few. By the ordinary observer the bird may readily be mistaken for others of its class which it closely resembles. Mr. Saunders has found it near London, and I have met with it now and then in the moist, secluded ravines by the shore of the Dundas Marsh, but it is by no means common.

Mr. White has collected one or two specimens at Ottawa, and Prof. Macoun got one at Lake Manitoba.

From its being found to the east, west and north it is most likely distributed throughout the Province, but just how rare or common it may be is hard to say, for there is nothing attractive in its appearance or manner, and in habits it is so retiring that it may readily be overlooked.

The difficulty in identifying these small Flycatchers is owing to
the close resemblance they bear to one another. In coloring the
present species is almost identical with the Least Flycatcher, but it
measures half an inch longer. It also resembles the Acadian Fly
catcher, but the present species measures a quarter of an inch less in
length, and is olive *brown*, while *acadicus* is olive *green*. This slight
difference in the shade of color, or in the size of a specimen, makes it
often difficult to say to which group it belongs.

EMPIDONAX MINIMUS Baird.

193. Least Flycatcher. (467)

Colors almost exactly as in *traillii*; usually, however, olive-gray rather
than olive-brown ; the wing markings, eye-ring and loral feathers, plain
grayish-white ; the whole anterior parts often with a slight ashy cast ; under
mandible, ordinarily dusky ; feet, black. It is a smaller bird than *traillii*, and
not so stoutly built ; the wing-tip projects only about ½ an inch beyond the
secondaries ; the fifth quill is but a little shorter than the fourth, the first apt
to be nearer the sixth than fifth ; the feet are differently proportioned, being
much as in *acadicus* ; the bill is obviously under ½ inch long. Length, 5-5.25 ;
wing, 2.60 *or less* ; tail, about 2.25.

Hab. —Eastern North America, south in winter to Central America ; breeds
from the Northern States northward

Nest, in the fork of a sapling or tree, composed of vegetable fibre and
wilted weeds, with a compact lining of plant down, horse-hair and fine grass.

Eggs, three or four. usually pure white, occasionally a set or part of a set are
found dotted with dusky.

The Least Flycatcher is very common throughout Ontario, and is ·
mentioned among the birds found by Prof. Macoun in the North-
West Territory. In the "Birds of Manitoba" it is mentioned as a
common summer resident, and many instances given of its capture
at different points. It arrives near Hamilton about the end of the
first week in May, soon after which its short, sharp call, "*Chebec*,"
is heard by the outer edge of the woods, and even in the city
orchards it takes its location and raises its family. As soon as the
young ones are able to fly, the birds disperse more generally over the
country, and are in no haste to retire, but linger till the cold weather
cuts off their supply of food.

As the correct identification of the small flycatchers is often a
puzzle to the amateur, and as the build of the nest and the markings

of the eggs are often strong points in the evidence, I shall give Dr. Coues' instructions, which may be of use in this connection :

" *E. acadicus*—Nest, in the trees, in horizontal forks, thin, saucer-shaped, open-work ; eggs, creamy-white, boldly spotted.

E. traillii—Nest, in trees, in upright crotch, deeply cupped, more or less compact walled ; eggs, creamy-white, boldly spotted.

E. minimus—Nest, in trees, in upright crotch, deeply cupped, compact walled ; eggs, immaculate white.

E. flaviventris—Nest, on the ground or near it, deeply cupped, thick and bulky ; eggs, white, spotted."

SUBORDER OSCINES. SONG BIRDS.

FAMILY ALAUDIDÆ. LARKS.

GENUS OTOCORIS BONAPARTE.

OTOCORIS ALPESTRIS (LINN.).

194. **Horned Lark.** (474)

Adult :—Above, brown, tinged with pinkish, brightest on the nape, lesser wing coverts and tail coverts ; other upper parts, gray, the centre of the feathers, dusky ; below, white, tinged with dusky on the sides, anteriorly with sulphur-yellow ; a large black area on the breast ; sides of the head and whole of the throat, sulphury-yellow, with a crescentic mark of black below each eye, and a black bar across the forehead, and thence along the side of the crown, prolonged into a tuft or "horn."

Middle tail feathers like the back, the others black, the outer web of the outer pair whitish; bill, blackish, livid blue at base below; feet, black. In winter, at which season it is observed in Southern Ontario, the colors are paler and much less decided.   Length, 7 to 7.50.   *Female:* –Smaller.

HAB.—North-eastern North America, Greenland and northern parts of the Old World, in winter, south in the eastern United States to the Carolinas, Illinois, etc.

Nest, a slight depression in the ground. lined with grass, horse-hair and feathers.

Eggs, four or five, grayish-white, marked with spots of brownish-purple.

The Shore Lark, when I became acquainted with it twenty-five years ago, was a rare winter visitor in Ontario, only a few being observed.   They usually are found in company with the snowbirds, and are thoroughly terrestrial in their habits, seldom alighting anywhere but on the ground.   While here they spend most of their time, during the short days of winter, searching for their daily fare on bare, gravelly patches, from which the snow has been blown away. Occasionally, toward the end of March, just before leaving, I have seen the male settle himself on a hillock and warble out a pleasing lark-like song, which is probably given with more power and pathos later in the season near his grassy home, with his mate for an audience.

This is the north-eastern type of the family, and it is believed to be identical with the British bird of the same name.   In Ontario it is as rare as formerly, its breeding place being far to the north and east, and its migratory course generally along the coast of the Atlantic. It breeds abundantly in the region around Hudson Bay, including Labrador, and has also been found in Greenland.   In the south and west it is represented by several varieties, differing somewhat in size and markings.

The present species, though rather irregular in its movements, is often very abundant along the shores of the eastern States.   It breeds in Newfoundland, and some are supposed to spend the summer in Maine, but the bulk of the species go farther north.

OTOCORIS ALPESTRIS PRATICOLA HENSH.

195.   Prairie Horned Lark.   (474b)

Adult male: In spring, posterior portion of the crown, occiput, nape, sides of the neck and breast, lesser wing coverts and shorter upper tail coverts, light vinaceous; back, scapulars and rump, grayish-brown; the feathers with darker centres, becoming darker and much more distinct on the rump; middle wing coverts, light vinaceous terminally, brownish-gray basally ; wings (except as described), grayish-brown, the feathers with paler edges, outer primaries with outer web chiefly white ; middle pair of tail feathers light brown (paler on edges), the general portion (longitudinally) much darker, approaching dusky ; remaining tail feathers uniform black, the outer pair with exterior webs broadly edged with white ; longer upper tail coverts, light brown, edged with whitish and marked with a broad lanceolate streak of dusky ; forehead (for about .15 of an inch) yellowish-white, this continued back in a broad superciliary stripe of nearly pure white ; fore part of crown (for about .35 of an inch) deep black, continued laterally back to and including the ear-like tufts ; lores, suborbital region, and broad patch on cheeks (with convex posterior outline) deep black, jugular crescent, also deep black, this extending to lower part of throat ; chin and throat, pale straw yellow, gradually fading into white on sides of fore-neck : anterior half of ear coverts white, posterior half drab-gray, each portion forming a crescent-shaped patch : lower parts posterior to the jugulum crescent pure white, the sides of the breast light vinaceous, the sides similar but brown and indistinctly streaked with darker : upper mandible, plumbeous black : lower, bluish-plumbeous ; iris, deep brown ; legs and feet, brownish-black. Size, slightly less than the preceding.

HAB. -Upper Mississippi Valley and the region of the Great Lakes.

Nest, a hollow in the ground, lined with grass, horse hair and feathers.

Eggs, four or five, dull olive, marked with spots and speckles of drab.

So far as I can remember, this species first appeared in Ontario about the year 1868. It was noticed at once as different from our winter visitor, being less in size and its plumage having the washed-out look peculiar to the prairie birds. Since that time it has increased annually, until it has become quite established. I think they do not all leave in the fall, but that a few remain over the winter. Great numbers appear in February or early in March, and should the season be late, they swarm in the road tracks and bare places everywhere, waiting for the disappearing of the snow, and even before it is quite gone many pairs commence building their nests. Soon the flocks separate, the birds scatter in pairs over the country, and are not again seen in such numbers until the following season. This species is very common in Manitoba, where they raise their young in suitable places all over the country.

Besides the original *alpestris* there are now seven different subspecies of the Genus *Otocoris*, described as being found in North America. They all have a strong family likeness, but differ sufficiently, in the eyes of the Committee, to warrant distinction, though several of the groups are of very recent formation. They are found mostly in the west and south-west, only one of the varieties having, till now, been observed in Ontario.

FAMILY CORVIDÆ. CROWS, JAYS, MAGPIES, ETC.

SUBFAMILY GARRULINÆ. MAGPIES AND JAYS.

GENUS PICA BRISSON.

PICA PICA HUDSONICA. (SAB.)

196. American Magpie. (475)

Bill, black ; head, neck, fore-part of the breast and back, black, glossed with green and blue ; middle of the back, grayish-white ; scapulars, white ; smaller wing coverts, black, secondary and primary coverts, glossed with green and blue ; primaries, black, glossed with green, their inner webs white except at the end ; secondaries bright blue changing to green, the inner webs greenish-black ; tail, glossed with green, changing to bluish-purple and dark green at the end ; breast and sides, pure white ; legs, abdomen, lower tail coverts, black. Length, 18-20 inches.

HAB. Northern and Western North America, casually east and south to Michigan (accidentally in Northern Illinois in winter) and the Plains, and in the Rocky Mountains to New Mexico and Arizona.

Nest, in a tree, ten or twelve feet or more from the ground, built of coarse sticks, plastered with mud and lined with hair, feathers and other soft materials.

Eggs, five or six, greenish, thickly shaded and dashed with purplish-brown.

The gaudy, garrulous Magpie is, on the American continent, peculiar to the north and west, and is mentioned as a bird of Ontario on the authority of Mr. C. J. Bampton, Registrar of the District of Algoma, who reports it as a rare winter visitor at Sault Ste. Marie. It has been seen by surveying parties along the northern tier of States, and is said to be possessed of all the accomplishments attributed to the British Magpie, whose history has been so often written. Mr. Trippe, who found it breeding in Colorado, describes the nest as being dome shaped, having two apertures, one at each side, so that when the bird enters by the front it leaves by the one at the back, and while sitting on the nest the long tail projects outside.

The Magpie is a gay, dashing fellow, whom we always like to see in his native haunts, and we should welcome him to the woods of Southern Ontario should his curiosity lead him this way. In Alaska he is common in certain districts, though not generally distributed. His long tail, showy colors, and cunning ways always gain him attention wherever he appears.

In the rural districts of Scotland these birds are regarded with suspicion, from the belief that they know more than birds ought to know. They are supposed to indicate future joy or sorrow to the wayfarer, according to the number he sees together, the idea being thus expressed in popular rhyme :

"One, mirth ;
Two, grief :
Three, a wedding :
Four, a death."

GENUS CYANOCITTA STRICKLAND.

CYANOCITTA CRISTATA. (LINN.).

197. Blue Jay. (477)

Purplish-blue ; below, pale gray, whitening on the throat, belly and crissum ; a black collar across the lower throat and up the sides of the neck and head behind the crest, and a black frontlet bordered with whitish ; wings and tail, pure rich blue, with black bars, the greater coverts, secondaries and tail feathers, except the central, broadly tipped with pure white ; tail, much rounded, the graduation over an inch. Length, 11-12 ; wing, 5½ ; tail, 5¾.

HAB.—Eastern North America to the Plains, and from the fur countries south to eastern Texas.

Nest, in trees or bushes, built of sticks, lined with weeds, grasses and other soft material.

Eggs, four or five, variable in color, usually clay color with brown spots.

18

This species is common throughout Ontario, and may be considered resident, for though the greater number migrate in the fall, a few always remain and are heard squalling among the evergreens any mild day in the depth of winter.

Notwithstanding his gaudy attire, the Jay is not a favorite, which is probably owing to his having many traits of character peculiar to the "bad boy," being always ready for sport or spoil. He frequently visits the farm-house for purposes of plunder, and when so engaged works silently and diligently till his object is attained. He then gets off to the woods as quickly as possible, where he may be heard chuckling to himself over his success.

There is a swampy spot in a clump of bush in West Flamboro' where a colony of Blue Jays has spent the winter for several seasons; and they seem to have lots of fun even in the severest weather. I have occasionally called in when passing, and have found amusement listening to their varied notes, issued in quite a colloquial strain. Sometimes the birds are on the ground, busily gathering nuts with which to replenish their storehouses, but if a scout arrives with some interesting intelligence, off goes the whole troop, each individual apparently knowing the object of the excursion. On the return, notes are compared, and I almost fancy I hear them laugh at their narrow escapes and ludicrous exploits. On such occasions I know I am often the subject of remark, but if I keep quiet they do not seem to object much to my presence.

They are somewhat gregarious in their habits, and even in the breeding season have a custom of going round in guerilla bands of four or five, visiting the farm-house in the early morning, seeking a chance to suck eggs; and woe betide the unlucky owl whom they happen to come across on any of these excursions : its peace for that day is done, for the excitement is often kept up till darkness forces the Jays to retire.

PERISOREUS CANADENSIS (Linn.).

198. Canada Jay. (484)

Upper parts, dull leaden-gray ; lower, dull yellowish-white ; forehead, yellowish-white ; hind part of the head and neck, grayish-black ; throat and band passing round the neck, grayish-white ; secondary quills and tail feathers narrowly tipped with white. *Young :-* Dull slate color, paler on the abdomen, darker on the head, the white tips of the wings and tail duller than in the adult. Length, 10-11 inches.

Hab. —Northern New England, Michigan and Canada, northward to Arctic America.

Nest, on the branch of an evergreen, composed of twigs and grass, lined with feathers.

Eggs, four or five, variable, usually grayish-white, marked with yellowish-brown.

The Indian name for this bird is *Wis-Ka-Tjan*, which pronounced by an English tongue sounds much like "Whiskey John." Through familiarity this has become "Whiskey Jack," the name by which the bird is best known in the districts he frequents. The Canada Jay is found in high latitudes, from Labrador to the Pacific coast. It is quite common in the District of Muskoka, where it breeds and is resident. I have also heard of one individual being taken at Oshawa, but have no record of its having been seen farther south in Ontario.

In the "Birds of the North-West," Dr. Coues, quoting from Mr. Trippe, says: "During the warmer months the Canada Jay frequents the darkest forests of spruce, occasionally flying a little way above the trees. It is quite tame, coming about the mining camps to pick up whatever is thrown out in the way of food, and evincing much of the curiosity that is characteristic of the family. In winter its supply of food is very precarious, and it is often reduced to mere skin and bones. At such times it will frequently weigh no more than a plump sparrow or snowbird, and undoubtedly it sometimes starves to death. During the latter part of the autumn, its hoarse croaking is almost the only sound to be heard in the cold, sombre forests which lie near the timber line."

This species is quite common in Manitoba, and has also been found in Northern Michigan and Minnesota, northward to Hudson Bay and the Arctic Regions, and also in Alaska. In the west it is rather less in size, and being slightly different in color also, it has been created into a subspecies, under the name of *fumifrons*.

In the coast region of Labrador is another form, somewhat darker than the others, which has been separated and named *nigricapillus*. But although their names have been changed, their natures remain the same. All have the habit of taking eggs and young birds from the nests of other birds, and devouring them in sight of the agonized parents. They become very familiar about the camps of lumbermen and others who have occasion to toil in the woods. This habit is no doubt to a great extent the result of hunger, which in the winter time, when deep snow covers the ground, must be the greatest evil with which the birds have to contend.

SUBFAMILY CORVINÆ. CROWS.

GENUS CORVUS LINNÆUS.

CORVUS CORAX SINUATUS (WAGL.).

199. **American Raven.** (486)

Entire lustrous black ; throat feathers acute, lengthened and disconnected.
Length, about 2 feet ; wing, 16-18 inches ; tail, 10.

HAB.—Continent of North America, from the Arctic Regions to Guatemala,
but local and not common in the United States east of the Mississippi River.

Nest, on high trees or inaccessible cliffs, built of sticks with a lining of
coarse grass, sea weed and wool.

Eggs, four or five, greenish, dotted, blotched and clouded with purplish and
blackish-brown.

Few birds are so widely distributed over the face of the earth, and
few have obtained so great a share of notoriety as the Raven, that
"grim, ungainly, ghastly, gaunt and ominous bird of yore." In
Southern Ontario it is now seldom seen. The specimen in my
collection was obtained at St. Clair Flats some years ago, where it
was reported as an occasional visitor in the fall. Wilson, speaking of
this species, says : "On the lakes, and particularly in the neighborhood of Niagara Falls, they are numerous, and it is a remarkable fact
that where they so abound the common Crow seldom appears. I had
an opportunity of observing this myself in a journey along the shores
of Lakes Erie and Ontario during the months of August and
September. The Ravens were seen every day, but I did not see
or hear a single Crow within several miles of the lakes." Since the
days of Wilson the case has been reversed, and any one travelling
now round the lakes named will see Crows in plenty, old and young,
but not a single Raven. They are said to be common in the rocky
region of Muskoka, where they probably nest on the cliffs. They are
believed to continue mated for life, and are often heard expressing
their feelings of conjugal attachment in what to human ears sounds
but a dismal croak.

The Raven appears so seldom in Ontario that we have little
opportunity for becoming acquainted with the habits of the bird.
We find, however, that they are tolerably common in Manitoba,
while in Alaska they seem to reach their centre of abundance.
Referring to them, Mr. Nelson says : "Everywhere throughout the
entire territory of Alaska, including the shores of the Arctic Ocean
and Behring Sea, this bird is a well-known resident. Here, as in

some more civilized regions, it bears a rather uncanny character, and many and strange are the shapes it assumes in the folklore of the natives of these far-off shores."

Eskimos and Indians unite in accusing this bird of pecking out the eyes of new-born reindeer and afterwards killing them. That they are notoriously mischievous in robbing traps of bait, is well known throughout the fur countries, though the thief sometimes pays the penalty by getting a foot in the trap.

Lütke tells us that "the Ravens are the brigands of Sitka. No poultry can be raised, as the Ravens devour the fledglings as fast as they appear, making only one mouthful of them. The porkers are too big to be overcome in the same manner, and the Ravens have to satisfy their greediness by keeping the pigs' tails close cropped. This is why the Sitka pigs have no tails." In spite of this ancient persecution, Mr. Dall assures us that the Sitka pigs of the present day have the caudal appendage of the usual length.

CORVUS AMERICANUS Aud.

200. American Crow. (488)

Color, uniform lustrous black, including the bill and feet ; nasal bristles about half as long as the bill ; throat feathers, oval and blended ; no naked space on cheeks. Length, 18-20 ; wing, 13-14 ; tail, about 8 ; bill, 1.75.

HAB.—North America, from the fur countries to Mexico.

Nest, in trees, built of sticks and twigs, lined with moss, strips of bark and fine grass.

Eggs, four to six, green, spotted and blotched with blackish-brown.

While the Raven prefers to frequent the uncleared parts of the country, the Crow delights in the cultivated districts, where, in the opinion of the farmer, his services could well be dispensed with. Though exposed to continued persecution, he knows the range of the gun accurately, and is wide awake to the intention of all sorts of ambuscades planned for his destruction, so that he thrives and increases in number as the country gets more thickly settled. The Crows mostly leave Ontario at the approach of cold weather, but should the carcase of a dead animal be exposed, even in the depth of winter, it is curious to observe how quickly it will be visited by a few individuals of this species, which are probably remaining in sheltered parts of the woods, and have some means of finding out where a feast is to be had.   Early in April the northern migration begins, and the birds may be seen daily, singly, in pairs, or in loose straggling flocks, passing toward the north-west.

Much has been said and written in regard to the Crow and his relation to the farmer, but so far no very satisfactory conclusion has been drawn.   It is matter for regret that the Crow was not included in the list of birds handed to Dr. Fisher to report upon, for we should then have had full details on the subject of his diet.   More than likely he would have appeared among those whose good deeds balance the evil, for, though he takes eggs, chickens, sprouting corn, fruits and vegetables, he destroys immense numbers of mice, moles, grubs, caterpillars and grasshoppers.   He is subjected to continued persecution, but he is well able to take care of himself, and that he does so is evident by the increase which from year to year takes place in his numbers.

In regions where he has not been molested, he comes about the dwellings much after the manner of the Canada Jay, and where the young are raised as pets they become quite familiar.   Perhaps a limited check is needed to keep the species in its proper position.

201. **Bobolink.** (494)

*Male :—*In spring, black ; cervix, buff ; scapulars, rump and upper tail coverts, ashy-white ; interscapulars streaked with black, buff and ashy ; outer quills, edged with yellowish : bill, blackish-horn : feet, brown. *Male* in fall, *female* and *young*, entirely different in color : yellowish-brown above, brownish-yellow below : crown and back conspicuously, nape, rump and sides less broadly streaked with black ; crown, with a median and lateral light stripe : wings and tail, blackish, pale edged : bill, brown. The *male* changing shows confused characters of both sexes. Length, $6\frac{1}{2}$-$7\frac{1}{2}$; wing, $3\frac{1}{2}$-4 ; tail, $2\frac{1}{2}$-3 : tarsus, about 1 : middle toe and claw, about $1\frac{1}{4}$.

Hab.—Eastern North America to the Great Plains, north to Southern Canada, south in winter to the West Indies and South America. Breeds from the Middle States northward, and winters south of the United States.

Nest, a cup-shaped hollow in the ground in a hay-field ; lined with withered grass.

Eggs, four or five, brownish-white, heavily blotched and clouded with chocolate-brown making the general appearance very dark.

In Southern Ontario the merry, rollicking Bobolink is well known to all who have occasion to pass by the clover fields, or moist meadows, in summer. He attracts attention then by his fantastic dress of black and white, as well as by his gay and festive manner, while he seeks to cheer and charm his modest helpmate, who, in humble garb of yellowish-brown, spends much of her time concealed among the grass. Toward the close of the season, the holiday dress and manners of the male are laid aside, and by the time the birds are ready to depart, male and female, young and old, are clad alike in uniform brownish-yellow. The merry, jingling notes are succeeded by a simple *chink* which serves to keep the flocks together, and is often heard overhead at night in the early part of September. In the south, where they get very fat, they are killed in great numbers for the table.

Genus MOLOTHRUS Swainson.

MOLOTHRUS ATER (Bodd.).

202. Cowbird. (495)

Male: Iridescent black ; head and neck purplish-brown. *Female :*—Smaller, an obscure-looking bird, nearly uniform dusky grayish-brown, but rather paler below, and appearing somewhat streaky, owing to darker shaft lines on nearly all the feathers ; bill and feet black in both sexes. Length, 7½-8 ; wing, over 4 ; tail, over 3.

Hab.—United States, from the Atlantic to the Pacific, north into Southern British America, south, in winter, into Mexico.

Nest, none.

Eggs, deposited in the nest of another bird, dull white, thickly dotted, and sometimes blotched with brown ; number uncertain.

In Southern Ontario nearly all the Cowbirds are migratory, but on two occasions I have seen them located here in winter. There were in each instance ten or a dozen birds which stayed by the farm-house they had selected for their winter residence, and roosted on the beams above the cattle in the cow-house. Early in April the migratory flocks arrive from the south, and soon they are seen in small solitary parties, chiefly in pasture-fields and by the banks of streams all over the country.

At this interesting season of the year, when all other birds are mated and are striving to make each other happy in the faithful discharge of their various domestic duties, the Cowbirds, despising all family relations, keep roving about, enjoying themselves after their own free-love fashion, with no preference for any locality save that where food is most easily obtained. The deportment of the male at this season is most ludicrous. With the view of pleasing his female associate of the hour, he puffs himself out to nearly double his usual size, and makes the most violent contortions while seeking to express his feelings in song, but like individuals of the human species whom we sometimes meet, he is "tongue-tied," and can only give utterance to a few spluttering notes.

As the time for laying draws near, the female leaves her associates, and, manifesting much uneasiness, seeks diligently for the nest of another bird to suit her purpose. This is usually that of a bird smaller than herself, which the owner has just finished and may have made therein a first deposit. Into such a nest the female Cowbird drops her egg, and leaving it, with evident feelings of satisfaction, joins her comrades and thinks no more about the matter. By the owners of the nest the intrusion is viewed with great dislike, and should it contain no eggs of their own it is frequently deserted. But another expedient to rid themselves of the incumbrance is sometimes resorted to which shows a higher degree of intelligence than we are accustomed to call ordinary instinct. Finding that their newly-finished cradle has been invaded, the birds build a floor over the obnoxious egg, leaving it to rot, while their own are hatched on the new floor in the usual way.

Should the owners of the nest have one or more eggs deposited before that of the Cowbird appears, the intrusion causes them much anxiety for an hour or two, but in the majority of cases the situation is accepted, and the young Cowbird being first hatched, the others do not come to maturity. The foster-parents are most attentive in supplying the wants of the youngster till he is fit to shift for himself, when he leaves them, apparently without thanks, and seeks the society of his own kindred, though how he recognizes them as such is something we have yet to learn.

Much speculation is indulged in regarding the cause of this apparent irregularity in the habits of the Cowbird, and different opinions are still held regarding it, but whatever other purpose it may serve in the economy of nature, it must cause a very large reduction in the number of the different species of birds on which it entails the care of its young. Some idea may be formed of the extent of this reduction by looking at the vast flocks of Cowbirds swarming in their favorite haunts in the fall, and considering that for each bird

in these flocks from three to four of a different species have been
prevented from coming to maturity.

The number of species imposed upon by the Cowbird is large,
including Warblers, Vireos, Sparrows, Thrushes, Bluebirds, etc., but
the one most frequently selected in this locality is the Summer
Yellowbird. On the prairies, where the Cowbirds are numerous, and
the number of foster-parents limited, it is said that in the month of
June nearly every available nest contains an egg of the Cowbird.

In Southern Ontario they disappear during July and August, but
usually return in vast flocks in September, when they frequent the
stubble fields and patches of wild rice by the edge of the marshes.

Genus XANTHOCEPHALUS Bonaparte.

XANTHOCEPHALUS XANTHOCEPHALUS (Bonap.).

203. Yellow-headed Blackbird. (497)

Male :—Black ; whole head (except lores), neck and upper breast, yellow,
and sometimes yellowish feathers on the belly and legs ; a large white patch
on the wing, formed by the primary and a few of the outer secondary coverts.
Female and *young* :—Brownish-black, with little or no white on the wing, the
yellow restricted or obscured.   *Female* much smaller than the *male*, about 9½.
Length, 10-11 ; wing, 5½ ; tail, 4½.

Hab. —Western North America, from Wisconsin, Illinois and Texas to the
Pacific coast.   Accidental in the Atlantic States (Massachusetts, South Carolina,
Florida).

Nest, composed of aquatic grasses fastened to the reeds.

Eggs, three to six, grayish-green spotted with reddish-brown.

A wanderer from the west, this handsome Blackbird has appeared
from time to time at different points in the Eastern States.   The
only record I have of its occurrence in Ontario is that given by Mr.
E. E. Thompson, in the *Auk* for October, 1885, as follows : "This
species has been taken a number of times in company with the
Red-winged Blackbirds by Mr. Wm. Loane, who describes it as the
Californian Blackbird.   The specimen I examined was taken near
Toronto by that gentleman, and it is now in the possession of Mr.
Jacobs, of Centre Street."

Though the Yellow-headed Blackbird is only a casual visitor, I
think it is quite probable we may yet see it as a summer resident
in the grassy meadows of Ontario.   At present it comes east as far as
Iowa, Minnesota, Illinois and Wisconsin, while in a northerly direc-

tion it extends its migrations to the interior of the Fur Countries, reaching the Saskatchewan about the 20th of May.

We should like to see him here, his yellow head making a bright spot among the sombre-plumaged Cowbirds and Grackles.

GENUS AGELAIUS VIEILLOT.

AGELAIUS PHŒNICEUS (LINN.).

204. Red-winged Blackbird. (498)

Male :— Uniform lustrous black ; lesser wing coverts, scarlet, broadly bordered by brownish-yellow or brownish-white, the middle row of coverts being entirely of this color, and sometimes the greater row likewise are similar, producing a patch on the wing nearly as large as the red one. Occasionally there are traces of red on the edge of the wing and below. *Female* :— Smaller, under 8 ; everywhere streaked ; above, blackish-brown with pale streaks, inclining on the head to form median and superciliary stripes ; below, whitish, with very many sharp dusky streaks ; the sides of the head, throat and the bend of the wing, tinged with reddish or fulvous. The *young male* at first like the *female*, but larger ; apt to have a general buffy or fulvous suffusion, and bright bay edgings of the feathers of the back, wings and tail, and soon showing black patches. Length, 8·9 ; wing, 4½-5 ; tail, 3½-4.

HAB.—North America in general, from Great Slave Lake south to Costa Rica.

Nest, large for the side of the bird, composed of rushes and sedges loosely put together and lined with grass and a few horse-hairs, usually fastened to the bulrushes, sometimes placed in a bush or tussock of grass near the ground.

Eggs, four or five, pale blue, curiously marked with brown.

This species is generally distributed and breeds in suitable places throughout the Province. It is very common near Hamilton, breeding abundantly in the Dundas marsh, and in the reedy inlets all around the shores of Hamilton Bay. As soon as the young broods are able to fly, old and young congregate in flocks, frequenting the stubble fields and moist meadows by day, and roosting at night among the reeds in the marsh. As the season advances the numbers are increased by others arriving from the north, and during October very large flocks are observed in the places they frequent. Towards the end of that month, if the weather gets cold, they all move off to the south. None have been observed here during the winter.

This species has a wide distribution, being found from Texas and Florida, as far north as 50° on the Atlantic coast, and 57° in the west, and breeding throughout its range. It is said also to occur on the Pacific slope, where several closely allied species are more or less common.

GENUS STURNELLA VIEILLOT.

STURNELLA MAGNA (LINN.).

205. Meadowlark. (501)

Above, the prevailing aspect brown. Each feather of the back blackish, with a terminal reddish-brown area, and sharp brownish-yellow borders ; neck similar, the pattern smaller ; crown, streaked with black and brown, and with a pale median and superciliary stripe ; a blackish line behind eye ; several lateral tail feathers white, the others with the inner quills and wing coverts barred or scolloped with black and brown or gray ; edge of wing, spot over eye, and under parts generally, bright yellow ; the sides and crissum, flaxen brown, with numerous sharp blackish streaks ; the breast, with a large black crescent (obscure in the young) ; bill, horn color ; feet, light brown. Length, 10-11 ; wing, 5 ; tail, 3½ ; bill, 1¼. *Female :*—Similar, smaller, 9½.

HAB.—Eastern United States and Southern Canada to the Plains.

Nest, on the ground, at the foot of a tuft of grass or weeds, lined with dry grass, and sometimes partly arched over.

Eggs, four to six, white, dotted and sprinkled with reddish-brown.

The Meadowlark is found in all suitable districts throughout Ontario, apparently preferring the south-west. In the southern portion of the Province, it is generally distributed throughout the agricultural districts, where its loud, clear, liquid notes are always associated in our minds with fields of clover and new-mown hay. Here it may be considered migratory, the greater number leaving us in October to return again in April, but it is no uncommon thing to find one or two remaining during the winter in sheltered situations. On the 7th of February, 1885, when the cold was intense and snow covered the ground, I noticed an individual of this species digging vigorously into a manure heap at Hamilton Beach. When examined he was found to be in very poor condition, and looked altogether as if he had been having a hard time. The present species is found north to Manitoba, where it is replaced by the Western Meadowlark, which resembles our eastern form so closely that it is doubtful if any one, judging by appearance only, could separate them with certainty. The song of the birds is so entirely different that, chiefly on this account, the western bird has been recorded as a sub-species under the name of *Sturnella magna neglecta*, or Western Meadowlark, the dry central plains forming the boundary between the two habitats.

Genus ICTERUS Brisson.

ICTERUS SPURIUS (Linn.).

206. Orchard Oriole. (506)

Male :—Black ; lower back, rump, lesser wing coverts, and all under parts from the throat, deep chestnut ; a whitish bar across the tips of greater wing coverts ; bill and feet, blue-black ; tail, graduated. Length, about 7 ; wing, 3¼ ; tail, 3. *Female :*—Smaller, plain yellowish-olive above, yellowish below ; wings, dusky ; tips of the coverts and edges of the inner quills, whitish ; known from the *female* of the other species by its smaller size and very slender bill. *Young male :*—At first like the *female*, afterwards showing confused characters of both sexes ; in a particular stage it has a black mask and throat.

Hab.—United States, west to the Plains, south, in winter, to Panama.

Nest, pensile, composed of grass and other stringy materials ingeniously woven together and lined with wool or plant down, rather less in size and not quite so deep in proportion to its width as that of the Baltimore.

Eggs, four to six, bluish-white, spotted and veined with brown.

On the 15th of May, 1865, I shot an immature male of this species in an orchard at Hamilton Beach, which was the first record for Ontario. I did not see or hear of it again till the summer of 1883, when they were observed breeding at different points around the city of Hamilton, but since that year they have not appeared near this place.

Mr. Saunders informs me that they breed regularly and in considerable numbers near London and west of that city, from which we infer that the species enters Ontario around the west end of Lake Erie, and does not come as far east as Hamilton. Most likely it does not at present extend its migrations in Ontario very far from the Lake Erie shore. The notes of the male are loud, clear and delivered with great energy, as he sits perched on the bough of an apple tree, or sails from one tree in the orchard to another. This species would be a desirable acquisition to our garden birds, both on account of his pleasing plumage of black and brown, and because of the havoc he makes among the insect pests which frequent our fruit trees.

I learn from Dr. Macallum that the Orchard Oriole breeds regularly in small numbers along the north shore of Lake Erie, near Dunnville, but it evidently does not proceed far north of our southern boundary. One wanderer, but only one, is reported by Dr. Coues as having appeared at Pembina.

ICTERUS GALBULA (LINN.).

207. Baltimore Oriole. (507)

Male : With head and neck all round, and the back, black ; rump, upper tail coverts. lesser wing coverts, most of the tail feathers, and all the under parts from the throat. fiery-orange, but of varying intensity according to age and season ; middle tail feathers, black ; the middle and greater coverts and inner quills, more or less edged and tipped with white, but the white on the coverts not forming a continuous patch : bill and feet, blue-black. Length, 7½-8 ; tail, 3. *Female :*—Smaller, and much paler, the black obscured by olive, sometimes entirely wanting. *Young :*—Entirely without the black on throat and head, otherwise colored nearly like the *female*.

HAB. —Eastern United States, west nearly to the Rocky Mountains.

Nest, purse shaped, pensile, about six inches deep, composed chiefly of vegetable fibre, with which is often intertwisted rags, paper, thread, twine and other foreign substances, usually suspended from the outer branches of a tree, most frequently an elm, at a height of ten to fifty feet from the ground.

Eggs, four to six, white, faintly tinged with blue, when blown, spotted, scrawled and streaked with lilac and brown mostly toward the larger end.

The gay, dashing, flashing Baltimore Oriole seems to court the admiration so generally bestowed on him, and is much more frequently seen among the ornamental trees in our parks and pleasure grounds than in the more retired parts of the country. He arrives from the south with wonderful regularity about the end of the first week in May, after which his clear flute-like notes are heard at all hours of the day till the early part of July, when with his wife and family he retires, probably to some shady region to avoid the extreme heat of summer. At all events they are not seen in Southern Ontario again till the beginning of September, when they pay us a passing visit while on their way to winter-quarters. The species seems to be well distributed in Ontario, for in the report of the "Ottawa Field Naturalists' Club," it is said to be common in that district, arriving about the 10th of May. It is also included in the list of birds observed at Moose Mountain in the North-West by Prof. Macoun.

Dr. Bell has a specimen which was taken at York Factory, but it prefers the west, being abundant throughout Manitoba, and Dr. Coues found it breeding at Pembina on the boundary south of Winnipeg.

Genus SCOLECOPHAGUS Swainson.

SCOLECOPHAGUS CAROLINUS (Müll.).

208. Rusty Blackbird. (509)

Male :—In summer, lustrous black, the reflections greenish, and not noticeably different on the head ; but not ordinarily found in this condition in the United States ; in general glossy black, nearly all the feathers skirted with warm brown above and brownish-yellow below, frequently continuous on the fore parts ; the *male* of the first season, like the *female*, is entirely rusty-brown above, the inner quills edged with the same ; a pale superciliary stripe ; below, mixed rusty and grayish-black, the primaries and tail above, black ; bill and feet, black at all times. Length, *male*, about 9; wing, 4½; tail, 3½; bill, ⅜ : *female*, smaller.

HAB.—Eastern North America, west to Alaska and the Plains. Breeds from Northern New England northward.

Nest, a coarse structure, resting on a layer of twigs, composed of grass mixed with mud; well formed inside and lined with fine grass and rootlets, usually placed in alder or similar bushes overhanging the water.

Eggs, four to six, grayish-white marked with brown.

During the last week in April or the first in May, according to the weather, the Rusty Grackles are seen in small flocks hurrying on to their breeding places farther north. Their stay at that time is very short, and the collectors have but little chance of securing a male in adult plumage, spring being the only season when such can be had here, and even then only a few in each flock have acquired their nuptial dress. They may yet be found breeding in Ontario, although, owing to the number of observers being small, the fact (so far as I know) has not yet been recorded. About the end of August, or early in September, they return in flocks of much greater dimensions than those which passed up in the spring, and in company with the Cowbirds and Redwings continue to frequent the plowed fields, cornfields and wet places till the weather gets cold in October, when they all move off to the south and are not seen again till spring.

This species goes farther north than any other of the Blackbirds, for it is found not only throughout Manitoba and the North-West, but is common in Alaska, where Mr. Nelson says : " It arrives in the British fur country, at Great Bear Lake, latitude 65° north, by the 3rd of May, and breeds throughout the northern extreme of the continental land, reaching the farthest limit of the wooded region on the Lower Anderson and Mackenzie Rivers. In Northern Alaska it reaches latitude 70 . On the Behring Sea and Arctic coast of this territory, from the mouth of the Kuskoquim River, the bird is a regular, but not numerous, summer resident wherever trees and bushes are found reaching the vicinity of the sea coast." It feeds largely on insects, but is also said to be fond of corn, though it leaves us too early in the spring and arrives too late in the fall to do much damage in Ontario.

In the *Auk*, Volume II., page 107, Mr. Banks, of St. John, N.B., gives an account of a nest of this species, which he found in a different position from that usually assigned to it.

It was placed in a large spruce, about 28 feet from the ground, and was a coarse, bulky nest, composed of dried vines of the honeysuckle, loosely entwined at the sides and fastened together by a solid mass of mud at the bottom.

There was no attempt at lining of any sort. It contained two eggs and two young birds.

GENUS QUISCALUS VIEILLOT.

SUBGENUS QUISCALUS.

QUISCALUS QUISCULA ÆNEUS (Ridgw.).

209. **Bronzed Grackle.** (511b)

Metallic tints, rich, deep and uniform : head and neck all round, rich, silky steel-blue, this strictly confined to these portions, and abruptly defined behind, varying in shade from an intense Prussian blue to brassy-greenish, the latter tint always, when present, most apparent on the neck, the head always more violaceous ; lores, velvety-black ; entire body, above and below, uniform continuous metallic brassy-olive, varying to burnished golden olivaceous-bronze, becoming gradually uniform metallic purplish or purplish-violet on wings and tail, the last more purplish ; primaries, violet-black ; bill, tarsus and toes, pure black ; iris, sulphur-yellow. Length, 12.50 to 13.50; wing, 6.00; tail, 6.00; culmen, 1.26 ; tarsus, 1.32 ; third and fourth quills, longest and equal ; first, shorter than fifth : projection of primaries beyond secondaries, 1.28 ; graduation of the tail, 1.48. (*Ridgway.*)

HAB. - From the Alleghanies and New England north and west to Hudson Bay and the Rocky Mountains.

Nest, coarse and bulky, composed of twigs and weeds, with a mixture of mud, often placed in a spruce or hemlock tree, sometimes in a bush overhanging the water, and occasionally in a hollow stub or deserted woodpecker's hole.

Eggs, four to six, smoky-blue with irregular dark brown blotches, lines and spots.

The Bronzed Grackle was christened by Mr. Ridgway at the Proceedings of the Academy of Natural Sciences of Philadelphia, in June, 1869. Prior to that date, Dr. Baird had separated one as peculiar to Florida, but all the others were supposed to belong to the species named by Linnæus, *Quiscalus quiscula*, or Purple Grackle. Mr. Ridgway, on comparing a large number of specimens from different points, found the group to contain two well-defined sub-species, and his decision has now been generally adopted. The original Purple Grackle is the most southern bird of the three, its habitat being given as "Atlantic States from Florida to Long Island," while our present form is said to extend from the Alleghanies and New England, north and west to Hudson Bay and the Rocky Mountains. Since giving my attention to this subject, I have made a point of examining all available mounted Crow Blackbirds in barber shops, country taverns, etc., and find that all belong to the Bronzed division.

It is quite possible that a few of the others may yet be found along our southern border, but unquestionably the Crow Blackbird of Ontario is the Bronzed Grackle. They like to be near water, and are very common in the town of Galt, breeding close to the houses along the banks of the river. There is a colony established at East Hamilton, where they breed in the Norway spruce trees near the residence of Mr. Barnes, who protects them from being molested, whether wisely or not is open to question, for there rests at their door the serious charge of robbing the nests of small birds and destroying the eggs and young, besides that of being very destructive to the sprouting corn in spring-time.

The Bronzed Grackle is very abundant throughout Manitoba and the North-West, and has been captured by Dr. Bell at York Factory. The males arrive in Southern Ontario a few days before the females, usually about the middle of April. It is noticed that the first to arrive are in the richest plumage, the fine purple bronze being fully developed. They are soon generally distributed over the country, in suitable places, where they cause much family affliction during the breeding season by carrying off the young of the smaller birds. During October they gather in flocks and retire to the south, where they spend the winter.

FAMILY FRINGILLIDÆ.   FINCHES, SPARROWS, ETC.

GENUS COCCOTHRAUSTES BRISSON.

SUBGENUS HESPERIPHONA BONAPARTE.

COCCOTHRAUSTES VESPERTINA (COOP.).

210. Evening Grosbeak. (514)

Dusky olivaceous, brighter behind ; forehead, line over the eye and under tail coverts, yellow ; crown, wings, tail and tibiæ, black ; the secondary quills. mostly white ; bill, greenish-yellow, of immense size, about ⅞ of an inch long and nearly as deep. Length, 7½-8½ ; wing, 4-4½ ; tail, 2½. The *female* and *young* differ somewhat, but cannot be mistaken.

HAB.—Western North America, east to Lake Superior, and casually to Ohio and Ontario, from the Fur Countries south into Mexico.

Nest, in a small tree or bush, a comparatively slight structure rather flat. composed of sticks and roots, lined with soft vegetable material.

Eggs, three, greenish ground color, blotched with brown.

This is a western species whose line of travel in the season of migration seems to be along the Mississippi Valley, casually coming as far east as Ontario.

I have heard of its being observed during the winter at St. Cloud, St. Paul and Minneapolis, and last winter I had a pair sent me by mail in the flesh, from Redwing, Minnesota. The first report of its appearance in Ontario was made by the late Dr. T. J. Cottle, of Woodstock, who in the month of May, 1866, observed a flock among the evergreens near his residence, and obtained one or two of them.

Again, in 1871, they were noticed near London about the same season, and several were procured, three of them coming into my possession. I did not hear of the species again till the 17th of March, 1883. When enjoying a sleigh ride along a road which runs through a swamp in West Flamboro', my son and I came unexpectedly upon two in the bush by the roadside and secured them both.

I have also heard of a female having been obtained by the Rev. Mr. Doel in Toronto, on the 25th of December, 1854, which completes the record for Ontario, so far as I know. The Evening Grosbeak is much prized by collectors on account of its rarity, its beauty, and the desire we have to know more of its history.

Dr. Coues speaks of it as "a bird of distinguished appearance, whose very name suggests the far-away land of the dipping sun, and the tuneful romance which the wild bird throws around the fading light of the day. Clothed in striking color contrasts of black, white and gold, he seems to represent the allegory of diurnal transmutation, for his sable pinions close around the brightness of his vesture as night encompasses the golden hues of sunset, while the clear white space enfolded in these tints foretells the dawn of the morrow." Thus the glowing words flow from the pen of an accurate observer and graceful writer, while to the mass of the people, the beauties of bird-life are a sealed book. By far the larger number of those who have the opportunity of observing our wild birds in their native haunts, belong to that practical class of which the representative is Peter Bell, of whom it is written :

> " A primrose by the river's brim
> A yellow primrose was to him,
> And it was nothing more."

I once directed the attention of a successful farmer, whose speech betrayed his nationality, to a fine mounted specimen of the bird I have been describing. I pointed out the beauty of its markings and related the interesting parts of its history, but failed to excite any enthusiasm regarding it. In fact the only remark elicited was that it was "unca thick i' the neb."

The first account we have of the nest and eggs of this species appears in the *Auk*, Vol. V., page 113. It is given by Mr. John Swinbourne, of Springerville, Arizona. Here is an extract: "On the 5th of June, 1884, while looking out for anything of ornithological interest in a thickly wooded cañon some fifteen miles west of the little town of Springerville, Apache County, Arizona, my attention was attracted by a bird which I did not know, flying off its nest in the top of a thick willow bush. Having climbed up to the nest and ascertained that it contained three eggs, I returned to the ranch. Next day I visited the cañon with my shotgun, and finding the number of eggs in the nest had not increased, concealed myself close by, and after a long wait succeeded in procuring the female as she flew from the nest. At that time I knew so little about American birds or their eggs that I took no eggs except when I could authenticate them by procuring the female bird.

"The nest was a comparatively slight structure, rather flat in shape, composed of small sticks and roots, lined with finer portions of the latter. The eggs, three in number, were of a clear greenish ground color, blotched with pale brown. They were fresh. The nest was placed about fifteen feet from the ground, in the extreme top of a thick willow bush. The slight cañon, with a few willow bushes in its centre bordering a small stream, lies in the midst of very dense pine timber, at an altitude of about 7,000 feet, as far as I can judge."

It will be noticed that the foregoing took place in 1884, but it was not published till 1888. In 1887, Mr. W. E. Bryant found a nest, of which he published an account in the "Bull. Cal. Acad. Sci." for that year. Thus, though Mr. Swinbourne was the first to find and identify the nest and eggs, Mr. Bryant's account of the one he found was first published.

The history of the species in Ontario remained as given above until December 19th, 1890, when a boy brought me a pair which he said he had shot from a flock he saw near the shore of Hamilton Bay. He described where he had found them, and I knew at once that it was a most likely place for the birds to be—the rough, steep bank of the bay, grown up with red cedars, close to the Roman Catholic cemetery, where many of the same trees were cultivated for ornament. In the afternoon K. C. McIlwraith visited the locality, and found a small flock feeding on the berries of the red cedar.

Sunday intervened, but on Monday I was there and was delighted to see a flock of twenty-five or thirty quite at home on the bank, some feeding and talking quietly to their companions, while others were down on the sandy shore, pecking gravel or dabbling themselves in the water. It was a clear, bright day, and they made a picture I would have travelled many miles to see; but there they were within a mile of my own residence, and my visit to them was repeated every day for over a month. I thought at first that the original flock had remained, but soon found that an easterly migration was going on, and that as one flock left another arrived, so that some were always present when the locality was visited, within the period named.

During February, few, if any, were observed here. In March, the return trip commenced, but was in all respects different from the easterly one. The birds were then fewer in number, and all seemed excited and desirous to go west with the least possible delay. Their food in this locality was pretty well exhausted, and they took any apples that still remained on the trees, using the seeds only. Small groups of four or five were seen going west till the middle of April, 1890, but since that time not one has been observed in this neighborhood.

The home of the Evening Grosbeak is in the coniferous forests of the North-West, though it has been found among the mountain tops as far south as Mexico. The first record we have of its capture is that of a specimen taken by Schoolcraft, in 1823, near Sault Ste. Marie. So much of its time is spent beyond the limits of civilization, that even now we are but little acquainted with its life history. It is known to be migratory, its usual route to the south being along the line of the Mississippi, but it is very irregular, often appearing at certain points one season where it may not again be seen for several years. Single birds and small groups have been observed in Canada for some years past, but never has it been known to make such an invasion as it did during the winter of 1889-90, already referred to. It is at all times difficult to account for the seeming irregularities which occasionally occur in the movements of the birds. The first flocks which start on the migratory journey take a course that the others usually follow, and if we knew just where the start was made, and under what circumstances, we might possibly be able to explain why the Grosbeaks came so far east of their usual limit as on the occasion mentioned.

It is quite possible that a strong west wind caused the leaders to diverge from their usual course till they found themselves east of Lake Superior, which they would not cross, but kept on along the eastern shore of Lake Huron till about opposite Detroit, where they took a turn due east towards the west end of Lake Ontario. I first heard of them from near Chatham ; then near London, and a few days later they appeared at Hamilton. Here the flock divided, some taking the south shore and others keeping close by the north, still moving eastward. They were observed at all points along the lake and the St. Lawrence River, to Montreal, a few even going as far east as Quebec. North of this point they did not seem to wander, as Mr. Goldie saw none at Guelph, neither did Mr. White observe them at Ottawa.

On the south side of the lake, Dr. Macallum saw them at Dunn-ville, and they were also common around Niagara Falls, at Buffalo, Oswego and other points, till, finally, Mr. Brewster told us in *Forest and Stream* that they had arrived at Eastern New Hampshire and Massachusetts.

Soon the return trip commenced, and again they were noticed as strangers at many different points on their route. They were greatly admired wherever they appeared, and many a wish was expressed that they might remain over the summer, but we have not heard of a single instance in which the wish was gratified. All that were left of them returned to the North-West.

GENUS PINICOLA VIEILLOT.

PINICOLA ENUCLEATOR (LINN.).

211. **Pine Grosbeak.** (515)

Male: Carmine red, paler or whitish on the belly, darker and streaked with dusky on the back ; wings and tail dusky, much edged with white, the former with two white bars. *Female:* Ashy-gray, paler below, marked with brownish-yellow on the head and rump. Length, 8.9 : wing, 4½ ; tail, 4.

HAB. Northern portions of the northern hemisphere, breeding far north, in winter, south, in North America, irregularly to the Northern United States. South in the Rocky Mountains to Colorado, and in the Sierra Nevada to California.

Nest, in a bush, four feet from the ground, composed entirely of coarse green moss.

Eggs, four, slate color, tinged with green, spotted and clouded with brown and purple.

In Southern Ontario the Pine Grosbeak is an irregular winter visitor, sometimes appearing in large flocks and again being entirely absent for several years in succession. During the winter of 1882-83, and also of 1883-84, they were quite common, and were observed throughout the country, wherever their favorite red cedar or mountain ash berries were to be found, but since that time not one has been seen. They are fine, robust birds of a most sociable, gentle disposition. I have often watched them feeding in flocks, sometimes in places where food was not over abundant, but never noticed a quarrel among them, all being willing to share alike.

Very many of the individuals which visit us are females, or young males, clad in a uniform of smoky-gray, more or less tinged with greenish-yellow, but in every flock of twenty or thirty there are two or three adult males in the showy crimson dress, which, when seen with a background of the sombre foliage of the Norway spruce, forms a most attractive object at this season of the year when the tide of bird-life is at its lowest ebb.

Our knowledge of the breeding habits of this species is as yet very imperfect, the description given of the nest and eggs being that of a *supposed* Grosbeak's nest which was found in Maine by Mr. Boardman, but the birds to which the nest belonged were not secured.

Mr. Trippe found them in Colorado in summer, living up near the timber line, and observed young birds fully feathered and shifting for themselves in June, which gives the impression that they must breed very early. I think it highly probable that they may yet be found breeding in Northern Ontario, for on the occasions already referred to they appeared early in January, and remained as late as April, so that they would not have time to travel far before engaging in their domestic duties.

They are reported as common winter visitors in Manitoba, while in Alaska Mr. Nelson says that "along the entire west and north-west coast of America, from Vancouver Island north to within the Arctic circle, these birds occur in greater or less abundance. I have frequently passed a pleasant half-hour on the wintry banks of the Yukon while making a mid-day halt and waiting for the natives to melt the snow for our tea, listening to the chirping and fluttering of these birds as they came trooping along the snow-laden woods in small parties. They withstand the severest cold in these forests, even within the Arctic circle, and appear to be about equally distributed throughout the wooded region."

Genus CARPODACUS Kaup.

CARPODACUS PURPUREUS (Gmel.).

212. Purple Finch. (517)

Male: Crimson, rosy or purplish-red, most intense on the crown, fading to white on the belly, mixed with dusky streaks on the back ; wings and tail, dusky, with reddish edgings, and the wing coverts tipped with the same ; lores and feathers all round the base of the bill, hoary. *Female* and *young* with no red ; olivaceous-brown, brighter on the rump, the feathers above all with paler edges, producing a streaked appearance ; below, white, thickly spotted and streaked with olive-brown, except on the middle of the belly and under tail coverts ; obscure whitish superciliary and maxillary lines. *Young males* show every gradation between these extremes in gradually assuming the *male* plumage, and are frequently brownish-yellow or bronzy below. Length, 5¾-6¼ ; wing, 3-3¼ ; tail, 2¼-2½.

Hab.—Eastern North America, from the Atlantic coast to the Plains. Breeds from the Middle States northward.

Nest, usually but not always in an evergreen, composed of weeds, grass, strips of bark, vegetable fibre, etc., lined with hair.

Eggs, four or five, pale green, scrawled and spotted with dark brown and lilac, chiefly toward the larger end.

In Southern Ontario the Purple Finch is most abundant during the month of May. At this season the few which have remained with us during the winter put on their brightest dress, and being joined by others which are daily arriving from the south, they make the orchards for a time quite lively with their sprightly song. Their presence, however, could well be dispensed with, for they are observed at this time to be very destructive to the buds and blossoms of fruit trees. As the season advances, they become generally distributed over the country and are not so often seen.

The male does not acquire the bright crimson dress till after the second season. The young male, in the garb of the female, being observed in full song has led to the belief that both sexes sing alike, but such is not the case. Crimson Finch would have been a more appropriate name for this bird than Purple Finch, for the color is certainly more crimson than purple.

It breeds sparingly in Southern Ontario, but many go farther north. It is common during summer in Manitoba, beyond which I have not heard of it having been observed. It remains quite late in the fall ; and, occasionally, I have seen stragglers in the depth of winter. In Ontario the species can at no time be said to be abundant.

In the proceedings of the Ornithological Sub-section of the Canadian Institute is an account of an interesting hybrid in which this species is concerned.

Mr. Wm. Cross says regarding it: "On January 22nd, 1890, a small-sized Finch, which was taken from a flock of Pine Grosbeaks, was brought into my store. As the bird was new to me, I had it submitted to J. A. Allen, who said: 'It is clearly a hybrid between the common Purple Finch and the Pine Grosbeak. It is certainly a most interesting capture, combining about equally the characteristics of both. It is just half-way between them in size and very nearly so in all other features.'"

GENUS LOXIA LINNÆUS.

LOXIA CURVIROSTRA MINOR (BREHM).

213. American Crossbill. (521)

Male :—Bricky-red ; wings, blackish, unmarked. *Female :*—Brownish-olive, streaked and speckled with dusky, the rump saffron. Immature males mottled with greenish and greenish-yellow. Length, about 6 ; wing, 3¼ ; tail, 2¼.

HAB.—Northern North America, resident sparingly south in the Eastern United States to Maryland and Tennessee, and in the Alleghanies, irregularly abundant in winter ; resident south in the Rocky Mountains to Colorado.

Nest, among the twigs of a spruce, composed of twigs, rootlets, lichens, etc., lined with hair and feathers.

Eggs, three or four, pale green, spotted toward the larger end with brown-purple and lilac or brown.

Throughout Ontario the Crossbills are very erratic in their movements, sometimes appearing unexpectedly in considerable numbers in sections of the country where, for several preceding years, they have been entirely absent. Their time of nesting is also unusual, the duties of incubation being performed while the ground is still covered with snow. Hence the young, being soon set at liberty, are often seen in flocks quite early in summer, and sometimes in the fall we hear their rattling call, and see them descend from the upper air to visit a patch of sunflowers, on the seeds of which they feast with evident relish. Early in spring, when food is less abundant, I have seen them alight on the ground and dig the seeds from a squash which had been left out during the winter.

Their favorite resorts, however, are the spruce and hemlock trees, whose dark green foliage forms a fine back-ground for the rich red color of the male as he swings about in every possible position, searching for food among the cones at the end of the slender branches.

They are chiefly found east of the Plains, but north and south they are widely distributed, there being records of their nesting at various points from Northern Georgia to Alaska. In Southern Ontario they occur chiefly as winter visitants.

— —

LOXIA LEUCOPTERA Gmel.

214. **White-winged Crossbill.** (522)

Wings in both sexes, with two conspicuous white bars. *Male:*—Rosy-red. *Female:*—Brownish-olive, streaked and speckled with dusky, the rump saffron. Length, about 6; wing, $3\frac{1}{4}$; tail, $2\frac{1}{4}$.

Hab.—Northern parts of North America, south into the United States in winter. Breeds from Northern New England northward.

Nest, similar to the preceding species.

Eggs, three or four, pale blue, dotted toward the larger end with lilac and purple.

This species resembles the preceding in its habits, but does not appear in such large numbers. They visit the same localities, sometimes in company, sometimes in separate flocks. Both are quite unsuspicious, and when eagerly searching for food among the pine cones, they admit of a very near approach without taking alarm. They vary much in plumage with age and sex, but the present species can at all times be identified by the white wing-bars.

It seems to be more northern in its habitat than the Red Crossbill.

In the "Birds of Alaska," Mr. Nelson says: "Although the Red Crossbill in the northern portion of the territory occurs only as an exceedingly rare visitant, the present species is found in the greatest abundance wherever there are trees enough to afford it shelter. It is in even greater numbers than the Pine Grosbeak, sharing its range with that species." It has also been found occasionally in Greenland, and is often taken on board of ships in the North Atlantic, far from land.

ACANTHIS HORNEMANNII.

215. Greenland Redpoll. (527)

Bill, regularly conic, only moderately compressed and acute, as high as long at the base; color, black or yellow according to the season; frontlet, black, overlaid with hoary, a recognizable light superciliary stripe reaching to the bill; crimson cap over nearly all the crown; upper parts, streaked with brownish-black and white, the latter edging and tipping the feathers, this white nearly pure, only slightly flaxen on the sides of the head and neck; wings and tail, as in the other species; rump and entire under parts, from the sooty throat, white, free from spots; the rump and breast, rosy. Length, 6; wing, 3.30; tail, 2.80.

HAB. Greenland and North-eastern North America, south irregularly in winter to New England, New York and Northern Illinois.

About the year 1863, a friend who used to join me in some of my local collecting trips was in the town of Galt, and seeing a small flock of large light-colored Redpolls, secured two of the lot and sent them to me in the flesh. I have neither before nor since met with any so large and hoary. One of them which I still have, mounted, seems to answer to the above description, but the country from which the Redpolls come is large enough to produce varying forms from different latitudes, and I think it is open to question whether or not it is wise to divide them into so many species.

GENUS ACANTHIS BECHSTEIN.

ACANTHIS HORNEMANNII EXILIPES (COUES).

216. Hoary Redpoll. (527*a*)

Colors pale, the flaxen of *linaria* bleaching to whitish ; rump, white or rosy, entirely unstreaked in the adults ; breast, pale rosy, and streaks on the sides small and sparse ; bill, very small with heavy plumules ; feet, small, the middle toe and claw hardly equal to the tarsus. Length, 5.50 ; extent, 9 ; wing, 3 ; tail, 2.50.

HAB.--Arctic America and North-eastern Asia.

Nest, somewhat bulky for the size of the bird, built of small twigs and grass, lined with fine grass and feathers.

Eggs, two to five, pale bluish-green, speckled toward the larger end with reddish-brown.

So few Redpolls are taken from the vast flocks which in some winters visit us from the north, that it is unsafe to say how rare or common any particular species may be. I have, however, observed a good many in different winters during the last thirty years, and have only seen one of this species. It was killed by K. C. McIlwraith at Hamilton Beach, on the 6th of April, 1885, and on being picked up at once elicited the exclamations which follow the capture of a rare bird. It was a male in fine plumage, the feathers being full and soft, and beautifully tinted with the rosy color peculiar to the race.

This species is said to inhabit the whole of boreal America, but

has seldom been found as far south as even the northern tier of
States.

Regarding this species, Mr. Nelson says in the " Birds of Alaska ":
" This is the prevailing species of the genus throughout Northern
Alaska, where it occurs in great numbers. Its habits and range are
shared by the Common Redpoll, and the two are almost indistinguishable, excepting for the differences in coloration. Their notes, nesting
sites, nests and eggs are indistinguishable.

"Both forms are summer and winter residents, making a partial
migration into the interior during the severe weather of winter. The
series obtained by me contains both forms from various points along
the Yukon from the sea to the British boundary line. They are
usually found in parties of varying size comprising both forms,
although owing to the greater abundance of *exilipes*, parties composed wholly of it are found at times.

"It is common on the Commander Islands in limited numbers
during the winter."

ACANTHIS LINARIA (LINN.).

217. **Redpoll.** (528)

Upper parts, streaked with dusky and flaxen in about equal amounts ; rump,
white or rosy, streaked with dusky ; below, streaked on the sides ; belly, dull
white ; bill, mostly yellow ; feet, blackish. Length, 5¼-5¾ ; wing, 2¾-3 ; tail,
2¼-2½.

HAB.—Northern portions of northern hemisphere, south irregularly in
winter, in North America, to the Middle United States (Washington, D.C.,
Kansas, South-eastern Oregon).

Nest, in a low tree or bush, composed of grass and moss, lined with plant
down or feathers.

Eggs, four or five, pale bluish-white, speckled with reddish-brown.

Like our other winter birds, the Redpolls are somewhat irregular in their visits, but are more frequently seen than either the Grosbeaks or Crossbills. Sometimes they appear in October and remain till late in March, while in other seasons only an occasional roving flock is seen during the winter, and again they are entirely absent. They are hardy, active little birds, and must consume a large quantity of seeds, which can well be spared from the weedy places the birds frequent.

Before leaving in spring, the breast of the male assumes a soft rosy tint, which adds greatly to his beauty when seen among the snow.

This is another of our winter visitors whose home is in the far north. Mr. Nelson says regarding it: "This species is found in Alaska in smaller numbers than the preceding. On the south-eastern coast of the territory, including the Kadiak and Sitkan region, the present bird is found to the exclusion of the other, as also to the south along the coast to Washington Territory and British Columbia, where Mr. Lord found it resident."

ACANTHIS LINARIA HOLBŒLLII (Brehm).

218. **Holbœll's Redpoll.** (528a)

Like *linaria*. Length, 6; wing, 3.25; tail, 2.75; bill, longer.
Hab.—Northern portions of northern hemisphere, near the sea coast.

I have occasionally found among the common Redpolls, individuals of large size which answer to the description given of this species. As they are never numerous, and have not been observed in flocks by themselves, those we see may be stragglers from the main body of their race, which is said to keep well up to the north and east.

Genus SPINUS Koch.

SPINUS TRISTIS (Linn.).

219. American Goldfinch. (529)

Male : In summer, rich yellow, changing to whitish on the tail coverts ; a black patch on the crown ; wings, black, more or less edged and barred with white ; lesser wing coverts, yellow ; tail, black, every feather with a white spot ; bill and feet, flesh-colored. In September the black cap disappears and the general plumage changes to a pale flaxen-brown above and whitey-brown below, with traces of the yellow, especially about the head ; this continues till the following April or May. *Female :*—Olivaceous, including the crown ; below, soiled yellowish ; wings and tail, dusky, whitish-edged ; *young* like the *female*. Length, about 4⅞ ; wing, 2¾ ; tail, 2.

Hab. Temperate North America generally, breeding southward to the middle districts of the United States (to about the Potomac and Ohio Rivers, Kansas and California), and wintering mostly south of the northern boundary of the United States.

Nest, a neat, strong structure, resembling that of the Summer Yellow-bird, composed of miscellaneous soft materials firmly felted together and lined with plant down, usually placed in the upright fork of a tree or bush, from six to twenty feet from the ground.

Eggs, four to six, pale bluish-white, unmarked.

In Southern Ontario the Goldfinch may be considered a resident species, for it nests throughout the country generally, and even in the depth of winter, is often met with unexpectedly in some favored locality where it finds food and shelter. In the severe winter of 1885-86, I came upon a colony of this kind in West Flamboro', where several hundreds of the birds were frequenting a grove of hemlock ; and, judging by the amount of *débris* on the snow underneath, they must have been there all winter. They were very lively, keeping up a continual chattering as they swayed to and fro on the slender branches, extracting the seeds from the cones. Occasionally, when cheered by the mild rays of the wintry sun, some of the males would come to the sunny side of the tree and warble out a few of their varied summer notes, but they spent most of the short wintry day in feeding and in dressing their plumage, retiring early to the thick shelter of the evergreens.

At other seasons of the year, they frequent the cultivated fields, orchards and gardens; and in the fall, when they are seen in greatest numbers, they do good service in consuming the seeds of the thistle and other noxious weeds. They are not in any great haste to begin the duties of housekeeping, and are seen in flocks till towards the end of May. About that time they pair off, and are actively engaged in their domestic duties till some time in August, when the males throw off their gaudy summer dress and join with the females and young in making up the flocks we see roving about the country in their own wild way.

Genus PASSER (Brisson).

PASSER DOMESTICUS (Linn.).

220. House Sparrow.

Form, stout and stumpy ; bill, stout, conical, bulging, longer than deep ; upper mandible longer than the lower. *Adult male* :—Lores, black, a narrow streak of white over each eye ; crown, nape and lower back, ash-gray ; region of the ear coverts, chestnut-brown, streaked with black ; wings, brown, with a bar of white on the middle coverts ; tail, dull brown ; throat and breast, black, sometimes suffused with bright chestnut checks, and sides of the neck white ; belly, dull white ; bill, bluish-black ; legs, pale brown. Length, 6 inches ; wing, 3 inches. In winter the colors are duller and the bill yellowish-brown. In the *female* the upper parts are striated dusky brown ; there is no black on the throat or gray on the pate, and the under parts are brownish-white.

Hab.—From the Atlantic coast to the Mississippi and from New Orleans to Saulte Ste. Marie ; many isolated colonies elsewhere throughout the country, some of which have originated by the birds being brought from a distance through being accidentally closed in empty grain cars.

Nest, about houses, under projecting cornices or in crevices in outhouses, also in trees ; large and clumsy outside, but deep and warmly lined with hair, feathers and other soft material.

Eggs, varying in number from six to nine, soiled white, speckled with brown.

The name English Sparrow is a misnomer as applied to this bird, for it is no more English than it is Scotch, Irish, French, or German. House Sparrow is the name it has gone by in Europe, whence it came, and it was no doubt bestowed on account of its persistent habit of nesting about dwelling-houses wherever it occurs.

Throughout Europe, in former years, when very many of the houses were covered with thatch, the constant habit of the sparrows was to pull out as much of the thatch as made a hole big enough for them to build their nest in. This, of course, led to leaky roofs ; and the result was a constant warfare between the outer and inner tenants, in which the former usually gained their object. Sometimes there would only be one ladder in a large district, and it could not

20

always be borrowed conveniently to stop the sparrow holes. In other cases, when a building bore a financial burden and repairs of any kind received but little attention, I have seen as many as twenty pairs of sparrows with their nests among the thatch of one roof.

In America the bird was unknown till 1850, when the Directors of the Brooklyn Institute imported eight pairs into Brooklyn, N.Y.

This lot did not thrive, and in 1852, a much larger consignment was brought from England by the same parties. These were kept in confinement over the winter, and in the spring of 1853 they were all let loose in the grounds of Greenwood Cemetery, where they did well and multiplied rapidly. There seems to have been a wild craze on this subject just then, for many other shipments were made by individuals in different parts of the country, all of which have no doubt helped to swell the grand army which is now looked upon as a scourge to the agriculturists, and one which is entirely beyond control.

It was, I think, in 1875 that the first sparrows were observed on the streets in Hamilton. The fact was made known to the City Council, who, being always ready to encourage desirable settlers, at once provided a handsome, commodious house for them in the Gore Park, and arrangements were made to have them regularly fed and made as comfortable as possible. But the sparrows, unused to so much kindness, seemed afraid that there was some trap about the house, or poison in the food, and betook themselves to the roadtracks for food, and found nesting-places of their own choice. They increased so rapidly in numbers that the house was taken down, and the birds were allowed to shift for themselves.

My own experience with these birds was similar to that of others, but may here be repeated as illustrating their introduction to the country. It was in the summer of 1874 that I first noticed a pair of these birds about the out-houses, and in a few days they became quite familiar, having evidently made up their minds to stay with us. I made them welcome for old acquaintance' sake, and thinking they would make good settlers, was about to put up a house for them; but before my well-meant intentions were carried out, it became apparent that they were providing for themselves in a manner quite characteristic.

On a peak of the stable was a box occupied by a pair of swallows, who were at that time engaged in rearing their young, and of this box the sparrows seemed determined to get possession. The swallows resisted their attacks with great spirit, and their outcries bringing a host of friends to their assistance, the intruders were for a time driven off, but only to return again with renewed energy and perseverance. The swallows were now sorely beset, for one had to remain on guard while the other went in search of supplies. Still they managed to hold the fort till the enemy, watching his opportunity, made a strategic movement from the rear and darted into the box more quickly than I can tell it. He emerged with a callow swallow hanging by the nape of the neck in his bill, and dropped it on the ground below. Another soon followed, amid the distressing cries of the swallows, who, seeing their hopes so completely blighted, sat mute and mournful on the ridge of the house for a short time, and then went away from the place, leaving the sparrows in undisputed possession of the box. There they remained and raised some young ones during the summer.

In the spring of the following year the numbers had increased, and they began to roost under the veranda around the house, which brought frequent complaints from the sanitary department, and a protest was made against their being allowed to remain there at all. Still, in view of the prospective riddance of insect pests from the garden, matters were arranged with the least possible disturbance to the birds, and we even stood by and saw them dislodge a pair of house wrens who had for years been in possession of a box fixed for them in an apple tree in the garden. So the second year wore on, no further notice being taken of the sparrows except that they were getting more numerous.

I had missed the sprightly song and lively manners of the wrens, and in the spring when they came round again seeking admission to their old home, I killed the sparrows, which were in possession, in order to give the wrens a chance, and they at once took advantage of it and commenced to carry up sticks in their usual industrious manner. They had enjoyed possession only for two days, however, when they were again dislodged. Again the intruders were killed off, and domestic felicity reigned for three days, when a third pair of sparrows came along bent on the same object, and, if possible, more overbearing and determined than their predecessors. This time I thought of a different mode of accomplishing the object in view, and, taking down the box at night, nailed a shingle over the end and worked it flush around the edges. With a centre-bit I then pierced a hole just large enough to admit the wrens, but too small for the sparrows, and put the box back into its place. Early in the morning the assault was renewed, but the wrens found at once that

they were masters of the situation, and never were two birds more
delighted. From his perch aloft, the male poured forth torrents of
scorn and ridicule, while the female inside the box fairly danced
with delight, and I almost fancied was making faces at their enemy,
as he struggled ineffectually to gain admission, or sullenly, but fruit-
lessly, tried to widen the aperture.

Shortly after this dispute was settled, I noticed ten or twelve
sparrows quietly at work at the grape vines, and feeling pleased at
the havoc they were apparently making among the insects, passed
on, speculating mentally on the probable increase of fruit I should
have. In the afternoon they had moved to another trellis, and I
thought, "Well, they are doing the work systematically, and no
doubt effectually." But shortly afterwards, while passing the trellis
where they commenced, I observed a slight *débris* of greenery along
under the vines. This led to an examination which showed, to my
intense mortification, that the heart had been eaten out of every
fruit bud where the birds had been, and nothing left but the outside
leaves. The report of firearms was heard several times in the garden
that afternoon : many dead and wounded sparrows were left to the
care of the cats, and every crevice where the birds were known to
breed was at once closed up.

Since then the wrens have kept possession of their box, and with
a little attention I can keep the sparrows out of the garden, for they
find plenty of provender round the stables ; but they are still on the
increase, and if this continues in the future, as in the past, the time
is not far distant when the streets and stable-yards will not furnish
food enough for the increased numbers, and there is no doubt that
then they will betake themselves to the fields and gardens and take
whatever suits them. This is the serious view of the subject which
has called for legislation in other countries, and may do so here
unless some unexpected check arises to prevent the necessity for it.

In the meantime it is well that all parties who have the opportunity should take notes of the movements and increase of the birds
for future consideration.

Since the above was written, more accounts have been published
about this Sparrow than there ever have been about any American
bird. Articles without number have appeared in the different magazines, pointing out the folly of our having imported a bird whose
character as a plague and a pest was established in the countries
whence he was brought. Sparrow Clubs have been organized for their
extinction, bounties have been offered and paid in different States

for their heads, traps have been advertised warranted to catch them
by the hundred, poison has been recommended and used for their
destruction, and at all agricultural meetings it has been voted that
the sparrow "must go." But he has not gone yet, except in small
colonies to occupy new territory, and he is on the increase wherever
established.

The Department of Agriculture at Washington has issued a most
exhaustive report of over 400 pages, giving the most ample and
circumstantial details of the history and habits of the bird in this
and other countries, all tending to show the injury it causes to agriculture in all its branches, and how much would be saved by our
being rid of it; but, taking my eyes from the report and looking at
the birds outside my window, there is something in their manner
which tells me they are here to stay, and we may as well make up
our minds to it. We have recently had occasion to change our ideas
regarding our relation to the hawks and owls, who, from being
looked upon as enemies, are now, with one or two exceptions,
believed to be our best friends.

Some such change may yet take place in our feelings toward the
Sparrow, but at present I see no indications of it.

SPINUS PINUS (Wils.).

221. Pine Siskin. (533)

Bill, extremely acute; continuously streaked above with dusky and
olivaceous-brown or flaxen; below, with dusky and whitish, the whole plumage
in the breeding season more or less suffused with yellowish, particularly
bright on the rump; the bases of the quills and tail feathers extensively
sulphury-yellow, and all these feathers more or less edged externally with
yellowish. Length, 4¾; wing, 2¾; tail, 1¾.

HAB.—North America generally, breeding mostly north of the United States
and in the Rocky Mountain region, in winter, south to the Gulf States and
Mexico.

Nest, placed high in an evergreen. It is composed of dry grass and pine
needles with a lining of feathers.

Eggs, pale greenish-blue, speckled with brown.

The Siskin, or Pine Linnet, is a more northern bird than the
Goldfinch, and as a winter visitor in Southern Ontario is sometimes
present and sometimes absent. Occasionally they appear in October
in large flocks, swarming on the rank weeds in waste places, and

hanging on the alder bushes by the banks of creeks and gullies. They are extremely restless, and in certain districts the twittering sound of their voices fills the air for days together, till they rise and pass away like a cloud of smoke, perhaps to be seen no more for the season. They are said to have been found nesting in New York State, and also in Massachusetts, but at present I have no record of their being found so engaged in Ontario. As the country becomes more explored, we shall have many such items to add to our present stock of knowledge of the birds.

Genus PLECTROPHENAX Stejneger.

PLECTROPHENAX NIVALIS (Linn.).

222. Snowflake. (534)

Bill, small, truly conic, ruffed at base; hind claw, decidedly curved. In breeding plumage pure white, the back, wings and tail variegated with black; bill and feet, black. As generally seen in Ontario, the white is clouded with warm, clear brown, and the bill is brownish. Length, about 7; wing, 4½; tail, 2¾.

Hab.—Northern parts of the northern hemisphere, breeding in the Arctic Regions, in North America south in winter into the Northern United States, irregularly to Georgia, southern Illinois and Kansas.

Nest, on the ground, composed of grass and moss, lined with feathers, concealed by a tuft of grass or projecting ledge of rock, cavity deep, sides warm and thick.

Eggs, four, pale greenish-white, scrawled and spotted with brown and lilac.

The Snowbirds are our most regular visitors from the north, and they come in greater numbers than any of the other species which descend from high latitudes to avoid the rigors of winter. As early as the 20th of October, their tinkling, icy notes may be heard, but more frequently the birds are first observed later in the season, driving with wild eccentric flight before the earliest flurry of snow. By the shores of the lakes, on bare sandy spots, thinly grown over with the *Andropogon scoparius*, on the seeds of which they freely feed, they may be found with tolerable certainty any time between the end of October and the first of April. Elsewhere throughout the country, they are frequently seen by the roadsides, examining the tall weeds which appear in waste places above the snow, or running in the road tracks searching hurriedly for their scanty fare. They are exceedingly restless, never remaining long in one place, and even when feeding, the flock will often arise without apparent cause of alarm and go off as if never to return, but not unfrequently they come swirling back and alight on the spot from which they have just arisen. There are one or two instances on record of their nests and eggs having been found among the highest mountain peaks in Massachusetts, but their breeding ground is within the Arctic circle, from which they descend over the northern portions of both continents, enlivening many a dreary region with their sprightly presence during the dull days of winter, till reminded by the lengthening days and rising temperature to return again to their northern home.

All our Canadian boys and girls are familiar with the Snowbird, which is frequently the only one they see when out for the first sleigh ride of the season, and often have I been interrogated as to where the birds go when they leave us in spring and where they have their nests. With the view of satisfying my juvenile readers on these points, I shall here make a short extract from Mr. Nelson's most interesting work on the "Birds of Alaska," which tells how he found them in that distant and dreary country:

"The Snowflake is a well-known summer bird in all the circumpolar regions, and none of the various Arctic expeditions have extended their explorations beyond the points where this handsome species is found. About Plover Bay, on the high mountains rising abruptly from the water, I found it common and breeding the last of June, 1881, and on June 24th, the same season, it was also found in fine breeding plumage at the south-west cape of St. Lawrence Island, where we landed from the "Corwin." At the base of this bluff were the ruined huts of the famine-stricken Eskimo, and the steep hill-side was studded with the glistening skulls of the victims. Every large depression held a heavy snow-bank, and the tops of the hills were gray with masses of lichen-colored rocks or a stunted vegetation.

"On the summit overlooking the desolate scene were two walrus-hide huts, sheltering a few survivors of a village which contained nearly 200 inhabitants a little over a year before. As we made our way up to these huts, we were greeted by the hard, rattling *chirr* of several Snow Buntings as they flew from place to place before us.

"Their note was different from any I ever heard them utter during their winter visit to the south, and was one of protest or alarm, as shown by the uneasiness of the birds as they flitted overhead.　We suspected they were nesting from their movements, and asked three or four native children, who ran to meet us, if they knew where the birds had their eggs.　In an instant a broad smile illuminated each grimy countenance, and away they scampered, each eager to be the first to reach the spot and gain the prize.

"Just back of the huts, about 100 yards distant on the hill-side, and sheltered by a slight tussock, was placed a warm closely-made structure of fine grass stems, interwoven throughout with feathers and the cottony seed-tops of plants.　The central depression was built uniformly like the rest of the nest, and the entire structure measured $2\frac{1}{4}$ inches high by $1\frac{1}{2}$ inches across the top, and $2\frac{1}{2}$ inches across the central depression, which latter was $1\frac{1}{4}$ inches deep and contained one fresh egg.

"The nest was taken and the female shot as she hovered restlessly about, uttering her sharp chirring note.

"As we returned slowly to the shore, the male flew about us continually, uttering a loud plaintive call note, the entire time of our stay in the neighborhood.　It exhibited the greatest distress, and appeared to be perfectly aware that we had the female in our possession, for the vicinity of the rifled nest was deserted, and it followed us over 100 yards, circling about and keeping close by, perfectly regardless of any danger to itself."

Genus CALCARIUS Bechstein.

CALCARIUS LAPPONICUS (Linn.).

223. Lapland Longspur. (536)

Bill, moderate, unruffed, but with a little tuft of feathers at the base of the rictus; hind claw, straightish, with its digit longer than the middle toe and claw. *Adult male:*—Whole head and throat, jet black, bordered with buffy or whitish, which forms a postocular line, separating the black of the crown from that of the sides of the head; a broad chestnut cervical collar; upper parts in general, blackish, streaked with buffy or whitish that edges all the feathers; below, whitish, the breast and sides black streaked; wings, dusky, the greater coverts and inner secondaries edged with dull bay; tail, dusky, with an oblique white area on the outer feathers; bill, yellowish, tipped with black; legs and feet, black. Winter *males* show less black on the head, and the cervical chestnut duller; the *female* and *young* have no continuous black on the head, and the crown is streaked like the back, and there are traces of the cervical collar. Length, 6-6½; wing, 3¼-3½; tail, 2¼-2¾.

HAB.—Northern portions of the northern hemisphere, breeding far north; in North America south, in winter, to the Northern United States, irregularly to the Middle States, accidentally to South Carolina, and abundantly in the interior to Kansas and Colorado.

Nest, under a tussock of grass, built of moss and fine dried grass, lined with a few feathers.

Eggs, four or five, greenish-gray, which color is nearly obscured by a heavy mottling of chocolate-brown.

Like the Snowflake, the present species is common to both continents. They come and go together and keep company while here : but at all times the Snowflakes far exceed the others in numbers.

The male Longspur, in full breeding plumage, is a very handsome bird. It is seldom found in Ontario in this dress, but some years ago, two young men who were collecting at Mitchell's Bay, met with quite a large flock in the month of May, and got some very fine specimens, several of which came into my possession. All those I have met with have been in winter dress, in which state the colors are obscured by the black feathers of the head and breast being tipped with yellowish-gray.

Here again we have pleasure in taking a summer chapter from the life of this interesting bird as observed in Alaska. "Like the preceding species, the Lapland Longspur is a widely-spread circumpolar bird, whose presence is recorded from nearly every point visited by the explorers along the shores of the Arctic coast. It is found breeding in Iceland, Greenland, and on nearly all those islands lying in the icy sea just north of the continental mainlands. In the territory covered by the present paper, it is an extremely abundant and familiar bird, found, perhaps, more numerously upon the mainland, but also known from the various islands of Behring Sea. Regarding its presence on the Seal Islands, Elliott tells us : 'This bird is the vocalist *par excellence* of the Pribilov group, singing all through the month of June in the most exquisite manner, rising high in the air and hovering on fluttering wings above its sitting mate. The song is so sweet that it is always too short.'"

GENUS POOCÆTES BAIRD.

POOCÆTES GRAMINEUS (GMEL.).

224. Vesper Sparrow. (540)

Thickly streaked everywhere above, on sides and across breast ; *no* yellow anywhere ; lesser wing coverts, *chestnut*, and one to three outer pairs of tail feathers partly or wholly white ; above, grayish-brown, the streaking dusky and brown with grayish-white ; below, white, usually buffy-tinged, the streaks very numerous on the fore part and sides ; wing coverts and inner quills much edged and tipped with bay ; crown, like back, without median stripe ; line over and ring round eye, whitish : feet, pale. Length, 5¾-6¼ ; wing, 2⅜-3¼ ; tail, 2¼-2¾.

Hab.—Eastern North America to the Plains, from Nova Scotia and Ontario southward, breeds from Virginia, Kentucky and Missouri northward.

Nest, a deep cup-shaped hollow in the ground, lined with grass and hair.

Eggs, four, grayish-white, heavily clouded with chocolate-brown.

This is one of the "Gray Birds," and the most abundant in Ontario of the several species to which this name is applied.

Its song is very sweet and plaintive, and being most frequently uttered in the evening about sundown, it has gained for the bird the appropriate name of Vesper Sparrow.

It is a summer resident, arriving in Southern Ontario toward the end of April, and soon becoming common all over the country. It does not penetrate far north in the Province, and in Manitoba it is replaced by the Western Vesper Sparrow, a pale gray form peculiar to the prairies.

The favorite perch of the male is the top of a fence post, and his nesting place among the grass close by. In the fall the birds get to be abundant before leaving, but from their habit of skulking among the rank weeds, they are not so conspicuous as the blackbirds and other species which keep in flocks on the wing. They move to the south in October, none having been observed during the winter.

Genus AMMODRAMUS Swainson.

Subgenus PASSERCULUS Bonaparte.

AMMODRAMUS SANDWICHENSIS SAVANNA (Wils.).

225. Savanna Sparrow. (542a)

Above, brownish-gray, streaked with blackish, whitish-gray and pale bay, the streaks largest on the inner scapulars, smallest on the cervix, the crown divided by an obscure whitish line ; superciliary line and edge of wing, yellowish ; sometimes an obscure yellowish suffusion about the head ; below, white, pure or with faint buffy shade, thickly streaked with dusky, the individual spots edged with brown, mostly arrow-shaped, running in chains along the sides, and often aggregated in an obscure blotch on the breast ; wings and tail, dusky, the wing coverts and inner secondaries black edged and tipped with bay. Length, 5¼-5¾ , wing, 2½-2¾ ; tail, 2-2¼.

HAB.—Eastern Province of North America, breeding from the Northern United States to Labrador and Hudson's Bay Territory.

Nest, composed of fine withered grass placed in a deep cup-shaped hole in the ground.

Eggs, four to six, ground color grayish, heavily clouded with chocolate-brown.

These quiet, unobtrusive little sparrows may be seen and heard in the moist meadows in spring and summer, but they are not very plentiful anywhere.

Towards the end of August they become abundant along the marshy shores of Hamilton Bay, where they evidently find food to suit their taste, and they continue to enjoy it until reminded by the cool nights in September that it is time to be off to the south. The specimens secured at this season are evidently northern-bred birds, being more fully developed in size and markings than those which breed with us. Often when picking one up, I fancy I have got the Ipswich Sparrow, but so far have not succeeded in doing so. I still think the latter species will be found near Hamilton, for there are several suitable resorts for it which will in future be carefully watched at the proper season.

The Savanna is chiefly an eastern species, going as far north as Labrador and Hudson Bay. In the North-West, it is replaced by the Western Savanna Sparrow, which differs from the present species in having the pale gray colors peculiar to the birds of the prairies.

SUBGENUS COTURNICULUS BONAPARTE.

AMMODRAMUS SAVANNARUM PASSERINUS (WILS.).

226. Grasshopper Sparrow. (546)

Edge of wing, conspicuously yellow ; lesser wing coverts and short line over the eye, yellowish ; below, not or not evidently streaked, but fore parts and sides buff, fading to dull white on the belly ; above, singularly variegated with black, gray, yellowish-brown, and a peculiar purplish bay in short streaks and specks, the crown being nearly black, with a sharp median brownish-yellow line, the middle of the back chiefly black, with bay and brownish-yellow edgings of the feathers, the cervical region and rump chiefly gray, mixed with bay ; wing coverts and inner quills, variegated like the back ; feet, pale. *Young:*—Similar, not so buffy below, and with pectoral and maxillary dusky spots. Length, $4\frac{1}{2}$-$5\frac{1}{4}$: wing, $2\frac{1}{4}$: tail, 2 or less, the outstretched feet reaching to or beyond its end.

Hab.—Eastern United States and Southern Canada to the Plains, south to Florida, Cuba, Porto Rico and coast of Central America.

Nest, a cup-shaped hole in the earth, lined with dry grass.

Eggs, four or five, crystal-white, speckled with reddish-brown.

So far as at present known, the Grasshopper Sparrow is of very rare occurrence in Ontario, the southern border seeming to be the northern limit of its distribution.

Many years ago I killed a male, who was squeezing out his wheezy notes from the top of a mullein stalk. Mr. Saunders mentions having taken one near London, but these two cases complete the record for Ontario.

It is named among the birds found in the North-West by Prof. Macoun, but is not found in Mr. Thompson's list of the "Birds of Western Manitoba." It is much given to concealing itself among the rank herbage, and *may* in some localities be a rare summer resident in Southern Ontario, but I do not expect to see it here, except as a casual visitor.

Genus CHONDESTES Swainson.

CHONDESTES GRAMMACUS (Say.).

227. Lark Sparrow. (552)

Head, curiously variegated with chestnut, black and white; crown, chestnut, blackening on the forehead, divided by a median stripe and bounded by two lateral stripes of white; a black line through and another below the eye, enclosing a white streak under the eye and the chestnut auriculars; next, a sharp black maxillary stripe, not quite reaching the bill, cutting off a white stripe from the white chin and throat; a black blotch on middle of breast; under parts, white, faintly shaded with grayish-brown; upper parts, grayish-brown; the middle of the back with fine black streaks; central tail feathers, like the back, the rest jet black, broadly tipped with pure white in diminishing amount from the lateral pair inward, and the outer web of outer pair entirely white. Length, 6½-7; wing, 3½; tail, 3.

Hab.—Mississippi Valley region, from Ohio, Illinois and Michigan to the Plains, south to Eastern Texas. Accidental near the Atlantic coast (Massachusetts, Long Island, New Jersey and Washington, D.C.).

Nest, on the ground, composed of dry grass.

Eggs, three to five, white, irregularly veined with dark.

In May, 1862, a pair of these birds was observed near Hamilton, and the male was obtained and shown to me shortly afterwards.

I did not hear of the species again till the publication of the

"List of Birds of Western Ontario," in 1882, in which it is mentioned as "breeding, but rare." More recently, Mr. Saunders informs me that it breeds regularly near London. In the spring of 1885, I saw several on the Beach near Hamilton, and it is also reported by Mr. Thompson as having been observed near Toronto.

It is evidently, like some others, making its way into Ontario around the west end of Lake Erie, and all lovers of birds will do well to encourage its progress, for it is a sweet songster and a handsome little bird of confiding, pleasing manners.

GENUS ZONOTRICHIA SWAINSON.

ZONOTRICHIA LEUCOPHRYS (FORST.).

228. White-crowned Sparrow. (554)

Adults:—Of both sexes with the crown pure white, enclosing on either side a broad black stripe that meets its fellow on the forehead and descends the lores to the level of the eyes, and bounded by another black stripe that starts behind the eye and curves around the side of the hind head, nearly meeting its fellow on the nape ; edge of under eyelid, white. Or, we may say, crown black, enclosing a median white stripe and two lateral white stripes, all confluent on the hind head. General color, a fine dark ash, paler below, whitening insensibly on the chin and belly, more brownish on the rump, changing to dull brownish on the flanks and crissum, the middle of the back streaked with dark purplish-bay and ashy-white. No bright bay like that of *albicollis* anywhere, except some edging on the wing covert and inner secondaries ; middle and greater coverts, tipped with white, forming two bars ; no yellow anywhere ; bill and feet, reddish. *Young :* Birds have the black of the head replaced by a very rich warm brown, the white of the head by pale brownish, and the general ash has a brownish suffusion and the back is more like *albicollis.* Length, 6.25-7 ; extent, 9.20-10.20 ; tail, 2.90-3.20.

HAB.—North America at large, breeding chiefly in the Rocky Mountain region (including Sierra Nevada) and north-east to Labrador.

Nest, on the ground among the bushes, composed of grass and weeds, intermixed with moss and lined with fine, hair-like grass and rootlets.

Eggs, four or five, ground color, greenish-blue, heavily clouded with chocolate-brown. Very variable in pattern.

The White-crowned Sparrow is a more northern bird than its white-throated relative, but it does not arrive so early in spring, seldom appearing along our southern border before the first week in May. During the two succeeding weeks, it is very common among the brambles and thorn bushes by the wayside.

They travel in small companies of ten or twelve, the individuals keeping each other in view, as they skulk from one brush pile to another to avoid being observed. By the 25th of May they have all gone north, apparently far north, for I have no record of their having been found breeding in Ontario.

In the fall they are again seen on the return trip, but not in such great numbers as in the spring, and none have been observed to winter within our limits.

ZONOTRICHIA ALBICOLLIS (Gmel.).

229. White-throated Sparrow. (558)

Adult male:—With the crown black, divided by a median white stripe, bounded by a white superciliary line and yellow spot from the nostril to the eye ; below this a black stripe through the eye ; below this a maxillary black stripe bounding the indefinitely pure white throat, sharply contrasted with the dark ash of the breast and sides of the neck and head ; edge of wing, yellow ; back, continuously streaked with black, chestnut and fulvous-white ; rump, ashy, unmarked ; wings, much edged with bay, the white tips of the median and greater coverts forming two conspicuous bars ; quills and tail feathers, dusky, with pale edges ; below, white, shaded with ashy-brown on sides, the ash deeper and purer on the breast ; bill, dark ; feet, pale. *Female:*—And immature birds with the black of head replaced by brown, the white of throat less conspicuously contrasted with the duller ash of surrounding parts, and frequently with obscure dusky streaks on the breast and sides. Length, 6½-7½ ; wings and tail, each about 3.

Hab.—Eastern North America, west to the Plains, north to Labrador and the Fur Countries. Breeds in Northern Michigan, Northern New York and Northern New England, and winters from the Middle States southward.

Nest, among the bushes, on or near the ground, composed of weeds, grass and moss, lined with fibre and thread-like rootlets.

Eggs, four to six, variable in color and pattern, usually greenish-blue, clouded and blotched with chocolate-brown.

These beautiful Sparrows make their appearance in Southern Ontario about the 20th of April, and till the middle of May are seen among the shrubbery and underbush, working their way in small flocks towards their summer residence to the north of us. Great numbers are said to go right on to the Fur Countries, but many no doubt find suitable nesting places in the intermediate districts. I first found them breeding near a retired pond surrounded by tamaracks, in the township of Dumfries, about thirty miles north-west of Hamilton. It was towards the close of a warm day in the early part of July, and the last rays of the sun were brightening the tops of the tamaracks, while, underneath, the still waters of the pond, enclosed in a deep natural basin, were shrouded in gloom. There was little to break the silence, till a bird, mounting to the topmost twig of one of the trees, his bill pointing upward, his tail hanging limp and motionless, and his whole attitude indicating languor and weariness, drawled out the plaintive, familiar "Old Tom Peabody, Peabody." This song harmonized so perfectly with the surroundings that I felt at once he was at home. The hour, the attitude, and above all the feeling of weariness expressed in the plaintive notes of the bird, reminded me strongly at the time of the Yellow-hammer of Britain.

Allan Brooks has also found this species breeding at Milton, a few miles north of the west end of Lake Ontario, but such cases are by no means common in this district. In the fall they are again seen in limited numbers, but at that season the plumage of the male has lost much of its brightness, and young and old, male and female, resemble each other in appearance.

Their food, which consists chiefly of seeds, is obtained on or near the ground. During October they are seen travelling from one brush pile to another, and by the end of that month they are gone for the season.

GENUS SPIZELLA BONAPARTE.

SPIZELLA MONTICOLA (GMEL.).

230. Tree Sparrow. (559)

Bill, black above, yellow below ; legs, brown : toes, black ; no black on forehead ; crown, chestnut (in winter specimens the feathers usually skirted with gray) bordered by a grayish-white superciliary and loral line, and some vague chestnut marks on sides of head ; below, impurely whitish, tinged with ashy anteriorly, washed with pale brownish posteriorly : the middle of the breast with an obscure dusky blotch ; middle of back boldly streaked with black, bay and flaxen ; middle and lesser wing coverts, black, edged with bay and tipped with white, forming two conspicuous cross bars ; inner secondaries similarly variegated : other quills and tail feathers, dusky, with pale edges. Length, 6 ; wing and tail, nearly 3.

HAB.—Eastern North America, westward to the Plains, and from the

Arctic Ocean south, in winter, to the Carolinas, Kentucky and Eastern Kansas. Breeds north of the United States, east of the Rocky Mountains.

Nest, indifferently on the ground or in a tree or bush, composed of grass, rootlets, mud, fine grass and hair.

Eggs, bluish-green, speckled and blotched with reddish-brown.

In Southern Ontario the Tree Sparrow is a regular winter visitor, arriving from the north during the month of October, and remaining over the winter in sheltered hollows or among the brush and weeds by the banks of streams. In appearance it does not look like a hardy bird, but while here it is exceedingly lively and cheerful, its silvery, tinkling notes being frequently heard during the coldest snaps in winter. At the approach of spring all the Tree Sparrows move off to the north, and none are observed during summer.

SPIZELLA SOCIALIS (Wils.).

231. **Chipping Sparrow.** (560)

Adult.—Bill, black; feet, pale; crown, chestnut, extreme forehead black, a grayish-white superciliary line, below this a blackish stripe through eye and over auriculars; below, a variable shade of pale ash, nearly uniform and entirely unmarked; back, streaked with black, dull bay and grayish-brown; inner secondaries and wing coverts, similarly variegated, the tips of the greater and lesser coverts forming whitish bars; rump, ashy, with slight blackish streaks; primaries and tail dusky, the bill pale brown, and the head lacking definite black. Length, 5-5½; wing, about 2¾; tail, rather less.

Hab.—Eastern North America, west to the Rocky Mountains, north to Great Slave Lake, and south to Eastern Mexico.

Nest, in a bush or among the vines, composed of rootlets and fine grass, lined with horse-hair.

Eggs, three or four, pale bluish-green, dotted, speckled or scrawled with dark brown.

Prior to the advent of the House Sparrow, the Chipper was the most familiar and best known bird around our dwellings, and though now in the minority, it still builds its nest in the garden, and comes familiarly near the door to pick up crumbs for the support of its family.

It is very generally distributed over Ontario, being found near the dwellings of rich and poor alike; in shade trees in the city as well as in weedy corners and thorn bushes in the pasture-field.

It arrives from the south about the end of April, and at once

21

begins building its nest. It is most diligent in the discharge of its varied domestic duties during the summer, and when the young are able to shift for themselves, old and young get together in flocks, and about the end of October all move off to the south.

SPIZELLA PUSILLA (WILS.).

232. Field Sparrow. (563)

Bill, pale reddish ; feet, very pale ; crown, dull chestnut ; no decided black or whitish about head ; below, white, unmarked, but much washed with pale brown on breast and sides ; sides of head and neck, with some ague brown markings ; all the ashy parts of *socialis* replaced by pale brownish ; back, bright bay, with black streaks and some pale flaxen edgings ; inner secondaries, similarly variegated ; tips of median and greater coverts forming decided whitish cross-bars. Size of *socialis*, but more nearly the colors of *monticola*. *Young :*—For a short time, streaked below as in *socialis*.

HAB.—Eastern United States and Southern Canada, west to the Plains.

Nest, on the ground, or near it, in a low bush, composed of grass and rootlets, lined with fine grass and hair.

Eggs, four or five, greenish-white, variously marked with reddish-brown.

The Field Sparrow is sparingly distributed in suitable places in Southern Ontario, which probably forms its northern limit. It arrives from the south during the first week in May, and soon makes its presence known by its pleasing ditty, which proceeds from the top of a low tree or bush in the pasture field. It resembles the Chipper in size, but is more like the Tree Sparrow in coloring. The cinnamon-tinted bill is always a ready mark by which to distinguish it from any other of the small sparrows.

It raises two broods in the season and retires to the south in September.

GENUS JUNCO WAGLER.

JUNCO HYEMALIS (LINN.).

233. Slate-colored Junco. (567)

Blackish-ash ; below, abruptly pure white from the breast ; two to three outer tail feathers, white ; bill, flesh colored. In the *female*, and in fact in most fall and winter specimens, the upper parts have a more grayish, or even a decidedly brownish cast, and the inner quills are edged with pale bay. Length, 6-6½ ; wing and tail, about 3.

Hab.—North America at large, but chiefly east of the Rocky Mountains, breeding from the higher parts of the Alleghanies and Northern New York and northern New England northward ; south in winter to the Gulf States.

Nest, on the ground, rarely in a bush above it, composed of strips of bark, grass and rootlets, lined with moss and hair.

Eggs, four or five, greenish-white, spotted and blotched with reddish-brown.

In Southern Ontario the "White Bill," as this species is familiarly called, may fairly be considered resident, for although it is most numerous in April and October, it breeds commonly throughout the country, and a few are always observed remaining during the winter.

It is a very familiar species, showing a marked partiality for rocky ravines, quarries and stone heaps. It is also common by the roadsides and in gullies and other uncultivated places, but in the dense bush we seldom see it, until we come to a spot where men and horses have been at work felling and hauling timber. In such a place at all seasons, its white tail feathers are almost sure to be noticed flirting about among the brush. The ordinary note of this species is a simple "*chip*," like the sound produced by striking two pebbles together, but in the spring the male has a rather pleasing little song, with which he cheers his mate while they are fitting up their home.

This species extends its migrations a long way to the north, where Mr. Nelson says : "This is one of the rarest sparrows visiting the coast of the Behring Sea. It is, however, much more numerous in the interior, and is found more or less common along the entire course of the Yukon, at the mouth of which it breeds."

GENUS MELOSPIZA BAIRD.

MELOSPIZA FASCIATA (GMEL.).

234. Song Sparrow. (581)

Below, white, slightly shaded with brownish on the flanks and crissum, breast and sides with numerous dusky streaks, with brown edges, coalescing to form a pectoral blotch, and maxillary stripes bounding the throat ; crown dull bay, with fine black streaks, divided and bounded on either side by ashy-whitish lines ; vague brown or dusky and whitish markings on the sides of the head ; the interscapular streaks black, with bay and ashy-white edgings ; rump and cervix, grayish-brown, with merely a few bay marks ; wings with dull bay edgings, the coverts and inner quills marked like the interscapulars ; tail, obviously longer than the wings, pale brown, with darker shaft lines on the middle feathers at least, and often with obsolete wavy markings. Length, 6-6½ ; wing, about 2½ : tail, about 3.

HAB.—Eastern United States to the Plains, breeding from Virginia and the northern portion of the Lake States northward.

Nest, on the ground, more rarely on a low tree or bush, composed of rootlets and leaves, lined with fine grass and occasionally some horse-hair.

Eggs, four or five, very variable in marking, usually grayish or greenish-white, blotched or spotted with brown, the shades of which differ greatly in different specimens.

This is an abundant summer resident, and one which seeks the society of man, being found wherever human habitations have been raised within its range. Large numbers pass on to the north in April, returning again in October on their way south, but they do not all leave us. While getting on or off the ice on Hamilton Bay in the depth of winter, I have several times been surprised by seeing a Song Sparrow rise from among the flags, which at that season have a roof of snow, and no doubt afford a comfortable shelter to the little birds. In the same locality, on a comparatively mild day in the middle of winter, I have seen a male of this species mount to the top of a bulrush and warble forth his pleasing, familiar notes, perhaps in appreciation of the rising temperature.

In the "Birds of Ohio," Dr. Wheaton mentions the following singular instance of the strong attachment which this species has for its nest. Some laborers, who were cutting grass on a railroad track near Columbus, found a nest of a Song Sparrow on the embankment, and though rather a delicate piece of work for this class of men to undertake, they moved it from its original site among the grass and placed it gently, but loosely, on the fork of a horizontal limb of a maple sapling, three feet from the trunk. Instead of deserting the nest, as many birds would have done, or attempting to secure it to the limb on which it was placed, the sparrows brought long stems of timothy grass, and twisted them together and around a limb extending over the nest at a distance of one and a half feet. The lower ends of these stems were firmly fastened into the rim of the nest, and other stems were woven in transversely, forming a complete basket. The whole structure resembled an inverted balloon, and in this remarkable construction the eggs were hatched and the young safely raised. After the nest was deserted, the guy ropes were found to be sufficiently strong to bear up the nest, after the limb on which it was placed had been removed.

The Song Sparrow is generally distributed throughout Ontario,

and extends its summer visit to the Fur Countries. It is not at any
season gregarious, each individual coming and going as suits itself.
It is so common everywhere, that should those frequenting any given
district all take wing at one time, the flock would be very large.

MELOSPIZA LINCOLNI (Aud.).

235. Lincoln's Sparrow. (583)

Below, white ; breast banded and sides often shaded with yellowish ; everywhere, except on the belly, thickly and sharply streaked with dusky ; above,
grayish-brown, crown and back with blackish, brownish and paler streaks ;
tail, grayish-brown, the feathers usually showing blackish shaft lines ; wings,
the same, the coverts and inner quills blackish, with bay and whitish edgings ;
no yellow on wings or head. Length, 5½ ; wing and tail, about 2¼.

HAB.—North America at large, breeding chiefly north of the United States
and in the higher parts of the Rocky Mountains, south in winter to Guatemala.

Nest, on the ground, composed of grass throughout, the finest used for
lining inside.

Eggs, four to six, greenish-white, clouded with brown.

Nest and eggs scarcely distinguishable from those of the Song Sparrow.

This quiet little Sparrow is almost unknown in the east, although
it has been found at a number of different points, and from its
retiring habits may be more common than we think it is.

Audubon found it first in Labrador, the young being able to fly
on 4th of July. It has occasionally been captured during the season
of migration, chiefly in Massachusetts and Connecticut, and there is
in the Bulletin of the Nuttal Club, 1878, an account of a nest being
found by Mr. Bagg in Hamilton County, N.Y.

Ontario was without a record of this species till the 23rd of May,
1885, when K. C. McIlwraith got into a bird wave which had been
stopped at Hamilton Beach by a head wind during the previous
night, and from a crowd, composed of different classes in large
numbers, picked out two Lincoln's Sparrows, and on the 25th he got
two more at the same place. Since that time Mr. George R. White
reports having taken one at Ottawa, and Mr. Saunders has also
secured one at London.

In the west the history of the species is entirely different. Mr.
Trippe, writing from Colorado, says : " Lincoln's Finch is abundant
and migratory. It breeds from about 9,500 or 10,000 feet up to the
timber line. It arrives at Idaho Springs early in May, and soon

becomes very common, haunting the thickets and brush heaps by the brooks, and behaving very much like the Song Sparrow. During the breeding season, it is most abundant among the bushes near and above timber line, nesting as high as it can find the shelter of willows and junipers. Reappearing in the valleys in October, it lingers by the streams for a few weeks and then disappears."

It is also said to be abundant in spring and fall in Iowa, and Mr. Ridgway reports it as wintering in great numbers in Southern Illinois.

It has been found breeding at Fort Yukon in Alaska, and also throughout the northern portions of British America to the Arctic Ocean.

MELOSPIZA GEORGIANA (LATH.).

236. Swamp Sparrow. (584)

Crown, bright bay or chestnut, blackening on the forehead, often with an obscure median ashy line and usually streaked with black ; cervix, sides of head and neck and the breast, strongly ashy, with vague dark auricular and maxillary markings, the latter bounding the whitish chin, the ashy of the breast obsoletely streaky : belly, whitish ; sides, flanks and crissum, strongly shaded with brown and faintly streaked : back and rump, brown, rather darker than the sides, boldly streaked with black and pale brown or grayish ; wings so strongly edged with bright bay as to appear almost uniformly of this color when viewed closed, but inner secondaries showing black with whitish edging ; tail, likewise strongly edged with bay and usually showing black shaft lines. Further distinguished from its allies by the emphasis of the black, bay and ash. Length, 5½-6 ; wing and tail, 2¼-2½.

HAB.—Eastern North America to the Plains, accidentally to Utah, north to the British Provinces, including Newfoundland and Labrador. Breeds from the Northern States northward, and winters in the Middle States and southward.

Nest, on the ground in a moist place, sometimes in a tussock of grass or low bush, composed of weeds, grass and rootlets, lined with fine fibrous substances.

Eggs, four to six, grayish-white, speckled with reddish-brown.

This is, perhaps, the least known of any of our common Sparrows, for it seldom comes within reach of the ordinary observer, and even by the collector it is apt to be overlooked, unless he knows its haunts and goes on purpose to seek it. It is very common by the shores of Hamilton Bay, where it may be seen skulking along the line where land and water meet, and if disturbed at once hides itself among the rank herbage of the marsh. Occasionally, during the excitement of the mating season, a male will mount a bulrush and warble out his not unpleasant song, but most of their time is spent in places which are difficult of access either by land or water, and therefore they are seldom seen.

This species is common during the breeding season throughout Ontario and Manitoba. It was found in Labrador by Audubon, and Richardson records its occurrence at Fort Simpson.

They arrive from the south early in May and leave again in October, none being observed during winter.

GENUS PASSERELLA SWAINSON.

PASSERELLA ILIACA (MERR.).

237. Fox Sparrow. (585)

General color, ferruginous or rusty-red, purest and brightest on the rump, tail and wings; on the other upper parts appearing as streaks laid on an ashy ground; below, white, variously but thickly marked, except on the belly and crissum, with rusty red; the markings anteriorly in the form of diffuse confluent blotches, on the breast and sides consisting chiefly of sharp sagittate spots and pointed streaks; tips of middle and greater coverts forming two whitish wing bars; under mandible, dark; lower, mostly yellow; feet, pale. Length, 6¾-7¼; wing and tail, each 3 or more.

HAB.—Eastern North America, west to the Plains and Alaska (Valley of
the Yukon to the Pacific), and from the Arctic Coast south to the Gulf States.
Breeds north of the United States, winters chiefly south of the Potomac and
Ohio Rivers.

Nest, indifferently on the ground or in a tree, composed of grass, moss and
fibrous roots, lined with hair and feathers.

Eggs, four to six, greenish-white, thickly spotted with rusty-brown.

This large and handsome Sparrow breeds in the north and winters
in the south, but by what particular route it passes between the two
points I am at a loss to determine, for in this part of Ontario it is
seldom seen.

In the London list it is mentioned as "rare during migration, four
or five specimens taken." In all my rambles I have only met with
it a few times, and but once have I heard it utter its rich, musical
notes, which are the admiration of all who hear them. Speaking
of this species, Dr. Coues, in his "Birds of the North-West," says:
"During the sunny days which precede their departure, the males
are fond of perching on the top of a small tree or bush to warble a
few exquisitely sweet notes, the overture of the joyous music which,
later in the year, enlivens the northern solitudes whither the birds
resort to breed." The nest has not been found within the limits of
the United States or Ontario, so far as I am aware, but in the list
of "Birds of Western Manitoba," Mr. Thompson mentions it as
breeding on Duck Mountain.

Audubon found it breeding in Labrador in July; it is known to
spend the summer in Newfoundland. Kennicott and Macfarlane
found it breeding in the Arctic Regions.

Of Alaska, Mr. Nelson says: "Along the coast of Norton Sound,
this bird is an abundant summer resident, sharing with the Tree
Sparrows the bushy shelter of the alder thickets on the hill-sides
and sheltered ravines. Wherever, along the northern coast, a fair-
sized alder patch occurs, this hardy species may confidently be looked
for. . . . . On pleasant frosty mornings at this season, the males take
their stand upon the roof of the highest building, or the cross upon
the Greek church, just back of the dwellings, whence they pour
forth their clear thrush-like whistle."

I have noticed it once or twice passing south in October, but none
have been observed during the winter.

GENUS **PIPILO** VIEILLOT.

GENUS **PIPILO** VIEILLOT.

PIPILO ERYTHROPHTHALMUS (LINN.).

238. **Towhee.** (587)

Adult male:—Black; belly, white sides, chestnut; crissum, fulvous-brown; primaries and inner secondaries with white touches on the outer webs; outer tail feathers with the outer web and nearly the terminal half of the inner web white. the next two or three with white spots, decreasing in size; bill, blackish; feet, pale brown; iris, red in the adult, white or creamy in the *young,* and generally in winter specimens. *Female:*—Rich warm brown where the male is black; otherwise similar. Very *young* birds are streaked brown and dusky above; below, whitish, tinged with brown and streaked with dusky. Length, male, 8½; wing, 3½; tail, 4; female rather less.

HAB.—Eastern United States and Southern Canada, west to the Plains.

Nest, on the ground, more rarely in a bush or sapling, a rude structure, composed of grape-vine bark, weed stalks, leaves and grass, lined with fine vegetable fibre.

Eggs, three or four, variable, usually white, thickly freckled with reddish-brown.

This species has a more northern range than we have been in the habit of attributing to it, for it was found by Prof. Macoun in the North-West Territory, and Mr. Thompson reports it as common in Southern Manitoba. In Southern Ontario it arrives from the south about the 1st of May, the males coming on a few days ahead of the females. Much of their time is spent on the ground, scratching and rustling about among the withered leaves in search of seeds and insects. During the pairing time, the male will frequently rise from the scrub bush to the lower branch of a tree, and sing his original song in his best style, accompanying the performance with many a jerk and flirt of his long handsome tail, which shows to advantage on these occasions. If we sit down to watch his motions for a little, we may be favored with a glimpse of the female stealing through the underbrush, but except under such circumstances she is rarely seen.

This is one of the species which apparently enters Ontario from the south-west, for on looking at the dates of its arrival at London and Chatham, we find it is always there before it reaches Hamilton, while at Ottawa Mr. White has not met with it at all.

During the heat of summer, the loud, ringing *Towhee,* which has given the birds their common name, is discontinued, and they spend their time quietly in the shade. In September it is again heard, perhaps as a bugle note to call the flocks together before starting for the south. We have no record of any being observed during winter.

Genus CARDINALIS Bonaparte.

CARDINALIS CARDINALIS (Linn.).

239. **Cardinal.** (593)

Male :—Rich vermilion or rosy-red, obscured with ashy on the back ; face, black : bill, reddish ; feet, brown. *Female* :—Ashy-brown, paler below, with evident traces of the red on the crest, wings, tail and under parts. Length, 8.9 ; wing, about 3⅜ ; tail, 4.

Hab.—Eastern United States, north to New Jersey and the Ohio Valley (casually farther), west to the Plains.

Nest, in a bush or low tree near the ground, usually not far from water, composed of bark, leaves, grass and rootlets rather loosely put together.

Eggs, three or four, greenish-white, thickly spotted with dark reddish-brown often wreathed round the larger end.

The Cardinal can only be regarded as a casual visitor along our south-western border. It is quite common in Ohio, and, as might be expected, a few occasionally cross the lake. Mr. Norval reports one or two being found at Port Rowan, and Dr. Macallum mentions that a few are seen every summer along the lake shore south of Dunnville, where they are supposed to breed among the evergreens. They make showy, interesting cage birds, on account of which great numbers are caught in trap cages and sold in the southern markets.

Dr. Coues describes the Cardinal as "a bird of striking appearance and brilliant vocal powers, resident and abundant from the Middle States southward ; inhabits thickets, tangle and undergrowth of all kinds, whence issue its rich, rolling notes, while the performer, brilliantly clad as he is, often eludes observation by his shyness, vigilance and activity."

Genus HABIA Reichenbach.

HABIA LUDOVICIANA (Linn.).

240. **Rose-breasted Grosbeak.** (595)

Adult male :—With head and neck all round and most of the upper parts, black ; the rump, upper tail coverts and under parts, white ; the breast and under wing coverts, exquisite carmine or rose-red ; wings and tail, black, variegated with white ; bill, pale ; feet, dark. *Female :*—Above, streaked with blackish and olive or flaxen brown with median white coronal and superciliary line ; below, white, more or less tinged with fulvous and streaked with dusky ; under wing coverts, saffron-yellow ; upper coverts and inner quills with a white spot at end ; bill, brown. *Young males :*—At first resemble the female. Length, 7¼-8½ ; wing, about 4 ; tail, about 3¼.

Hab.—Eastern United States and Southern Canada, west to the eastern border of the Plains, south in winter to Cuba, Central America and Northern South America.

Nest, in a low tree, composed of twigs, vegetable fibre and grass, rather loosely put together.

Eggs, three or four, greenish-blue, thickly spotted with reddish-brown.

This robust and gaily attired songster arrives from the south about the 10th of May, and soon its rich, rolling song is heard in the trees and thickets where it spends the summer. It breeds regularly along the southern border of Ontario, and has also been found in Labrador and in the Red River Valley. Its favorite haunts are along the wooded banks of streams, where, even at noonday, when most other birds are silent, the male, in the shade of the luxuriant foliage, cheers his mate during the tedious hours of incubation with the song she loves to hear. The food of the species consists of seeds, buds and berries, but it also takes a variety of insects, and is one of the few birds which visit the potato patch and snap up the potato bugs. On this account alone, it is entitled to our protection, but it is also one of the most attractive birds which visit the shrubbery, and should be most welcome if it could be taught to consider itself protected and come nearer to our dwellings.

Before retiring in the fall, the males lose the greater portion of their black, but retain the carmine on the breast and under wing coverts.

--- --- ---

GENUS PASSERINA VIEILLOT.

PASSERINA CYANEA (LINN.).

241. Indigo Bunting. (598)

Adult male :—Blue, tinged with ultra-marine on the head and throat, elsewhere with verdigris-green ; wings and tail, blackish, glossed with green ; feathers around base of bill, black ; bill, dark above, rather paler below, with a curious black stripe along the gonys. *Female :*—Above, plain warm brown ; below, whitey-brown, obsoletely streaky on the breast and sides ; wing coverts and inner quills, pale edged, but not whitish : upper mandible, blackish, lower pale, with the black stripe just mentioned. *Young male :*—Is like the female, but soon shows blue traces, and afterwards is blue, with white variegation below. Length, $5\frac{1}{2}$; wing, $2\frac{3}{4}$; tail, $2\frac{1}{2}$.

HAB.—Eastern United States, south in winter to Veragua.

Nest, in a bush, composed of leaves, grass and weed stalks, lined with finer material of the same kind.

Eggs, four or five, white, tinged with blue, sometimes speckled with reddish-brown.

About the 15th of May, the Indigo arrives from the south, and at once commences to deliver his musical message, such as it is, with considerable animation. While so engaged, he is usually perched on the upper twig of a dead limb, within hearing of the female, who is of retiring habits and seeks to elude observation among the briars and underbrush.

It is rather a tender species, and probably does not penetrate far north into Ontario. It is not mentioned either by Prof. Macoun or by Mr. Thompson as having been seen in the North-West, and by the middle of September it has disappeared from Southern Ontario. The rich plumage and lively manners of the male make him quite conspicuous while here. Individuals vary considerably in the regularity of their coloring and in the intensity of the blue, but a male in rich spring plumage is a very handsome little bird.

A favorite resort of the species near Hamilton is about the railroad track, near the waterworks reservoir.

Genus SPIZA Bonaparte.

SPIZA AMERICANA (Gmel.).

242. Dickcissel. (604)

Male :—Above, grayish-brown ; the middle of the back, streaked with black ; the hind neck, ashy, becoming on the crown yellowish-olive, with black touches ; a yellow superciliary line and maxillary touch of the same ; eyelid, white ; ear coverts, ashy ; chin, white ; throat, with a large jet black patch ; under parts in general, white, shaded on the sides, extensively tinged with yellow on the breast and belly ; edge of wing, yellow ; lesser and middle coverts, rich chestnut, the other coverts and inner secondaries edged with paler ; bill, dark horn blue ; feet, brown. *Female :*—Smaller ; above, like the male, but head and neck plainer ; below, less tinged with yellow, the black throat patch wanting and replaced by sparse sharp maxillary and pectoral streaks. Length, 6½-7 ; wing, 2¼ ; tail, 2¾.

Hab.—Eastern United States to the Rocky Mountains, north to Massachusetts, New York, Wisconsin and Minnesota, and south in winter through Central America to Northern South America.

Nest, on the ground or in a low bush, built of leaves and fine grass.

Eggs, four or five, greenish-white, sometimes speckled with reddish-brown.

The only record we have at present of the Black-throated Bunting as a bird of Ontario is that furnished by Mr. Saunders, in the *Auk*, for July, 1885, page 307. The writer describes finding the species in June, 1884, at Point Pelee, at the west end of Lake Erie. The birds were tolerably common and evidently breeding, one or two pairs being in every field within a limited district, but it was only after considerable waiting and watching that the party succeeded in discovering a nest with five fresh eggs.

It is just possible that "Dickcissel," like some others, having reached the north shore of Lake Erie, may come along as far as Lake Ontario, but it is rather a weakly, tender species, and we hardly expect to see it much north of the present limit, although there are several records of its capture in Massachusetts and Connecticut.

FAMILY TANAGRIDÆ. TANAGERS.

GENUS PIRANGA VIEILLOT.

PIRANGA ERYTHROMELAS (VIEILL.).

243. Scarlet Tanager. (608)

*Male :—*Scarlet, with black wings and tail ; bill and feet, dark. *Female :—*
Clear olive green ; below, clear greenish-yellow ; wings and tail, dusky, edged
with olive. *Young male :—*At first like the female, afterwards variegated with
red, green and black. Length. 7-7½ ; wing. 4 ; tail, 3.

HAB.—Eastern United States, west to the Plains and north to Southern
Canada ; in winter, the West Indies, Central America and Northern South
America.

Nest, on the horizontal limb of a low tree on the outskirts of the bush, a
shallow, saucer-shaped structure, composed of vine-bark, rootlets and leaves,
lined with vegetable fibre.

Eggs, three to five, dull greenish-blue, spotted with reddish-brown and
lilac.

The Scarlet Tanager is one of our most brilliantly colored birds,
but his rich plumage is all he has to commend him to popular
favor, for he is neither handsome in form nor eloquent in tongue.
Still he sings his song as well as he can, and it probably pleases
the female for whose gratification it is intended, so we let him pass.
In Ontario the species is peculiar to the south and makes but a
short stay, arriving about the 10th of May and leaving again about
the middle of September.

In the fall the bright scarlet of the male's plumage is replaced by
green, but he retains the black on wings and tail.

The food of the species consists chiefly of insects, in the capture
of which considerable dexterity is exhibited. In the fall, when the
wild berries are ripe, the Tanagers take to them with evident relish,
and though they usually keep to the retired parts of the woods,
sometimes at this season they visit the farmer's raspberry patch in
such numbers that they leave but little fruit for household use.

In Southern Ontario they are generally distributed but nowhere
abundant.

In Manitoba they occur only as stragglers during summer. In
the history of American birds, it is stated that at least three years is
required before the male assumes the perfect plumage. In the first
year the young male is like the female, but has the black wings and
tail, while in the following year the red predominates in patches.

PIRANGA RUBRA (Linn.).

244. Summer Tanager. (610)

Male :—Rich rose-red or vermilion, including wings and tail; the wings, however, dusky on the inner webs; bill, rather pale; feet, darker. *Female :*—Dull brownish-olive; below, dull brownish-yellow. *Young male :*—Like the female; the male changing plumage, shows red and green confused in irregular patches, but no black. The female, with general resemblance to female *erythromelas* is distinguished by the dull brownish, ochre or buffy tinge, the greenish and yellowish of *erythromelas* being much purer; the bill and feet also are generally much paler in *rubra*. Size of *erythromelas* or rather larger.

Hab.—Eastern United States to the Plains, north to Southern New Jersey and Southern Illinois, casually north to Connecticut and Ontario, and accidentally to Nova Scotia; in winter, Cuba, Central America and Northern South America.

Nest, on the horizontal bough of a tree, composed of strips of bark, rootlets and grass, lined with fine grass and fibre.

Eggs, three to five, light green, spotted with reddish-brown and lilac.

We sometimes meet in the humbler walks of life people with little education, who, from a natural love of the subject, are wonderfully correct in their observation of the birds.

A man of this class, who, at the time of which I speak, lived near a clump of bush on the "mountain," above the reservoir, three miles east of Hamilton, told me that one summer while he lived there a number of red birds, which had not black wings and tail like the common kind, bred near his house. I felt sure he was describing the Summer Red Bird, and looked through that bush with interest every subsequent spring, but it was not till May, 1885, that I found the first and only individual of the species I ever saw in Canada. It was a female in fine adult plumage, and was among a group of Scarlet Tanagers, which apparently had just arrived from the south, and were enjoying the last rays of the setting sun that gilded the topmost twigs of a dead tree in the bush already referred to. Individuals have been found straggling as far north as Massachusetts and Connecticut, but the home of the species is farther south, and the above is the only record for Ontario.

Mr. Ridgway says that the male requires several years to attain the full plumage, immature individuals showing a mixture of red and yellow in relative proportions, according to the age.

Family HIRUNDINIDÆ. Swallows.

Genus PROGNE Boie.

PROGNE SUBIS (Linn.).

245. Purple Martin. (611)

Lustrous blue-black ; the *female* and *young* are much duller above, and more or less white below, streaked with gray. Length, 7 or more ; wing, nearly 6 ; tail, 3½, simply forked.

Hab.—Temperate North America, south to Mexico.

Nest, of hay, straw, bits of twine and paper, lined with feathers.

Eggs, four or five, pure white, glossy, oblong, pointed at one end.

The Purple Martin arrives in Southern Ontario about the 10th of May, and though generally distributed is nowhere abundant. Its original nesting place was in a knot-hole or other hollow in a tree, but now, seeking the society of men, it raises its young in boxes put up for its accommodation, or in the interstices of the Gothic architecture of our city buildings.

Its flight is rapid and its aerial evolutions often extremely graceful, while at other times it may be seen sailing, hawk-like, with very little action of the wings.

The Martins are general favorites in town and country, and are made welcome everywhere. Before leaving in the fall they have a grand gathering, which is thus described by Dr. Wheaton in the " Birds of Ohio":

"After the breeding season is over, these birds congregate towards night in large flocks, and having selected a suitable cornice on some high building, make preparations for spending the night. The retiring ceremony is very complicated and formal, to judge from the number of times they alight and rise again, all the while keeping up a noisy chatter. It is not until twilight deepens into evening that all are huddled together in silence and slumber, and their slumbers are often disturbed by some youngster who falls out of bed, amid the derisive laughter of his neighbors, which is changed to petulant scolding as he clambers over them to his perch, tumbling others down. All at once the scene of last night's disturbance is quiet and deserted, for the birds have flown to unknown southern lands, where they find less crowded beds and shorter, warmer nights."

It occurs in Manitoba as a rare summer resident, and that seems to be the northern limit of the species.

Genus PETROCHELIDON Cabanis.

PETROCHELIDON LUNIFRONS (Say.).

246. Cliff Swallow. (612)

Lustrous steel-blue; forehead, whitish or brown; rump, rufous; chin, throat and sides of head, chestnut; a steel-blue spot on the throat; breast, sides and generally a cervical collar, rusty-gray, whitening on the belly. *Young:*—Sufficiently similar. Length, 5; wing, 4½; tail, 2¼.

Hab.—North America at large, and south to Brazil and Paraguay.

Nest, a flask-shaped building of mud, lined with wool, feathers and bits of straw.

Eggs, four or five, white, spotted with reddish-brown.

Early in May, the Cliff Swallow crosses the southern border of Ontario, and gradually works its way up to the far north, breeding in colonies in suitable places all over the country. In towns and villages, the nests are placed under the eaves of outhouses; in the country, they are fastened under projecting ledges of rock and hard embankments. The birds are of an amiable, sociable disposition, as many as fifty families being sometimes observed in a colony without the slightest sign of quarrelling. Two broods are raised in the season, and by the end of August they begin to move off and are seen no more till spring. They are somewhat fastidious in their choice of a nesting place, and on this account are not equally abundant at all points, but still they are very numerous throughout the Province, passing along to the North-West, where Mr. Thompson reports them as breeding abundantly in Manitoba. On the boundary, at Pembina, Dr. Coues noticed them as the most abundant of the family, and he traced them all along the line westward to the Rockies.

In Alaska, Mr. Dall states that he found the species nesting at Nullato, "about the trading stations, and was told by the natives that it nested on the faces of the sandstone cliffs along the Yukon, before the advent of the white man placed at its disposal the convenient shelter of the trading-post. The birds were quick to take advantage of the hospitality offered them, and to change from their primitive nesting sites to civilized domiciles."

GENUS CHELIDON FORSTER.

CHELIDON ERYTHROGASTER (BODD.).

247. Barn Swallow. (613)

Lustrous steel-blue; below, rufous or pale chestnut of varying shade; forehead, chin and throat, deep chestnut; breast, with an imperfect steel-blue collar; tail, with white spots on the inner web of all the feathers, except the inner pair. Sexes alike, *young*, less lustrous, much paler below; tail, simply forked. Wing, 4½-4¾; tail, 2½-5.

HAB.—North America in general, from the Fur Countries southward to the West Indies, Central America and South America.

Nest, in a barn or other outbuilding, composed of pellets of mud and bits of straw, and lined with feathers.

Eggs, four or five, white, spotted with reddish-brown.

While the Cliff Swallow chooses to fix its nest *outside* the building under the eaves, the present species prefers the *inside*, where its dwelling is seen attached to the beams and rafters.

They, too, are to some extent gregarious, as many as twenty or thirty pairs being often found nesting together in the same outhouse.

The Swallows, as a class, from their great rapidity of flight and graceful aerial evolutions, are the most easily recognized of all our birds, and this species is perhaps the most accomplished of the group. It is seen skimming over the fields and meadows at a rate which leaves the "lightning express" far behind, and suddenly checking its course it will dart, with surprising rapidity, to right or left in pursuit of some passing insect. It likes to be near a still, sheltered pond, where it can drink and bathe while on the wing. Beautiful it is, on a still summer evening, to see these birds take their plunge bath, and, almost without checking their speed, rise gracefully from the surface of the water, shake sparkling drops from their burnished backs, and continue their airy gambols till the fading light calls them to their humble home. They arrive in Ontario early in May, and are generally distributed over the country during summer, but about the end of August they begin to move toward the south, and soon all have disappeared.

During the summer this species seems to be somewhat irregularly distributed, for in Manitoba Mr. Thompson speaks of it as a rare spring visitant, and Dr. Coues says it is a very rare summer resident at Moose River and various other points along the boundary line, while in Alaska Mr. Nelson says that it is the most common and widely distributed of all the swallows throughout the north.

In Alaska it is found along the south-eastern coast, extending thence over nearly the entire Aleutian chain, and north along the coast of the mainland to Kotzebue Sound, and thence throughout the territory wherever suitable situations occur.

Genus TACHYCINETA Cabanis.

TACHYCINETA BICOLOR (Vieill.).

248. Tree Swallow. (614)

Lustrous green ; below, pure white. *Young :*—Similar, not so glossy. Length, 6-6½ ; wing, 5; tail, 2½.

HAB.—North America at large, from the Fur Countries southward, in winter to the West Indies and Central America.

Nest, of leaves and grass, lined with down and feathers.

Eggs, four to seven, white, unspotted.

A common summer resident, arriving early in May and leaving about the middle of September.

The White-bellied Swallows must at times have considerable trouble in finding suitable places for their summer abode, but it may be that, like people who move often, they have come to enjoy the occupation of house-hunting. The original nesting place was a hole in a tree or stub near water, but as the birds are incapable of making such an excavation themselves, they had to search for a natural aperture, or the deserted hole of a woodpecker to suit their purpose, the finding of which must have been to some extent accidental. As the country became settled, and the swallow trees were cleared away, the birds betook themselves to breeding in boxes, which in the east were put up in great numbers for their accommodation. On the advent of the English Sparrow, many pairs of swallows were summarily ejected from their boxes, and were obliged to retire to the remote parts of the country, and resume their primitive habit of nesting in trees. On this account they are not so common in towns and villages as they were some years ago, but are more generally distributed throughout the country. In Southern Ontario they are seen in greatest numbers during the season of migration.

This species is reported as common in Manitoba and the North-West, and reaches its northern limit in Alaska, where its distribution is limited to those portions where proper accommodation is found for its nesting.

It still retains its ancient habit of occupying holes in trees or banks, and where these are absent the birds are unknown.

In the fall, it is the first of the swallows to leave for winter-quarters.

GENUS CLIVICOLA FORSTER.

CLIVICOLA RIPARIA (LINN.).

249. Bank Swallow. (616)

Lustreless gray, with a pectoral band of the same; other under parts, white. Sexes exactly alike. *Young:*—Similar, the feathers often skirted with rusty or whitish. Length, 4½-4¾; wing, 3⅞-4; tail, 2.

HAB.—Northern hemisphere, in America, south to the West Indies, Central America and Northern South America.

Nest, a few bits of straw and some feathers placed at the end of a tunnel, two to four feet deep, dug by the birds in a sand bank.

Eggs, four to six, pure white.

A common summer resident is the Bank Swallow, breeding abundantly in suitable places all over the country. It arrives about the end of April and leaves in September, both dates being dependent, to some extent, on the weather.

Near Hamilton this species is very abundant, a favorite nesting place being in the gravel bank which is cut through to form the canal to Dundas. There are also many sand banks around the Bay shore, perforated to an extent which shows that flocks of young ones are raised there every summer.

Dr. Wheaton, in the "Birds of Ohio," mentions that this species, from being a common summer resident in the immediate vicinity of Columbus, has become only a passing migrant in spring and fall. This he attributes partly to the frequent disturbance of the nesting places by freshets, and partly to the advent of the Rough-winged Swallow, which, though comparatively a new species at Columbus, is rapidly increasing in numbers. The Bank Swallows are sprightly little birds, greatly attached to their homes, and we hope that nothing will happen here to cause them to change their residence.

This is another species which is common to Ontario and the North-West, and is also found in Alaska. In the latter territory it is rare along the sea coast, but on the river courses in the interior it is one of the most abundant species of Swallow. There, as elsewhere, it raises its brood in holes dug in a sand bank. It arrives at the mouth of the Yukon from the 20th to the 25th of May, and leaves for the south about the end of August.

GENUS STELGIDOPTERYX BAIRD.

STELGIDOPTERYX SERRIPENNIS (AUD.).

250. Rough-winged Swallow. (617)

Lustreless brownish-gray, paler below, whitening on the belly ; rather larger than the last. Hooklets on outer web of outer primary wanting, or much weaker in the female.

HAB.—United States at large (in the Eastern States north to Connecticut), south to Guatemala.

Nest, in holes dug by the birds in the sandy banks of creeks and rivers, a few straws and feathers at the end of the excavation representing the nest.

Eggs, five or six, pure white.

This species seems to be gradually advancing from the west to the east, for we hear every now and then of it being observed at points in the Eastern States where it has not before been noticed. I have no record of it from any part of Ontario except from London, where Mr. Saunders has found it breeding for the last year or two. It is not so decidedly attached to the sand or gravel bank for a breeding place as the Bank Swallow, the nests having been found in crevices of rocks, on beams under bridges, and even in a hole in a brick wall.

It bears a very close resemblance to the Bank Swallow, and as there are not many of them killed, it is possible the Rough-winged species may be more common than we imagine. When closely examined, the curious little hooklets on the outer web of the first primary, which are most fully developed in the male, are always sufficient to identify the species.

FAMILY AMPELIDÆ. WAXWINGS, ETC.

SUBFAMILY AMPELINÆ. WAXWINGS.

GENUS AMPELIS LINNÆUS.

AMPELIS GARRULUS (LINN.).

251. Bohemian Waxwing. (618)

General color, brownish-ash, shading insensibly from the clear ash of the
tail and its upper coverts and rump into a reddish-tinged ash anteriorly, this
peculiar tint heightening on the head, especially on the forehead and sides of
the head, into orange-brown. A narrow frontal line, and broader bar through
the eye, with the chin and throat sooty-black, not sharply bordered with
white ; no yellowish on belly ; under tail coverts, orange-brown or chestnut ;
tail, ash, deepening to blackish-ash towards the end, broadly tipped with rich
yellow ; wings, ashy-blackish : primaries tipped (chiefly on the outer webs)
with sharp spaces of yellow or white, or both ; secondaries with white spaces
at the ends of the outer webs, the shafts usually ending with enlarged, horny,
red appendages ; primary coverts, tipped with white : bill, blackish-plumbeous,
often paler at base below ; feet, black ; sexes alike. Length, 7 or 8 inches ;
wing, about 4¾ ; tail, 2½.

HAB.—Northern parts of the northern hemisphere ; in North America,
south, in winter, irregularly, to the Northern United States.

Nest and eggs, similar to those of the cedar bird.

This handsome, eccentric, garrulous wanderer is common to the
high latitudes of both continents, often appearing unexpectedly in
very large flocks, and disappearing quite as mysteriously, not to be
seen again for many years in succession.

The Ontario records are mostly of small flocks which occasionally
visit us during the winter, and feed on the berries of the red cedar
or the mountain ash. Sometimes they move by themselves, sometimes in company with the Pine Grosbeaks ; the Waxwings taking
the pulpy part of the berries and the Grosbeaks preferring the hard
seeds. The nest of this species was found by Mr Kennicott on the
Yukon, and by Mr. Macfarlane on the Anderson River, but when
we read the accounts of the vast flocks which have been seen by
travellers, we have to admit that it is little we know of their summer
haunts and homes.

I have always had a great admiration for these northern strangers,
as they appear from time to time during the winter, but those I
had captured became insignificant when compared with a few adult
specimens sent to me by Mr. Allan Brooks from British Columbia,

in which all the points of beauty peculiar to the species were most fully developed.

The nest and eggs of this species are still a *desiderata* in American collections, the only specimens we have for reference being those taken by Kennicott at Fort Yukon, July 4th, 1861. The nest was placed on the side of a branch of a small spruce growing on the edge of a clump on low ground. It was large, the base being made of dry spruce twigs, and the nest itself composed of fine grass and moose hair, lined with large feathers. The female was shot as she left the nest. The one egg obtained had a stone-colored ground, spotted with dark brown of a violet shade.

AMPELIS CEDRORUM (VIEILL.).

252. Cedar Waxwing. (619)

General color as in *garrulus*; under tail coverts, whitish; little or no orange-brown about head; no white on wings; chin, black, shading gradually into the color of the throat; a black frontal, loral and transocular stripe as in *garrulus*, but this bordered on the forehead with whitish; a white touch on lower eyelid; feathers on side of lower jaw, white; abdomen, soiled yellowish; tail, tipped with yellow. Length, 7-7½; wing, about 3⅞.

HAB.—North America at large, from the Fur Countries southward. In winter, south to Guatemala and the West Indies.

Nest, large, built in the orchard or in a low tree in the bush; composed of twigs, bark, leaves, rootlets, etc., lined with fine grass, hair or wool.

Eggs, three to five, slaty blue or stone color, spotted and blotched with brownish-black.

The Cedar Bird is generally distributed throughout Ontario. It is a resident species, being here both in summer and winter, but it is so uncertain in its movements that its presence at a particular point at a given time cannot be counted upon with any degree of certainty. The birds do not begin housekeeping until quite late in the season, and may be seen visiting the orchard in flocks up to the end of May. At this season their food consists chiefly of insects, some kinds of which they cleverly capture on the wing. They are also accredited with the destruction of large numbers of canker-worms and other noxious insects. As the season advances, they show a great liking for fruit, especially cherries, with which they often cram themselves till they can hardly maintain their balance on the branches. In the fall and winter the berries of the poke weed, red cedar and mountain ash afford them a bountiful supply of food. Their voice is heard only in a weak call note, easily recognized but difficult to describe.

In many individuals the secondaries finish with a hard horny appendage, having the appearance of red sealing-wax. This is not indicative of age or sex, but is most frequently found in the adult male. In some instances the tail feathers are similarly tipped. The use of these appendages is unknown to us.

FAMILY LANIIDÆ.   SHRIKES.

GENUS LANIUS LINNÆUS.

LANIUS BOREALIS (VIEILL.).

253. Northern Shrike. (621)

Clear bluish-ash, blanching on the rump and scapulars ; below, white, always vermiculated with fine wavy blackish lines ; a black bar along side of the head, not meeting its fellow across forehead, interrupted by a white crescent on under eyelid, and bordered above by hoary white that also occupies the extreme forehead ; wings and tail, black, the former with a large spot near base of primaries ; and the tips of most of the quills, white, the latter with nearly all the feathers broadly tipped with white, and with concealed white bars ; bill and feet, black. Length, 9-10 ; wing, 4½ ; tail, rather more. The *young* are similar, but none of the colors are so fine or so intense ; the entire plumage has a brownish suffusion, and the bill is flesh-colored at base.

HAB.—Northern North America, south, in winter, to the middle portions of the United States (Washington, D.C., Kentucky, Kansas, Colorado, Arizona, Northern California).

Nest, rested on a platform of sticks and twigs in a low tree or bush; composed of weeds, rootlets, bark strips, moss, fine grass and feathers.

Eggs, four to six. The ground color is greenish-gray, but this is almost hidden by the profuse markings of purple and reddish-brown.

In Southern Ontario a few individuals of this species are seen every winter. They arrive from the north in October, and remain with us if the weather be mild, but if it becomes severe about the end of the year, they disappear and are not observed again until March. They like the open country, usually taking their position, sentinel-like, on the topmost twig of a low tree or bush, from which they notice all that moves within a certain radius.

I once saw a pair of these birds unite to hunt down an unfortunate Junco. It took shelter in a patch of scrubby brush, and the Shrikes, not being able to clutch it as a hawk would have done, sought to wear it out by fright and fatigue. As there were two of them taking the work by turns, they would probably have succeeded had I not stopped the proceedings by collecting the two Shrikes, and so saving the life of the Junco. They, no doubt, breed in the northern portion of the Province, but in the south I have not heard of their being found so engaged.

They are common in Manitoba and the North-West, also throughout the interior of Alaska, and Dr. Bell mentions having seen them on the western side of Hudson's Bay. From these distant regions they advance through Ontario in the fall, finding an abundant supply of game awaiting them in *passer domesticus*, whose ranks are thus thinned without apparent reduction in their numbers.

LANIUS LUDOVICIANUS (Linn.).

254. Loggerhead Shrike. (622)

Slate-colored, slightly whitish on the rump and scapulars; below, white, with a few obscure wavy black lines, or none; black bar on one side of the head, meeting its fellow across the forehead, not interrupted by white on under eyelid, and scarcely or not bordered above by hoary white; otherwise like *borealis* in color, but smaller; 8-8½; wing, about 4; tail, rather more.

HAB.—Eastern United States, north to Maine, west to the Prairies of the Upper Mississippi Valley.

Nest, in a tree or bush not often more than fifteen feet from the ground, the middle of a thorn being often selected.

The eggs cannot with certainty be distinguished from those of the White-rumped Shrike.

This and the next species resemble each other so closely that a doubt is raised in the minds of many whether or not they should ever have been separated. Dr. Coues, in his "New Key," says on this subject : "Extreme examples of *excubitorides* look very different from *ludovicianus* proper, but the two are observed to melt into each other when many specimens are compared, so that no specific characters can be assigned." All those I have found near Hamilton agree best with the description given of *excubitorides*, but there are other observers who think we have both kinds, and some believe we have *ludovicianus* only. As a guide to a proper understanding of the matter, I have given the technical descriptions of both, but hold my own opinion that of the two, only *excubitorides* has been found in Ontario.

LANIUS LUDOVICIANUS EXCUBITORIDES (Swains.).

255. White-rumped Shrike. (622*a*)

With the size and essential characters of head stripe of var. *ludovicianus*, and the under parts, as in that species, not or not obviously waved, but with the clear light ashy upper parts and hoary whitish superciliary line, scapulars and rump of *borealis*.

HAB.—Western United States, east to the Middle and New England, breeding as far north as Northern New York and Northern New England and Ontario. Rare or local east of the Alleghanies.

Nest, in a tree or bush, seldom more than ten feet from the ground, exteriorly built of prickly twigs, interwoven with strips of bark, rags, twine and rootlets, lined with fine grass and pieces of cotton waste picked up on the railroad track.

Eggs, four to six, light grayish color, spotted with yellowish-brown.

Besides the great northern Butcher Bird (*Lanius borealis*), there are two Shrikes, smaller in size, described as North American. One is the Loggerhead Shrike of the south-east, and the other the White-rumped Shrike, which was originally described as a western species, but has of late years been extending its territory to the eastward, north of the Loggerhead's range. Taking examples from the south-east to compare with those from the north-west, the difference is seen at once, but as they approach each other in habitat, they also approach each other so closely in appearance that we are almost brought to the conclusion that they are simply different races of the same bird which should not have been separated. Those found in Ontario are of the western race. They were first observed about 1860, and have since become quite common, extending north to the banks of the Saskatchewan, where they were observed by Prof. Macoun. The species is also included in Mr. Thompson's list of the " Birds of Western Manitoba," and is said to be "abundant all over" from May till September.

In Southern Ontario the little Shrike is not found in the city nor in the dense bush, its favorite haunts being along the roadsides in the open country, where it may often be seen on a fence-post or on the telegraph wire by the railway track. My first acquaintance with this bird at its home was made on one of my Saturday after-noon excursions, shortly after its first appearance in this part of the country. While driving along a back road east of the city, my attention was attracted by an ancient negro, who, with a table fork fastened to the end of a fishing pole, was poking vigorously into the centre of a very large, dense thorn-bush near his shanty. Getting over the fence to find out what he was doing, I was informed that a little Chicken Hawk had its nest in there, and that it had killed two of his young chickens. Looking along the pole, I saw in the heart of the dense bush a Shrike's nest with some young ones, which one of the old birds was valiantly defending, biting at the end of the fork when it came too near the youngsters. Taking the pole from the man's hand I worked it into the bush, but it broke before I got it out, and this put an end to hostilities for the time. I tried to con-vince my colored friend that he was mistaken about the bird having killed his chickens, that this kind lived only on grasshoppers and crickets, but he insisted that it was a Chicken Hawk, giving em-phasis to the name by the use of several profane adjectives, and vowing he would have him out before night, even if he should have to burn him out. The appearance of the bush the next time I passed that way indicated that he had carried out his threat.

My opinion regarding the food of this species, which I gave in good faith at the time, I have since had occasion to change, and to believe that after all Sambo was probably right on the subject. For twenty-five years no one could have gone a few miles into the country in any direction near Hamilton, during June, July or August, without seeing one or more pairs of these birds in suitable places, until the year 1886, during which not one was observed.

Trusting that the exodus was only local and temporary, we watched for the return of the birds with interest.

The species evidently overlooked its Hamilton connection in 1886, but it has since then been as common as formerly, watching silently near the roadside for its favorite fare of beetles, mice, or small birds. It is very common in the west, going as far north as Manitoba. In the east it seems more rare, Mr. White not having yet observed it at Ottawa, though he has looked for it during several seasons.

FAMILY VIREONIDÆ, VIREOS.

GENUS VIREO VIEILLOT.

SUBGENUS VIREOSYLVA BONAPARTE.

VIREO OLIVACEUS (LINN.).

256. **Red-eyed Vireo.** (624)

Above, olive-green ; crown, ash, edged on each side with a blackish line, below this a white superciliary line, below this again a dusky stripe through the eye : under parts, white, faintly shaded with olive along sides, and tinged with olive on under wing and tail coverts ; wings and tail, dusky, edged with olive outside, with whitish inside ; bill, dusky, pale below ; feet, leaden olive ; eyes, red ; no spurious quill. Length, 5¾-6¼ ; wing, 3¼-3¼ ; tail, 2¼-2½ ; bill, about ⅜ ; tarsus, ¾.

HAB.—Eastern North America, to the Rocky Mountains, north to the Arctic Regions.

Nest, pensile, fastened by the rim to a horizontal fork, ten to twenty-five feet from the ground ; a thin light structure, composed of bark strips, pine needles, wasp's nest, paper and fine grass, felted and apparently pasted together.

Eggs, three to five, pure white, marked with fine dark reddish-brown spots toward the larger end.

A very common summer resident is the Red-eyed Vireo, and his loud, clear notes are heard in the outskirts of the woods at all hours of the day. Even during the sultry month of July, when most other songsters sing only in the morning or evening, the Red-eye keeps on all day with tireless energy. In Ontario it is the most numerous species of the family, arriving early in May and leaving in September. In the early part of the season its food consists entirely of insects, which it is at all times ready to capture, either on the wing or otherwise. In the fall it partakes of raspberries, and the berries of the poke weed and of other wild plants, with the juice of which its plumage is often found to be stained. It is frequently imposed upon by the Cowbird, whose young ones it rears as tenderly as if they were its own.

In the east it is said to travel as far north as Anticosti. Richardson found it at Fort Simpson, and according to Mr. Fannin, it is a summer resident along the southern border of British Columbia. A great many spend the winter in the Gulf States, and go even farther south.

VIREO PHILADELPHICUS (Cass.).

257. Philadelphia Vireo. (626)

Above, dull olive-green, brightening on the rump, fading insensibly into ashy on the crown, which is not bordered with blackish; a dull white superciliary line; below, palest possible yellowish, whitening on throat and belly, slightly olive-shaded on sides; sometimes a slight creamy or buffy shade throughout the under parts; no obvious wing bars; no spurious quill. Length, 4½-5¼; wing, about 2⅜; tail, about 2¼; bill, hardly or about ½; tarsus ¾.

HAB.—Eastern North America, north to Hudson's Bay, south, in winter, to Costa Rica.

The only record of the nest and eggs of this species I have ever seen is published by Mr. Ernest E. Thompson in the *Auk*, for July, 1885. He says: "On the 9th of June, 1884, while camped near Duck Mountain, I found a nest of this species. It hung from a forked twig, about eight feet from the ground, in a willow which was the reverse of dense, as it grew in the shade of a poplar grove. The nest was pensile, as is usual with the genus; formed of fine grass and birch bark. The eggs were four in number, and presented no obvious difference from those of the Red-eyed Vireo, but unfortunately they were destroyed by an accident before they were measured."

The owners were not secured.

Very many of the more recent additions to the list of our American birds have been made by the discovery that within certain well-known groups were individuals differing in some respects from the others. If these differences were found to be uncertain and irregular they received only a passing notice, but if they were found to be constant they were made the basis on which to build a new species.

Thus, although the American Vireos had passed in review before many distinguished ornithologists, it was not until 1842 that John Cassin found one closely resembling several of the others, but differing in some respects from all of them.

In 1851, he published a description of the bird he had found,

pointed out its peculiarities, claimed for it specific distinction, and
named it after the city near which he first observed it. For many
succeeding years it was again lost sight of, most likely because no one
was looking for it, but as the number of collectors increased and rare
birds were sought after, the species was again observed, and at far
distant points, giving it an extensive range from north to south, and
west to the middle of the continent. How relatively rare it is it
would be unsafe to say, for it is difficult to identify it without
close inspection, to accomplish which might require the slaughter of
Warbling Vireos enough to excite the ire of the Audubon Club.

Some time in the early part of 1883, I took up casually the
Bulletin of the Nuttall Ornithological Club, and read therein a
charming article by Wm. Brewster on the distribution of this Vireo
in the Eastern States. It showed the little bird to be more common
and more widely distributed than was first supposed, and left on my
mind the impression that it must pass through Ontario.

In May, when the Vireos began to arrive, I scrutinized them
closely, and the first I shot on suspicion proved to be of this species,
and was, I believe, the first record for Ontario. When seen in the
woods it looked rather smaller than the Warbling Vireo, was more
solid and compact in the plumage, and was noticeably tinged underneath with yellow. From these features in its general appearance,
I have since recognized it both in spring and fall. I have also heard
of its being taken at other points in Ontario, but have no record of
its having been found breeding within the Province.

Since the above was written, it has been found by Mr. White at
Ottawa, by Mr. Chamberlain at Edmonton, N.B., by Mr. Boardman
at St. Stephens, N.B., and by Mr. Drexler at Moose Factory, H.B.T.
These records show that it is widely distributed, but how abundant
it may be, relatively, is still an uncertain point in its history.

VIREO GILVUS (Vieill.)

258. Warbling Vireo. (627)

Primaries, ten, the exposed portion of the first of which is one-third or less
of the second, no obvious wing bars, no blackish stripe along the side of the
crown, and no abrupt contrast between color of back and crown; upper parts,
greenish, with an ashy shade, rather brighter on the rump and edgings of the
wings and tail, anteriorly shading insensibly into ashy on the crown. Ash of
crown bordered immediately by a whitish superciliary and loral line; region
immediately before and behind the eye, dusky ash; below, sordid white with
faint yellowish (sometimes creamy or buffy) tinge, more obviously shaded

along the sides with a dilution of the color of the back ; quills and tail feathers, fuscous, with narrow external edgings as above said, and broader whitish edging of the inner webs ; the wing coverts without obvious whitish tipping : bill, dark horn color above, paler below; feet, plumbeous ; iris, brown. Length, 5 inches, or rather more ; wing, 2.80 ; tail, 2.25 ; bill, .40 ; tarsus, .67.

Hab.—North America in general, from the Fur Countries to Mexico.

Nest and eggs, closely resembling those of the Red-eye, but usually placed at a greater distance from the ground.

This amiable little songster is very common in Southern Ontario, from the end of the first week in May till the beginning of September. Although less abundant than the Red-eye, it is probably known to a greater number of people, owing to the preference it shows for isolated ornamental trees in parks and gardens and the shade trees in cities. Its song is soft, subdued and flowing, like the murmuring of "a hidden brook in the leafy month of June."

It has little excitement in its nature, and keeps its usual composure under circumstances which would drive most other birds off in alarm. I once saw one warbling forth its pleasing ditty in a shade tree, quite close to which a large fire was in progress. Firemen, engines, and crowds of people were all around, but the bird was to windward of the blaze and seemed to be commenting on the unnecessary excitement which prevailed.

Subgenus LANIVIREO Baird.

VIREO FLAVIFRONS (Vieill.).

259. **Yellow-throated Vireo.** (628)

Above, rich olive-green ; crown, the same or even brighter; rump, insensibly shading into bluish-ash : below, bright yellow ; belly and crissum, abruptly white; sides, anteriorly shaded with olive, posteriorly with plumbeous; extreme forehead, superciliary line and ring around eye, yellow ; lores, dusky ; wings, dusky, with the inner secondaries broadly white-edged, and two broad white bars across tips of greater and median coverts ; tail, dusky, nearly all the feathers completely encircled with white edging ; bill and feet, dark leaden blue : no spurious quill. Length, 5¾-6 ; wing, about 3 ; tail, only about 2¼.

Hab.—Eastern United States, south, in winter, to Costa Rica.

The position and frame-work of the nest of this species is similar to that of the Red-eye, but its appearance and comfort are greatly increased by an artistic outside coating of gray moss, intertwisted with the silk of caterpillars.

The eggs are not with certainty distinguishable from those of the Red-eye.

This is a summer resident in Southern Ontario, but it is by no means common. It seems partial to the beech woods, and being

more retiring than the preceding and less noisy than the Red-eye, it
is not much observed. It is by some considered the handsomest of
all our Vireos, and a male in full spring plumage is pleasing to look
at, but I prefer the succeeding species. The Yellow-throated Vireo,
though not abundant, seems to be generally distributed throughout
Ontario. It has been found at Ottawa by Mr. White; at London
Mr. Saunders reports it as a common summer resident; and it is
also included in Mr. Thompson's " List of Birds of Manitoba," where,
however, it is mentioned as being rare or accidental. The female
differs but little in plumage from the male, the colors being less
decided.

VIREO SOLITARIUS (WILS.).

260. Blue-headed Vireo. (629)

Above, olive-green; crown and sides of head, bluish-ash in marked contrast;
a broad white line from nostrils to and around eye and a dusky loral line;
below, white; flanks washed with olivaceous, and auxiliaries and crissum, pale
yellow; wings and tail, dusky, most of the feathers edged with white or
whitish, and two conspicuous bars of the same across tips of middle and great
coverts; bill and feet, blackish horn color. Length, 5¼-5¾; wing, 2¾-3; tail,
2¼-2½; spurious quill, ½-¾, about ¼ as long as second.

HAB.—Eastern United States to the Plains. In winter, south to Mexico
and Guatemala.

Nest and eggs, similar to those of the other Vireos, resembling those of the
Yellow-throat more than either of the others.

This is a stout, hardy-looking bird, apparently better adapted to
live in the north than any other member of the family. It arrives
from the south with the earliest of the Warblers, and in some years
is quite common during the first half of May, after which it is not
seen again till the fall. While here it is much among the evergreens,
leisurely seeking its food, and is usually silent, but when at home it
is said to have a very pleasant song.

Some of the specimens procured in spring are beautiful birds; the
plumage is soft and silky, and the different shades of color delicate,
but in others it is worn and ragged, as if they had been roughing it
during the winter. Quite a number of these birds cross our southern
border in spring and fall, but I have no record of their nesting in the
Province. It is more of an eastern species, being rare in Michigan
and Manitoba, but, according to Mr. Ridgway, "its known range
during the breeding season extends from Eastern Massachusetts and
the region along the northern border of the Great Lakes northward
nearly to the limit of the tree growth."

FAMILY MNIOTILTIDÆ.   WOOD-WARBLERS.

GENUS MNIOTILTA VIEILLOT.

MNIOTILTA VARIA (LINN.).

261. Black and White Warbler.   (636)

Entirely white and black, in streaks except on the belly ; tail, white, spotted ; wings, white barred,   Length, about 5 ; wing, 2½ ; tail, 2¼.

HAB.—Eastern United States to the Plains, north to Fort Simpson, south, in winter, to Central America and the West Indies.

Nest, on the ground, built of bark fibre, grass and leaves, lined with plant down or hair.

Eggs, four to six, creamy white, spotted and sprinkled with reddish-brown.

This dainty little bird, formerly known as the Black and White *Creeper*, has now been named the Black and White *Warbler*, but as it is much more given to *creeping* than to *warbling*, it is likely that with the ordinary observer it will retain its former name as long as it retains its creeping habit.   It arrives in Southern Ontario during the last days of April, and even before the leaves are expanded, its neat, decided attire of black and white is observed in striking contrast to the dull colored bark of the trees, around which it goes creeping with wonderful celerity in search of its favorite insect food. It becomes very common during the first half of May, after which the numbers again decrease, many having passed farther north, and only a few remain to spend the summer and raise their young in Southern Ontario.   The note of the male is sharp and penetrating, resembling the sound made in sharpening a fine saw.

The Black and White Warbler is a typical representative of the family of Wood-Warblers, which is remarkable for the number of its

23

members, as well as for the richness and variety of their dress. There is, perhaps, no group of small birds which so much interests the collector, or furnishes so many attractive specimens to his cabinet, as that we are now about to consider.

Some of the members of this family are so rare that the capture of one is the event of a lifetime. To get any of them in perfect plumage they must be collected during the spring migration, and that season is so short and uncertain that if a chance is missed in May, another may not occur for a year.

Game birds are followed by sportsmen with much enthusiasm and varying success, though Ruffed Grouse, Woodcock and Quail are now so scarce in the more settled parts of the country that it is hardly worth while searching for them.

Our inland lakes and rivers are, at certain seasons, visited by crowds of water-fowl, and the hunter, hidden behind his screen of rushes in the marsh, delights to hear the hoarse honking of an old gander as he leads on his ∧-shaped flock of geese, or to see the flocks of ducks wheeling around and pitching down into the open water beside his decoys. At Long Point and other shooting places where the ducks have been protected, the number killed in a day is often very large. Dull, windy weather with light showers of rain is preferred. If the hunter is fortunate in choosing a good point at which to screen his boat among the rushes, he may remain there all day, and if the ducks are moving about he needs only to load as quickly as he is able and kill as many as he can, the proof of his success being the number he brings home at night.

Not so with the field ornithologist, whose pursuits I have always felt to be more refined and elevating than those of the ordinary sportsman. As soon as the winter of our northern clime relaxes its grasp, and the season of flowers and brighter skies returns, he enters the woods as if by appointment, and hears among the expanding buds the familiar voice of many a feathered friend just returned from winter-quarters. The meeting is pleasant and the birds pass on. The walk is enjoyable, the bush is fragrant and freckled with early spring flowers. The loud warning note of the Great Crested Fly-catcher is heard in the tree-tops ; Tanagers, Rose-breasted Grosbeaks, etc., are there in brilliant plumage and full of life, but a note is heard or a glimpse is seen of something rare, and that is the time for the collector to exercise his skill. He must not fire when the bird is too close or he will destroy it. He must not let it get out of reach or he may lose it. He must not be flurried or he may miss it, and if he brings it down he must notice particularly the spot where it fell and get there as quickly as he can, for if the bird is only wounded it may flutter away and hide itself, and even if it falls dead it may be covered with a leaf and not seen again, unless the spot where it fell is carefully marked.

All seasons have their attractions, but the month of May above all others is enjoyed by the collector, and bright and rare are the feathered gems he then brings from the woods to enrich his cabinet.

--

Genus PROTONOTARIA Baird.

PROTONOTARIA CITREA (Bodd.) Bd.

262. **Prothonotary Warbler.** (637)

Golden yellow, paler on the belly, changing to olivaceous on the back, thence to bluish ashy on the rump, wings and tail ; most of the tail feathers largely white on the inner webs ; bill, black. Length, 5½ ; wing, 2¾-3 ; tail, 2¼.

HAB.—South Atlantic and Gulf States, north to Ohio, Illinois, Missouri and Kansas. Accidental in Maine and New Brunswick.

Nest, in a hole in a tree or stump, lined with moss, leaves and grass. If the hole is too deep at first, it is filled to within four or five inches of the top.

Eggs, five or six, creamy-white or buff, spotted with rich chestnut-red.

The only record I have of this species in Ontario is that of a female taken by K. C. McIlwraith, which was reported at the time in the *Auk* as follows : "While collecting Warblers near Hamilton, on the morning of the 23rd of May, 1888, I met a group which had evidently just arrived from some favored point in the South, their plumage being particularly fresh and bright, and such rare species as the Mourning and Connecticut Warblers and the Green Black-cap being conspicuous. Presently I noticed, on a willow overhanging the water, one which seemed to be a compromise between the Summer Yellow Bird and the Yellow-throated Vireo. On picking it up, I was greatly pleased to find I had got a specimen of the Prothonotary Warbler—a female in the ordinary plumage of the season.

"It is the first record of the species for Ontario, and the second for Canada, the first being that of a specimen which was found at St. Stephens, New Brunswick, by Mr. Boardman, in October, 1862."

Along the Atlantic coast it is rare or accidental, throughout the Gulf States it is common, but its centre of abundance in the breeding

season is probably reached in the States lying about the junction of
the Ohio and Mississippi Rivers. So it has been reported by Mr.
Brewster, who spent several weeks in that region in the spring of
1878, and as one result of the trip, has given us in the Nutt. Bull.,
Vol. 3, page 153, by far the best account we have of the life habits
of this species. We have not again seen or heard of its being taken
in Ontario, or anywhere near the boundary, so that the instance
recorded above must have been only accidental.

GENUS HELMINTHOPHILA RIDGWAY.

HELMINTHOPHILA CHRYSOPTERA (LINN.).

263. Golden-winged Warbler. (642)

Male :—In spring, slaty-blue, paler or whitish below where frequently tinged
with yellowish ; crown and two wing bars, rich yellow ; broad stripe on side of
head through eye, and large patch on the throat, black ; both these bordered
with white ; several tail feathers, white blotched ; bill, black ; back and wings,
frequently glossed with yellowish-olive in young birds in which the black
markings are somewhat obscure. Length, 4.75; extent, 7.50; wing, 2.40;
tail, 2.00.

HAB.   Eastern United States, Central America in winter.

Nest, on the ground, built of dry leaves and grape-vine bark, lined with
fine grass and horse-hair.

Eggs, four to six, pure white, spotted with reddish-brown or lilac.

A trim and beautifully marked species, very seldom seen in
Ontario and not abundant anywhere. The Golden-winged is spoken
of as one of the rarer Warblers in the Eastern States. Westward
it is said to be common in one district in Indiana. Mr. Saunders
meets with it regularly near London, from which I infer that it is
one of those birds which enter Ontario at the south-west corner, and
having crossed the boundary do not care to penetrate farther into
the Dominion. I have met with it on two occasions near Hamilton ;
have also heard of its being noticed at Port Rowan. Dr. Macallum
sees it every spring and summer near his home at Dunnville. It is
an exceedingly active, restless species, and is most frequently found
among the low shrubbery on the moist ground near some creek or
marshy inlet, where under a broad leaf or tuft of grass the nest is
usually found.

HELMINTHOPHILA RUFICAPILLA (Wils.).

264. Nashville Warbler. (645)

Above, olive-green, brighter on the rump, changing to pure ash on the head ; below, bright yellow, paler on the belly, olive-shaded on the sides ; crown, with a more or less concealed chestnut patch ; lores and ring round the eye, pale ; no superciliary line. *Female :*—And autumnal specimens have the head glossed with olive, and the crown patch may be wanting. Length, 4½-4¾ ; wing, 2⅓-2½ ; tail, 1¾-2.

Hab.—Eastern North America to the Plains, north to the Fur Countries, breeding from the Northern United States northward ; Mexico in winter.

Nest, on the ground, composed of withered leaves and strips of bark, lined with fine grass, pine needles or hair.

Eggs, four or five, white, speckled with lilac or reddish-brown.

The Nashville Warbler, although an abundant species, is not very regular in his visits to this part of Ontario, being sometimes with us in considerable numbers during the season of migration, and again being almost or altogether absent. When they pass this way in the spring, a few pairs usually remain over the summer with us, but the greater number go on farther north. In the fall they are again seen in limited numbers, working their way southward in company with their young, which are distinguished by the absence of the crown patch. In this part of Ontario, we never see so great a number of Warblers in the fall as we do in spring. Either they are less conspicuous on account of the time of their migration extending over a longer period, or they have some other return route by which the majority find their way south.

HELMINTHOPHILA CELATA (Say.).

265. Orange-crowned Warbler. (646)

Above, olive-green, rather brighter on the rump, never ashy on the head ; below, greenish-yellow, washed with olive on the sides ; crown, with a more or less concealed orange-brown patch (sometimes wanting) ; eye ring and obscure superciliary line wanting. Length, 4.80-5.20 ; extent, 7.40-7.75 ; wing, 2.30-2.50.

Hab.—Eastern North America (rare, however, in the North-eastern United States), breeding as far northward as the Yukon and Mackenzie River districts and southward through the Rocky Mountains, wintering in the South Atlantic and Gulf States and Mexico.

Nest, on the ground, composed of leaves, bark fibre and fine grass.

Eggs, four to six, white, marked with spots and blotches of reddish-brown.

The range of this species is chiefly along the west coast or middle district of the continent. In the east it occurs rarely. As a straggler, I have met with it only on two occasions, the latter being on the 11th of May, 1886, when a specimen was taken at the Beach by K. C. McIlwraith. Mr. Saunders mentions having obtained two near London, and Mr. Allan Brooks got one at Milton.

It is a very plainly attired species, and may readily be overlooked, for there is nothing in its dress or manner to attract attention, but on close examination the color of the crown patch is a distinguishing mark not likely to be mistaken. The sexes closely resemble each other, and the young are like them, except that they do not always have the brown crown patch till after the first year.

They are by no means conspicuous birds, but they have their own route to follow, and keep by it regularly every season.

Province, county, township, territory are passed quietly over till they reach the far distant land of Alaska, where Mr. Nelson says they have been observed throughout the wooded region of the north from the British boundary line, west to the shores of Behring Sea, and north within the Arctic circle as far as the tree limit. They breed throughout the interior, and Kennicott secured a set of eggs on the 10th of June and another on the 15th of the same month.

HELMINTHOPHILA PEREGRINA (Wils.).

266. Tennessee Warbler. (647)

Olive-green, brighter behind, but never quite yellow on the tail coverts, more or less ashy towards and on the head; no crown patch; below, white, often glossed with yellowish, but never quite yellow; a ring round the eye and superciliary line, whitish, frequently an obscure whitish spot on outer tail feathers; lores, dusky; in the *female* and *young*, the olivaceous glosses the whole upper parts. Length, $4\frac{1}{2}$-$4\frac{3}{4}$; wing, about $2\frac{4}{7}$; tail, 2 or less.

This comparative length of wing and tail, with other characters, probably always distinguishes this species from the foregoing.

Hab.—Eastern North America, breeding from Northern New York and Northern New England northward to Hudson's Bay Territory; Central America in winter.

Nest, on or near the ground, built of grasses, mosses and bark strips, lined with fine grass and hair.

Eggs, four, white, with markings of reddish-brown about the larger end.

The Tennessee Warbler breeds in the Hudson's Bay Territory, where it is by no means rare, but the line of its migration seems to be along the Mississippi Valley, so that in the east it is seldom seen. I have only met with it twice, once in spring and once in fall. It is probable that a few visit us with the migratory birds every season, but like one or two other species, it may owe its safety to its plain attire, being allowed to pass where one of more gaudy plumage would be stopped.

GENUS COMPSOTHLYPIS CABANIS.

COMPSOTHLYPIS AMERICANA (LINN.).

267. Parula Warbler. (648)

Male in spring:—Above, blue, back with a golden-brown patch, throat and breast yellow, with a rich brown or blackish patch, the former sometimes extending along the sides ; belly, eyelids, two wing bars and several tail spots, white ; lores, black : upper mandible, black ; lower, flesh-colored. *Female in spring:*—With the blue less bright, back and throat patches not so well defined. *Young:*—With these patches obscure or wanting, but always recognizable by the other marks and very small size. Length, 4¼-4¾ ; wing, 2¼ ; tail, 1¾.

HAB.—Eastern United States, west to the Plains, north to Canada, and south, in winter, to the West Indies and Central America.

Nest, globular, with a hole in the side, suspended from the end of a bough, often twenty feet or more from the ground, composed of hanging mosses, so as often to look like an excavation made in the side of a bunch of moss.

Eggs, four or five, creamy-white, with spots of lilac and brown.

This small and neatly dressed species is very common during the spring migration, when it may be seen in the tops of the tallest trees, often hanging back downward like a Titmouse, searching for insects among the opening leaves. In winter it withdraws entirely from Canada, and even from the United States, great numbers being at that season observed in the West Indies.

On the return trip in spring a few pairs stop by the way, but the majority pass on still farther north to breed. I have not heard of the nest being found in Ontario, but I have the impression that this and many others of the same family will yet be found breeding in the picturesque District of Muskoka, between Georgian Bay and the Ottawa River.

Genus DENDROICA Gray.

Subgenus PERISSOGLOSSA Baird.

DENDROICA TIGRINA (Gmel.).

268. Cape May Warbler. (650)

Male in spring:—Back, yellowish-olive with dark spots; crown, blackish, more or less interrupted with brownish ; ear patch, orange-brown ; chin, throat and posterior portion of a yellowish superciliary line, tinged with the same; a black loral line, rump and under parts rich yellow, paler on belly and crissum, the breast and sides streaked with black ; wing bars, fused into a large whitish patch ; tail blotches large, on three pairs of retrices : bill and feet, black. *Female in spring:* -Somewhat similar, but lacks the distinctive head markings ; the under parts are paler and less streaked ; the tail spots small or obscure ; the white on the wing less. *Young:* -An insignificant-looking bird, resembling an overgrown Ruby-crowned Kinglet without its crest ; obscure greenish-olive above ; rump, olive-yellow ; under parts, yellowish-white ; breast and sides with the streaks obscure or obsolete : little or no white on wings, which are edged with yellowish ; tail spots very small.   Length, 5-5¼; wing, 2¾; tail, 2¼.

Hab.--Eastern North America, north to Hudson's Bay Territory, west to the Plains. Breeds from Northern New England northward and also in Jamaica; winters in the West Indies.

Nest, fastened to the outermost twigs of a cedar bough about three feet from the ground, composed of minute twigs of dried spruce, grasses and strawberry vines woven together with spider webs. The rim is neatly formed and the lining is entirely of horse-hair.

Eggs, three to five, creamy-white, marked with lilac and reddish-brown.

This rare and beautiful Warbler is peculiar to the east, not yet having been found west of the Mississippi. In the Eastern States it is occasionally obtained, but is so rare that it is always regarded as a prize, and the collector who recognizes in the woods the orange ear-coverts and striped breast of this species is not likely soon to forget the tingling sensation which passes up to his finger ends at the time.

I have altogether found six in Ontario, but the occasions of their capture extended over a good many years. The above description of the nest and eggs is condensed from an account given by Montague Chamberlain in the *Auk* for January, 1885, of the finding of a nest on the northern boundary of New Brunswick in the summer of 1882.

SUBGENUS DENDROICA GRAY.

DENDROICA ÆSTIVA (GMEL.).

269. **Yellow Warbler.** (652)

Golden-yellow ; back, olive-yellow, frequently with obsolete brownish streaks ; breast and sides, streaked with orange-brown, which sometimes tinges the crown ; wings and tail, dusky, the latter marked with yellow blotches ; bill, dark blue. *Female* and *young*, paler ; less or not streaked below. Length, 5¼ ; wing, 2⅗ ; tail, 2¼.

HAB.—North America at large, south, in winter, to Central America and Northern South America.

Nest, placed in the crotch of a small tree or bush, composed of a variety of soft, elastic materials, including wool, hair, moss, bark fibre and plant down, closely felted together.

Eggs, four or five, greenish-white, spotted and blotched with different shades of reddish-brown.

This is, perhaps, the best known of all the Warbler family, its nest being more frequently found in a lilac bush in the garden than in any more retired situation. About the 10th of May it arrives from the south, and soon makes its presence known by its sprightly notes, the males being in full song at the time of their arrival.

It spends much of its time picking small caterpillars off the foliage of the willows, and is a general favorite on account of its sociable disposition and confiding manners. Unfortunately for its domestic comfort, it is often reluctantly compelled to become the foster-parent of a young Cowbird, but it does not always accept the situation. After the obnoxious egg has been deposited, it has been known to raise the sides of the nest an inch higher, build a second bottom over the top of the egg, and raise its own brood above, leaving the Cowbird egg to rot in the basement.

This ambitious little bird, not satisfied with the United States and Canada as a breeding ground, has extended the area to Alaska, where Mr. Nelson says it is perhaps the most abundant Warbler throughout the territory. It is found everywhere in the wooded interior, on the bushy borders of the watercourses, or among the clumps of stunted alders on the shores of Behring Sea, and the coast of the Arctic about Kotzebue Sound on the south-east coast of the territory. Richardson reports its arrival at Fort Franklin in latitude 60°, on the 15th of May.

This species and the Green Black-capped Flycatcher are the only two of this class which breed in the alder thickets in the vicinity of St. Michael's.

DENDROICA CÆRULESCENS (Gmel.).

270. Black-throated Blue Warbler. (654)

Male in spring:—Above, uniform slaty-blue, the perfect continuity of which is only interrupted, in very high plumages, by a few black dorsal streaks ; below, pure white ; the sides of the head to above the eyes, the chin, throat and whole sides of the body continuously jet black ; wing bars wanting (the coverts being black, edged with blue), but a large white spot at the base of the primaries ; quill feathers, blackish, outwardly edged with bluish, the inner ones mostly white on their inner webs ; tail, with the ordinary white blotches, the central feathers edged with bluish ; bill, black ; feet, dark. *Young male:*—Similar, but the blue glossed with olivaceous, and the black interrupted and restricted. *Female:*—Entirely different : dull olive-greenish with faint bluish shade, below pale, soiled yellowish, recognizable by the white spot at the base of the primaries, which, though it may be reduced to a mere speck, is always evident, at least on pushing aside the primary coverts ; tail blotches, small or obscure ; feet, rather pale. Length, about 5 ; wing, 2½ ; tail, 2¼.

HAB. —Eastern North America to the Plains, breeding from Northern New England and Northern New York northward, and in the Alleghanies to Northern Georgia ; West Indies in winter.

Nest, placed in the fork of a bush near the ground, composed of grape-vine bark and rootlets, lined with vegetable fibre and horse-hair.

Eggs, three to five, creamy-white, with a few spots of reddish-brown toward the larger end.

During the spring migration this species is always fairly represented, and some seasons it exceeds in numbers any other group of the family to which it belongs. It arrives about the 10th of May, and continues common till the 25th, by which time those bound for the north have disappeared. I have heard of individuals being seen in the woods in summer, and think it quite likely that a few pairs breed in suitable places in the southern part of the Province, but the majority unquestionably go farther north. While here the favorite haunt of the species is in the open woods, but it also visits the orchard, and is often seen among the lilac bushes in search of its insect food. In the fall it is in the woods during the greater part of September, after which it disappears and is seen no more till the following spring.

Since the above was written I find the following notice in Davie's "Nests and Eggs of American Birds": "Mr. Wm. L. Kells found the Black-throated Blue Warbler breeding in the thick underbrush of the high timbered land near Listowel, Ontario, in June, and on the 5th of that month, 1886, discovered a compactly built nest of this species in a small maple. On the 9th it contained three eggs of the Warbler and one of the Cowbird."

DENDROICA CORONATA (Linn.).

271. Myrtle Warbler. (655)

Male:—In spring, slaty-blue streaked with black; breast and sides, mostly black; throat and belly, pure white, immaculate; rump, central crown patch and sides of breast, sharply yellow, there being thus four definite yellow places: sides of head, black; eyelids and superciliary line, white; ordinary white wing bars and tail blotches; bill and feet, black; *male* in winter and *female* in summer similar, but slate color less pure or quite brownish. *Young:*—Quite brown above, obscurely streaked below. Length, 5½-5¾; wing, 3; tail, 2¼.

Hab. —Eastern North America chiefly, straggling more or less commonly westward to the Pacific; breeds from the Northern United States northward, and winters from the Middle States and the Ohio Valley southward to the West Indies and Central America.

Nest, in a low tree or bush, composed chiefly of hemlock twigs and soft vegetable fibres, lined with feathers.

Eggs, three to five, creamy-white, marked with brownish-purple.

The familiar Yellow-rump is the first of the family to arrive in spring, often appearing early in April, and for a time it is the one most frequently met with in the woods, where it is observed passing in loose flocks among the upper branches of the trees.

By the middle of May, they have mostly disappeared, and are not again seen in Southern Ontario till the end of September. No doubt many of the warblers spend the summer in the thinly settled, uncultivated tracts of Ontario, but their haunts are so seldom visited by anyone interested in the birds, that it is only occasionally we hear of them.

Quite recently Mr. W. L. Kells found this species breeding near Listowel. The locality in which the nest was found was a clump of black ash, intermingled with cedars and balsams.

Macfarlane found this species nesting on the Anderson River. There its nest was often placed on the ground, but that might be a necessity, for trees or bushes are not always available in that northern region.

They linger late in the fall, as if unwilling to leave, and many probably do not go much beyond our southern boundary, though none have been known to remain here over the winter. On the Pacific coast, this species has been replaced by *Dendroica auduboni* (Audubon's Warbler). These two species resemble each other very closely, the principal difference being that in the western species the throat is yellow, while in ours it is white. Our eastern species has frequently been found on the Pacific coast, but in the east the western one has only once been observed, the record being of a specimen taken near Cambridge, Mass., on the 15th November, 1876.

DENDROICA MACULOSA (GMEL.).

272. Magnolia Warbler. (657)

Male in spring: —Back, black, the feathers more or less skirted with olive, rump, yellow ; crown, clear ash, bordered by black in front to the eyes, behind the eyes by a white stripe ; forehead and sides of the head, black, continuous with that of the back, enclosing the white under eyelid ; entire under parts (except white under tail coverts). rich yellow. thickly streaked across the breast and along the sides with black, the pectoral streaks crowded and cutting off the definitely bounded immaculate yellow throat from the yellow of the other under parts ; wing bars, white, generally fused into one patch ; tail spots small, rectangular, at the middle of the tail and on all the feathers except the central pair ; bill, black ; feet, brown. *Female in spring:* —Quite similar ; black of back reduced to spots in the grayish-olive ; ash of head washed with olive ; other head markings obscure ; black streaks below smaller and fewer. *Young:* —Quite different ; upper parts, ashy-olive ; no head markings whatever, and streaks below wanting or confined to a few small ones along the sides, but always known by the yellow rump in connection with extensively or completely yellow under parts (except white under tail coverts) and small tail spots near the middle of all the feathers except the central. Small, 5 inches or less ; wing, 2½ ; tail, 2.

HAB. —Eastern North America to the base of the Rocky Mountains, breeding from Northern New England, Northern New York and Northern Michigan to Hudson's Bay Territory. In winter, Bahamas, Cuba and Central America.

Nest, usually placed in a low spruce or hemlock, a few feet above the ground, sometimes ten to fifteen feet up in a young hemlock, composed of twigs, rootlets and grass, and lined with horse-hair.

Eggs, four or five, dull white, marked with lilac and brown.

This by many is considered the most gaily dressed of the Warbler family. In Southern Ontario it is a migrant in spring and fall, and usually quite numerous. From its remaining near Hamilton till late in May, and appearing again about the end of August, we may infer that some of the numbers which pass in spring breed at no great distance. Mr. C. J. Young, of the Collegiate Institute, Perth, mentions having found a nest of this species in his neighborhood on the 1st July, 1885. The description of the nest, its position and the four eggs it contained, correspond exactly with that given by others who have seen them elsewhere. So far as I have observed, this is not one of the high fliers, being seldom seen among the tree tops, but mostly in young woods, particularly evergreens, where its colors show to advantage against the back-ground of dark foliage.

When seen in spring, flitting from bush to bush, one is apt to suppose that it will not travel much farther before settling for the summer, but Macoun reports finding it at Lake Mistasini on the 25th May, 1885, and Richardson met with it on the banks of the Saskatchewan on the 26th May, 1827.

DENDROICA CÆRULEA (Wils.).

273. Cerulean Warbler. (658)

Male in spring:—Azure-blue, with black streaks; below, pure white; breast and sides with blue or blue-black streaks; two white wing bars; tail blotches small, but occupying every feather, except, perhaps, the central pair; bill, black; feet, dark. *Female* and *young* with the blue strongly glossed with greenish, and the white soiled with yellowish; a yellowish eye ring and superciliary line. Length, 4-4½.

HAB.—Eastern United States and Southern Canada to the Plains. Rare or casual east of Central New York and the Alleghanies. Cuba (rare) and Central America in winter.

Nest, in the outer fork of a branch, twenty to fifty feet from the ground, composed of bark strips, grass and rootlets, and lined with fine grass and fibre; outside are many pieces of gray moss fastened with spider silk.

Eggs, four, greenish-white, blotched with brown and lilac at the larger end.

The Cerulean Warbler is, I think, a regular summer resident in Southern Ontario, but is somewhat local in its distribution. One spring I searched for it carefully near Hamilton without seeing a single individual, while across the Bay, four miles off, Mr. Dickson reported it as quite common, and breeding in the woods near the Waterdown station of the Grand Trunk railway. Its home and haunts are among the upper branches of the trees, and, except on a blustering, rainy day, it is seldom seen among the lower branches. Its song is almost identical with that of the Parula Warbler, but in the latter species it rises to a slightly higher key at the close, while the Cerulean's ditty is uniform throughout. The colors of the bird are very pleasing when it is seen in a good light, fluttering among the topmost twigs of a beech or maple, the azure-blue and silvery-white seeming like a shred wafted from the drapery of the sky. Dr. Wheaton mentions the species as abundant in Ohio, but generally it is considered rare.

DENDROICA PENSYLVANICA (Linn.)

274. Chestnut-sided Warbler. (659)

Male in spring:—Back, streaked with black and pale yellow (sometimes ashy or whitish), whole crown pure yellow, immediately bordered with white, then enclosed in black ; sides of head and neck and whole under parts, pure white, the former with an irregular black crescent before the eye, one horn extending backward over the eye to border the yellow crown and be dissipated on the sides of the nape, the other reaching downward and backward to connect with a chain of pure chestnut streaks that run the whole length of the body, the under eyelid and auriculars being left white ; wing bands, generally fused into one large patch, and like the edging of the inner secondaries, much tinged with yellow ; tail spots white as usual ; bill, blackish ; feet, brown. *Female in spring:*—Quite similar ; colors less pure ; black loral crescent obscure or wanting ; chestnut streaks thinner. *Young:* Above, including the crown, clear yellowish-green, perfectly uniform or back with slight dusky touches ; no distinct head markings ; below, entirely white from bill to tail, or else showing a trace of chestnut streaks on the sides ; wing bands, clear yellow, as in the adult ; this is a diagnostic feature, shared by no other species, taken in connection with the continuously white under parts ; bill, light colored below. Length, 5-5¼ ; wing, 2½ ; tail, 2.

HAB. –Eastern United States and Southern Canada, west to the Plains, breeding southward to Central Illinois and in the Appalachian Highlands, probably to Northern Georgia. Visits the Bahamas and Central America in winter.

Nest, in the fork of a bush or sapling, three to eight feet from the ground, composed of bark strips and grass, and lined with plant down and hair.

Eggs, four or five creamy-white with reddish-brown marking.

The Chestnut-sided is a common summer resident, breeding in suitable places near the city and throughout the country, and raising two broods in the season. It is very partial to briar patches, but sometimes goes gleaning for insects among the trees, when the blending of its varied plumage with the fresh spring foliage produces a very pleasing effect. It arrives from the south about the 10th of May, and departs early in September.

Although it seems quite at home in Southern Ontario, many individuals must take a wider range, for Dr. Coues reports it as a common summer resident in the woodlands near Pembina, and Kennicott found it at the Lake of the Woods on the 25th of May. The song, when heard in its haunts in the early spring, is pleasing, and delivered with much spirit. Any of these birds whose notes I have once or twice heard I readily recognize again, and trust to the *ear* more than to the *eye* to tell what birds are about me in the bush, but only in a few exceptional cases could I convey to my readers in words or letters any idea of the songs of birds, and I must admit that I make but poor progress in trying to follow others who think they have succeeded in doing so.

DENDROICA CASTANEA (Wils.).

275. Bay-breasted Warbler. (660)

Male in spring:—Back, thickly streaked with black and grayish-olive; forehead and sides of head, black, enclosing a large deep chestnut patch; a duller chestnut (exactly like a Bluebird's breast) occupies the whole chin and throat, and extends, more or less interrupted, along the entire sides of the body; rest of under parts, ochrey or buffy whitish, a similar buffy area behind the ears; wing bars and tail spots, ordinary; bill and feet, blackish. *Female in spring:*—Is more olivaceous than the *male*, with the markings less pronounced, but always shows evident chestnut coloration, and probably traces of it persist in all *adult* birds in the fall. The *young*, however, so closely resemble young *striata* that it is sometimes impossible to distinguish them with certainty. *Castanea* is, however, tinged with buffy or ochrey below, instead of the clear pale yellowish of *striata;* moreover, *castanea* is usually not streaked on the sides at all. Size of *striata*.

HAB.—Eastern North America, north to Hudson's Bay. Breeds from Northern New England and Northern Michigan northward; winters in Central America.

Nest, in a hemlock tree, fifteen or twenty feet from the ground, composed of larch twigs and moss, woven together with spider silk, and lined with fibrous roots.

Eggs, four, bluish-green, thickly spotted with lilac and brown at the larger end.

My observations of this species agree with what has been published regarding it by those who have observed it in the Eastern States. I have found it abundant in spring some years, and in others rare, or entirely wanting, while in the fall it is always scarce, if it is seen at all. This has led to the belief that the species does not always follow the same line of migration in spring, and that in the fall the return trip is made along a line to the west of us, the few we see being only stragglers from the main body. It is a late comer, being seldom seen till after the middle of May, and it is less active in its movements than other members of the family. It is seldom seen on the ground or near it, usually keeping among the upper branches of the trees.

The only time I ever saw more than three or four together was in the spring of 1885, when I observed a flock of fifty or more feeding in a clump of willows overhanging an inlet of the Hamilton Bay.

Listowel seems a favorite locality with the Warblers, and Mr. Kells evidently gives them some attention, for this is another species which he found breeding in a low, swampy, mixed bush not far from his home. Mr. Kells found a nest placed between a slender limb and the trunk of a small cedar about five feet up. Another was found in a hemlock at an elevation of fourteen feet. The nests were built as described above, of rather small size, the interior being only about two inches in diameter by one in depth.

DENDROICA STRIATA (Forst.).

276. Black-poll Warbler. (661)

Male in spring: Upper parts thickly streaked with black and olivaceous-ash; whole crown, pure black; head below the level of the eyes and whole under parts, white, the sides thickly marked with black streaks crowding forward on the sides of the neck to form two stripes that converge to meet at base of the bill, cutting off the white of the cheeks from that of the throat; wing bars and tail blotches, white; inner secondaries, white-edged; primaries usually edged externally with olive; feet and under mandible, flesh color or pale yellowish; upper mandible, black. *Female in spring:* Upper parts, including the crown, greenish-olive, both thickly and rather sharply black streaked; white of under parts soiled anteriorly with very pale olivaceous-yellow, the streaks smaller and not so crowded as in the male. *Young:* Closely resembling the adult female, but a brighter and more greenish-olive above with fewer streaks, often obsolete on the crown; below, more or less tinged with pale greenish-yellow, the streaks very obscure, sometimes altogether wanting; under tail coverts, usually pure white; a yellowish superciliary line; wing bars, tinged with the same color. Length, 5¼-5¾; wing, 2¾-3; tail, 2-2¼.

Hab. Eastern North America to the Rocky Mountains, north to Greenland, the Barren Grounds and Alaska, breeding from Northern New England northward; south, in winter, to Northern South America.

Nest, in an evergreen, eight or ten feet from the ground, built of larch twigs woven together with moss and grass, and lined with fine grass.

Eggs, four or five, variable, usually white, spotted with purple and reddish-brown.

The Black-poll is a regular visitor in Southern Ontario in spring and fall. It is the last of the family to arrive from the south, being seldom seen before the 20th of May. Its stay at that time is of short duration, and when it goes the collector considers the Warbler season is over. In the fall these birds are again seen in increased numbers, many being in the young plumage, and not in such haste to depart, although none remain over the winter.

The musical powers, if they have any, are not exercised in this latitude, the birds while here being mostly silent. They feed largely on winged insects, which are never plentiful till the end of May, and this may account for the Black-polls being late in arriving in spring.

Viewed from the ordinary traveller's standpoint, one would expect that the birds which go farthest north would be the first to start on the journey, but such is not the case. This species, which is the very latest to arrive from the south, keeps moving on the northern route, passing many which left the winter rendezvous before them, and though some may drop off by the way, large numbers keep on till they reach Alaska, and they are found even in Greenland.

Macfarlane noticed this species breeding on the Anderson River, and its eggs were taken at Fort Yukon on the 8th and 10th of June.

DENDROICA BLACKBURNIÆ. (GMEL.).

277. Blackburnian Warbler. (662)

Male in spring: Back, black, more or less interrupted with yellowish; crown, black, with a central orange spot; a broad black stripe through the eye, enclosing the orange under eyelid; rest of head, with whole throat, most brilliant orange or flame color; other under parts, whitish, more or less tinged with yellow, and sides streaked with black; wing bars, fused into a large white patch; tail blotches, white, occupying nearly all the outer feathers; bill and feet, dark. *Female and young male:*—Upper parts, olive and black, streaked; superciliary line and throat, clear yellow, fading insensibly on the breast; lower eyelid, yellow, confined in the dusky ear-patch; wing patch, resolved into two bars; tail blotches, nearly as extensive as in the adult male, the outer feathers showing white on the outer webs at base. Length, 5¼; wing, 2⅝; tail, 2¼.

HAB.—Eastern North America to the Plains, breeding from the northern and more elevated parts of the Eastern United States northward; in winter, south to the Bahamas, Central America and Northern South America.

Nest, in an evergreen, twenty feet from the ground; built of twigs, grass and moss, and lined with fine fibre, hair and feathers.

Eggs, three or four, bluish-green, speckled toward the larger end with reddish-brown and lilac.

This "flying gem," clad in black, and orange of the richest shade, is by many regarded as the most gaily attired of all the Warblers.

24

It is a regular visitor in spring and fall, and though not abundant is very generally distributed.

A few have been observed in Manitoba, and it has also been reported as a straggler in Labrador, but from its lingering late in the spring in Southern Ontario, and appearing again early in September, the bulk of the species probably does not go much farther north to spend the summer. In the *Auk*, Vol. II., page 103, Dr. Merriam gives an account of a nest of this species which he found in Lewis County, N.Y., on the 2nd of June, 1879. It was saddled on a limb of a large pine, eighty-four feet from the ground, and contained four fresh eggs of the owners of the nest and one of the Cowbird.

One of the few errors made by Wilson was his description of the young of this bird as a separate species, which he named the Hemlock Warbler. In this he was for a time followed by other writers, till further observations brought out the truth.

Like most of its class, this species crosses the southern frontier in May, and is again seen passing south in September.

DENDROICA VIRENS (GMEL.).

278. Black-throated Green Warbler. (667)

Male in spring: Back and crown, clear yellow-olive: forehead, superciliary line, sides of head, rich yellow (in very high plumage, middle of back with dusky marks, and dusky or dark olive lines through eyes, auriculars, and even bordering the crown); chin, throat and breast, jet black, prolonged behind in streaks on the sides; other under parts, white, usually yellow-tinged; wings and tail, dusky, the former with two white bars and much white edging, the latter with outer feathers nearly all white; bill and feet, blackish; *male* in the fall and *female* in the spring similar, but black restricted, interrupted or veiled with yellow. *Young:* Similar to the female, but the black more restricted or wanting altogether, except a few streaks along the sides. Length, about 5; wing, 2½; tail, 2¼.

Hab.—Eastern North America to the Plains, north to Hudson's Bay Territory, breeding from the Northern United States northward. In winter, south to Cuba and Panama. Accidental in Greenland and Europe.

Nest, small, neat, compact, placed in a fork of a pine tree, near the end of a branch, often twenty to fifty feet from the ground; composed of twigs, strips of vine bark and dried grass, and lined with vegetable fibre and horse hair.

Eggs, three or four, creamy-white, marked with reddish-brown and lilac, mostly toward the larger end.

The Black-throated Green Warbler is a regular visitor in spring and fall. It appears a few days earlier in spring than some others of its class, and soon announces its arrival by the frequent utterance of its characteristic notes, which are readily recognized when heard in the woods, but are difficult to translate into our language. When the Warblers are on their migratory journey, they use trees of all kinds as resting places, but while seeking food this species evidently prefers the pines, and is most frequently seen among the higher branches. In the fall they are as active as ever in their movements, but are mostly silent, except in the utterance of a simple chirp to advise each other of their whereabouts.

DENDROICA VIGORSII (Aud.).

279. Pine Warbler. (671)

Uniform yellowish-olive above, yellow below, paler or white on belly and under tail coverts, shaded and sometimes obsoletely streaked with darker on the sides; superciliary line, yellow; wing bars, white; tail blotches confined to two outer pairs of feathers, large, oblique. *Female* and *young:* —Similar, duller; sometimes merely olive-gray above and sordid-whitish below. The variations in precise shade are interminable, but the species may always be known by the lack of any special sharp markings whatever, except the superciliary line, and by the combination of white wing bars with large oblique tail spots confined to the two outer pairs of feathers. Length, $5\frac{1}{4}$ to nearly 6 inches.

Hab. —Eastern United States to the Plains, north to Ontario and New Brunswick, wintering in the South Atlantic and Gulf States, and the Bahamas.

Nest, in a pine tree, well up from the ground, built of strips of bark, rootlets and grass, and lined with plant down, hair and feathers.

Eggs, usually four, white, tinged with pink and spotted with reddish-brown and lilac toward the larger end.

The Pine-creeping Warbler is remarkable neither for the gaiety of dress nor the activity of movement which distinguish most of the others of its class. It is a large, quiet Warbler, yellowish-green above, and greenish-yellow below, and is most frequently observed creeping on the trunks or branches of the pine trees, searching for insects among the crevices of the bark. It does not seem to be generally distributed, for Dr. Wheaton speaks of it as being rare in Ohio, and Mr. Saunders has not met with it near London, while at Hamilton it is rather a common species, and raises its young near the city every season.

It arrives from the south quite early in spring, and for a time is quiet, but as the weather gets warmer the male indulges in a rather pleasant little song, resembling that of the Chipping Sparrow. In the fall they disappear about the middle of September.

DENDROICA PALMARUM (Gmel.).

280. Palm Warbler. (672)

Adult male:— In spring, beneath, yellowish-white, tinged with yellow, the throat and crissum deepening into gamboge; sides of the neck, sides and entire breast streaked with umber-brown, tinged with rusty, the shafts of the feathers darker; a distinct superciliary stripe of clear yellow; pileum, uniform rich chestnut, darker next the bill, when divided medially by a short and indistinct streak of yellow; upper parts in general, olive-gray, deepening into yellowish olive-green on the upper tail coverts; tail feathers, dusky, edged externally with pale olive-yellowish, the two outer pairs with their inner webs broadly tipped with white; wings, dusky, the remiges edged like the tail feathers with yellowish olive-green, both rows of coverts tipped with pale grayish-buff, forming rather distinct indications of two bands. Wing, 2.55; tail, 2.30.

Hab.—Northern interior to Great Slave Lake; in winter and in migrations, Mississippi Valley and Gulf States, including Western and Southern Florida and the West Indies. Casual in the Atlantic States.

Nest, on the ground under a bush or tussock of grass, composed of grass, lined with rootlets, hair, caterpillar silk and moss.

Eggs, three, rosy white, spotted with brown and reddish at the larger end.

From the way in which western birds creep up into Ontario around the west end of Lake Erie, I think it highly probable that this species will be found here. I have noticed some individuals less bright in the yellow than others, but at present the number of specimens available for comparison is so small that I cannot say positively that we have both species of Palm Warbler, and have some doubts as to whether or not the recognized authorities have acted wisely in making the separation.

DENDROICA PALMARUM HYPOCHRYSEA (Ridgw.).

281. Yellow Palm Warbler. (672a)

Adult male:—In spring, entire lower parts and a conspicuous superciliary stripe bright yellow, entirely continuous and uniform beneath; entire sides marked with broad streaks of deep chestnut, these most distinct on the sides

of the breast; auricular mixed olive and chestnut (the latter prevailing),
somewhat darker immediately behind the eye; lore, with an indistinct dusky
streak; entire pileum, rich chestnut, becoming darker next the bill when
divided medially by a short and rather indistinct yellow streak; rest of the
upper parts, olive, tinged with brown on the back and brightening into
yellowish olive-green on the rump and upper tail coverts, the latter having
shaft streaks of reddish-chestnut; tail feathers, dusky, edged externally with
yellowish-olive, the inner webs of the two outer feathers broadly tipped with
white; wings, dusky, all the feathers edged with pale brownish-olive, this
edging rather widest on the ends of the middle and greater coverts, where,
however, they do not form any indication of bands. Wing, 2.65; tail, 2.50.

HAB.—Atlantic States north to Hudson's Bay. Breeds from New Brunswick and Nova Scotia northward; winters in the South Atlantic and Gulf
States.

Nest, on the ground, built of bark fibre, grass and moss, and lined with
hair and feathers.

Eggs, four, creamy-white, blotched with reddish-brown and lilac at the
larger end.

As this interesting bird is said to be abundant in the Eastern
States as far west as the Plains, we should expect to find it also
plentiful in Ontario, but I have not so observed it near Hamilton.
Occasionally, late in the fall or early in spring, it is seen running
about on the ground, by the roadsides or in bare weedy fields, but
it is not at any time abundant, and sometimes is altogether absent.
It is very different, in many respects, from the other members of
the group in which it has been placed; the building of its nest on
the ground and the jerky motions of its tail suggesting relationship
with the Tit Lark.

Some ten years ago, while examining a large series of specimens
of this species in the National Museum, Mr. Ridgway observed a
wide difference in the intensity of the coloring of different individuals
in the group. An examination, as to the localities from which they
had been obtained, showed that the highly colored individuals were
from the east of the Alleghanies, while those in plainer attire were
all from farther west. A comparison of specimens in the possession
of different collectors in these districts, showed that the differences
referred to were constant, and this has led to the variety we are
now considering being described as a sub-species, under the name of
Dendroica palmarum hypochrysea; the original *Dendroica palmarum*
of Gmelin being supposed to be the plain colored form observed in
the west. I have described both, so that collectors may satisfy
themselves as to whether we have here the eastern form, or the
western, or both.

Genus SEIURUS Swainson.

SEIURUS AUROCAPILLUS (Linn.).

282. Oven Bird. (674)

Crown, orange-brown, bordered with two black stripes, no superciliary line; above, bright olive-green; below, pure white, thickly spotted with dusky on breast and sides; a narrow maxillary line of blackish; under wing coverts, tinged with yellow; a white eye ring; legs, flesh color. Sexes alike. *Young:*— Similar. Length, 5½-6½; wing, 3; tail, 2¾.

Hab. -Eastern North America, north to Hudson's Bay Territory and Alaska, breeding from Kansas, the Ohio Valley and Virginia northward. In winter, Southern Florida, the West Indies and Central America.

Nest, on the ground, usually on a sloping bank, frequently roofed over with an entrance at the side; composed of twigs, leaves and moss, and lined with fine grass and hair.

Eggs, four or five, creamy-white, spotted with reddish-brown and lilac-gray.

The Oven Bird, so called from its habit of building its nest somewhat in the form of an oven, is a summer resident in Ontario, and is very generally distributed, being found in suitable places all over the country, from the early part of May till the beginning of September. To see it walking gingerly on the ground, jerking its tail after the manner of the Tit Lark, conveys the impression that it is a quiet, retiring little bird, with clear, handsome markings; but should it mount to one of the middle branches of a tree, it is astonishing to observe with what emphasis and energy it delivers its notes. With a little help from the imagination, its song resembles the word *teacher*, frequently repeated with increasing emphasis. This loud, clear call may often be heard in the moist woods during the month of May, but the bird is said to have also another song, more soft and musical, which must be reserved for special occasions, for I have not met with anyone who has heard it.

In Alaska it is known to breed from Fort Yukon some distance down the river, where the natives call it the Grandfather of the Ruby-crowned Kinglet.

SEIURUS NOVEBORACENSIS (Gmel.).

283. Water-thrush. (675)

Entire upper parts, deep olivaceous-brown; conspicuous superciliary line, yellowish; below, white, more or less tinged with pale yellowish, thickly and sharply spotted with the color of the back, except on lower belly and crissum; feet, dark. Length, 5½-6; wing, 2¾; tail, 2¼; bill, about ½.

Hab. —Eastern United States to Illinois, and northward to Arctic America, breeding from the Northern United States northward; south, in winter, to the West Indies and Northern South America.

Nest, on the ground, often under the exposed roots of a tree, built of leaves, moss and grasses, and lined with fine grass and rootlets.

Eggs, four to six, of crystalline whiteness, marked with reddish-brown or lilac.

This inhabitant of the moist woods and swampy thickets is found in all suitable places throughout the country, but it has not the loud decided notes of the Oven Bird, and is therefore less known, though quite as abundant. It is terrestrial in its habits, being often seen walking with careful steps by the edge of the pools, or along wet logs, nervously jerking its tail, after the manner of the Teeter Snipe.

In appearance it closely resembles the next species, with which it has often been confounded, but the distinction, once clearly understood, is afterwards readily recognized. In the present species the throat and breast are streaked from the bill downwards, while in the Louisiana the throat is always unstreaked.

This species also reaches Alaska, where Mr. Nelson says of it, after quoting the observations of others: "My own experience at the Yukon mouth proves the bird to be one of the commonest species breeding at that place. Its favorite haunts, in the midst of dense thickets, shelter it from the observation of one not accustomed to its song, which, however, is one of the most striking that reaches the ear of the traveller in that region; but the songster, perched on some low branch, is quick to take alarm, and skulks away beyond the sight of one penetrating its haunts."

SEIURUS MOTACILLA (Vieill.).

284. Louisiana Water-thrush. (676)

Very similar to the last; rather larger, averaging about 6, with the wing 3: bill, especially longer and stouter, over ½, and tarsus nearly 1; under parts, white, only faintly tinged, and chiefly on the flanks and crissum, with buffy-yellow; the streaks sparse, pale and not very sharp; throat, as well as belly and crissum, unmarked: legs, pale.

Hab.—Eastern United States, north to Southern New England and Michigan, west to the Plains. In winter, West Indies, Southern Mexico and Central America.

Nest, on the ground, composed of twigs, moss and leaves, and lined with fine grass and the fur of some quadruped.

Eggs, four or five, white, tinged with rose color and lightly marked with reddish-brown, chestnut or lilac gray.

Southern Ontario is perhaps the northern limit of this species, and even here it is not generally distributed. My first acquaintance with it was made early on a bright May morning, a good many years ago. I had gone out under the mountain, west of Hamilton, and was crossing a deep ravine, which there cut through the escarpment, when I heard farther up the glen the clear, rich, liquid notes of a bird that was then entirely new to me. Following with some difficulty the course of the stream, which was heard trickling beneath the moss-grown rocks at the bottom of the ravine, I came, at length, in sight of the musician. He was on the prostrate trunk of a tree, which, years before, had fallen and bridged over the chasm, and was then moss-grown and going to decay. On this carpeted platform the bird moved about with mincing steps, often turning around with a jerk of the tail, and uttering his characteristic notes with such energy that, for a time, the whole ravine seemed filled with the sound. I have seen the species many times since then, but the recollection of our first meeting has lingered long in my memory, and this particular bird still occupies a prominent place in my collection.

The Large-billed, or Louisiana Water-thrush, as it is now called, is by no means so common a bird in Ontario as the preceding species, but along the southern border of the Province, wherever there is a rocky ravine, its loud, clear notes are almost sure to be heard in the spring, mingling with the sound of the falling water. It arrives from the south early in May and leaves in September.

GENUS GEOTHLYPIS CABANIS.

SUBGENUS OPORORNIS BAIRD.

GEOTHLYPIS AGILIS (WILS.).

285. **Connecticut Warbler.** (678)

Above, olive-green, becoming ashy on the head; below, from the breast, yellow, olive-shaded on the sides; chin, throat and breast, grayish-ash; a whitish ring round eye; wings and tail, unmarked, glossed with olive; under mandible and feet, pale; no decided markings anywhere. Length, 5½; wing, 2⅞; tail, 2.

Hab.—Eastern North America, breeding north of the United States.

Nest, sunk in the ground level with the surface, composed entirely of dry grass.

Eggs, four, creamy-white, a few spots of lilac-brown and black, inclining to form a ring round the larger end.

The Connecticut Warbler was discovered by Wilson, and named by him after the State in which he found it. It is a widely distributed species, but is nowhere abundant, though it seems to be more common in the west than in the eastern portion of its habitat. It is of shy, retiring habits, frequenting low, swampy places and keeping near the ground.

On one or two occasions I have met with the adults in spring, and have seen them again in the fall, accompanied by their young. In their haunts and habits they closely resemble the Mourning Warbler, and in certain stages of plumage they are also like the latter in appearance, but the present species can always be recognized by its wings, which are longer and more pointed.

This species undoubtedly breeds in Ontario, and as the nest and eggs are still unknown to naturalists, they are a prize which our Canadian boys should try to secure. I found the young in August, and they certainly looked as if they had not travelled far.

Since the above was written, the nest of this species has been found at Duck Mountain, in Manitoba, by Mr. Ernest E. Thompson, who in the *Auk*, for April, 1884, gives an interesting account of the occurrence which happened on June 21st, 1883. The life history of this graceful species has so long continued in obscurity, that it was with exceptional pleasure Mr. Thompson found himself enabled to cast light upon several of the most important of its habits. He had the good fortune to find the nest, the first of its kind known to science, and it was subsequently sent to the Smithsonian Institute, where the identification was confirmed, and the nest finally deposited.

GEOTHLYPIS PHILADELPHIA (Wils.).

286. Mourning Warbler. (679)

Bright olive; below, clear yellow; on the head, the olive passes insensibly into ash; in high plumage, the throat and breast are black, but are generally ash, showing black traces, the feathers being black, skirted with ash, producing a peculiar appearance, suggestive of the birds wearing crape; wings and tail, unmarked, glossed with olive; under mandible and feet, flesh color; no white

about eyes. Young birds have little or no ashy on the head and no black on
the throat. thus nearly resembling the *Oporornis agilis*. Length, 5¼-5½; wing
and tail, each about 2¼.

HAB. –Eastern North America to the Plains, breeding from the mountainous portions of Pennsylvania, New England and New York, and Northern
Michigan northward. Central America and northern South America in winter.

Nest. on or near the ground, built of leaves and weed stalks, and lined with
fine black rootlets.

Eggs. three: "light flesh color, uniformly speckled with fine brown specks."

Very little is yet known of the nest and eggs of the Mourning
Warbler. The above description is given by John Burroughs of a
nest found by him in New York State, which is farther south than
these birds usually spend the summer.

Some years ago, while waiting for the train at a way station on
the Kincardine branch of the Grand Trunk Railway, I strolled into
the neighboring woods to pass the time. Sitting on a prostrate log
on the sunny side of a ravine, birds of many kinds fluttered about,
and a pair of Mourning Warblers soon attracted my attention by
the displeasure and anxiety they manifested at being disturbed. I
changed my position, and the female moved cautiously towards the
place I had left. A few minutes more and I certainly should have
seen the nest, but the engine whistle sounded, and being some
distance from the station I had to leave. Next day, as the train
slowly passed the place, the male was again observed singing on his
former perch.

Any one who has given attention to the movements of the birds
for a number of years, must have been surprised at the persistent
regularity with which certain species appear at particular places at
a given time, especially in spring.

For many years after I commenced collecting birds, I considered
the Mourning Warbler only a straggler in this part of Ontario,
having met with it on but two occasions. More recently, I have
carefully studied the topographical aspect of the neighborhood with
special reference to the habits of the birds, and have calculated
where certain species should be found at certain dates. One result
of this was, that on two visits made to a particular place in May,
1885, K. C. McIlwraith obtained nine Mourning Warblers in a very
short time. In the spring of 1886 they were again observed at the
same place, but were not molested. The name Mourning does not
refer to the manners of the bird, for it sings with much spirit and
is quite lively in its movements, but was suggested by the ashy tips

to the black feathers of the throat, resembling the effect produced by wearing crape.

Mr. Thompson found this species quite common in Manitoba, but difficult to collect, owing to its habit of keeping back in the underbrush.

GEOTHLYPIS TRICHAS (LINN.).

287. Maryland Yellow-throat. (681)

Male in spring: Olive green, rather grayer anteriorly; forehead and a broad band through the eye to the neck, pure black, bordered above with hoary ash; chin, throat, breast, under tail coverts and edge of wing, rich yellow, fading into whitish on the belly; wings and tail, unmarked, glossed with olive; bill, black; feet, flesh colored. *Female in spring:*—Without the definite black and ash of the head; the crown, generally brownish, the yellow pale and restricted. The *young* in general resemble the female, at any rate lacking the head markings of the male, but are sometimes buffy-brownish below, sometimes almost entirely clear yellow. Length, $4\frac{3}{4}$-5; wing and tail, $1\frac{3}{4}$-$2\frac{1}{4}$.

HAB.—Eastern United States, mainly east of the Alleghanies, north to Ontario and Nova Scotia, breeding from Georgia northward. In winter, South Atlantic and Gulf States, and the West Indies.

Nest, on the ground, composed of leaves and grass, and lined with fine withered grass, sometimes partly roofed over.

Eggs, four to six, white, thickly sprinkled with reddish-brown, chiefly at the larger end.

The Maryland Yellow-throat is widely but somewhat irregularly distributed. I have heard its familiar notes on the banks of the St. Lawrence, near Quebec; by the marshy ponds between Galt and Paris I have found it breeding abundantly; but near Hamilton, where there are places which to us appear equally suitable for its summer residence, it is only observed as a casual migrant in spring and fall. It is a very lively little bird, and makes its summer haunts ring with its loud, clear *whit-ti-tee*, often repeated, which once heard is not soon forgotten by any one who has an ear for bird music. It arrives during the first week in May, and disappears about the end of August.

It is very common in Manitoba, but beyond that to the north, I have not heard of its having being observed.

GENUS ICTERIA Vieillot.

ICTERIA VIRENS (Linn.).

288. Yellow-breasted Chat. (683)

Bright olive-green; below, golden-yellow; belly, abruptly white; lore, black, isolating the white under eyelid from a white superciliary line above and a short maxillary line below; wings and tail, unmarked, glossed with olive; bill and feet, blue-black. *Female* and *young*:— Similar, colors less bright. Length, 7-7½; wing, about 3; tail, about 3¼.

Hab.— Eastern United States to the Plains, north to Ontario and Southern New England, south in winter to Eastern Mexico and Guatemala.

Nest, in a thicket, in the upright fork of a sapling, three to six feet from the ground; composed of leaves, strips of grape vine bark and grass, lined with fine withered grass and fibre.

Eggs, three or four, very smooth, white, spotted and blotched with several shades of reddish-brown, heaviest toward the larger end.

Bird collecting is attended with all the excitement of other speculations. The very uncertainty, as to the amount of success attainable, tends to increase the feeling.

Laying aside accidents by gun, boat or buggy, much time and labor are sometimes expended with very slim results, while on the other hand the prizes are often obtained quite unexpectedly. On the 16th of May, 1884, I went for a short excursion to the woods, impressed with the idea that I had lately spent too much time collecting common species which I already had, and that, by a more careful inspection of the birds I came across, I should have a better chance of finding something new. I observed quite a number that afternoon, but came back without a specimen of any kind, and, as it began to rain, I got home thoroughly damped, and unhitched my horse, firm in the belief that the subject was unworthy the attention I was giving to it. Just then I noticed an olive-backed bird lying dead on the ground close by, and on picking it up found it to be the decaying body of a Yellow-breasted Chat, that had probably been killed by flying against the telegraph wire which crossed above the spot where it was found. It had evidently been there for two or three days, and I must have passed close to it several times daily. It was too far gone for preservation, so I had to console myself with its being the first record of the species in Canada. A week or so afterwards, when visiting Mr. Dickson, who is station-master on the G. T. R. at Waterdown, he pointed out to me an old, unused mill-race, grown up with briars and brambles, where the day before he had seen a pair of Chats mated. Mr. Dickson was collecting at the time, and was greatly surprised at their sudden appearance within ten feet of where he was standing, but on his moving backward, with a view of getting to a safer shooting distance, they disappeared in the thicket and did not again become visible, though they kept up their scolding as long as he remained near the place.

A pair of this species was also found by Mr. Saunders breeding on the north shore of Lake Erie, near Point Pelee, which completes the record for Ontario, so far as I have heard.

GENUS SYLVANIA NUTTALL.

SYLVANIA MITRATA (GMEL.).

289. **Hooded Warbler.** (684)

Clear yellow-olive; below, rich yellow shaded along the sides; whole head and neck, pure black, enclosing a broad golden mask across forehead and through eyes; wings, unmarked, glossed with olive; tail, with large white blotches on the two outer pairs of feathers; bill, black; feet, flesh color. *Female:*—With no black on the head; that of the crown replaced by olive, that of the throat by yellow. *Young male:*—With the black much restricted and interrupted, if not wholly wanting, as in the female. Length, 5-5¼; wing, about 2¾; tail, about 2¼.

HAB.—Eastern United States, west to the Plains, north and east to Michigan, Southern New York and Southern New England. In winter. West Indies, Eastern Mexico and Central America.

Nest, in a low bush or tree, a few feet from the ground; built of leaves and coarse grasses, and lined with fine grass and horse-hair.

Eggs, four, white, tinged with flesh color and marked with reddish-brown.

The Hooded Warbler is a southern species which rarely crosses our southern border. Mr. Norval reports finding it, occasionally, at Port Rowan, on the north shore of Lake Erie, and I once found a young male near Hamilton. It was toward the end of May, when there had been a big bird-wave during the previous night, and this one had apparently got carried away in the crowd. It is a most expert fly-catcher, very active on the wing, and has the habit of flirting its tail after the manner of the Redstart. Its favorite haunts are in thick, briary patches and among underbrush, where it finds food and shelter for itself and family.

There has been considerable discussion regarding the plumage of the *female* of this species, which has apparently subsided into the belief that in mature adult birds the sexes are nearly alike, but that the female is longer in acquiring the black of the head and throat, and is sometimes found with it imperfectly developed or entirely wanting.

SYLVANIA PUSILLA (Wils.).

290. Wilson's Warbler. (685)

Clear yellow-olive; crown, glossy blue-black; forehead, sides of head and entire under parts, clear yellow; wings and tail, plain, glossed with olive; upper mandible, dark; under, pale; feet, brown. *Female* and *young:*—Similar; colors not so bright, the black cap obscure.   Small, 4¾-5; wing, about 2¼; tail, about 2.

HAB.—Eastern North America, west to and including the Rocky Mountains, north to Hudson's Bay Territory and Alaska.   Breeds chiefly north of the United States, migrating south to Eastern Mexico and Central America.

Nest, a hollow in the ground, lined with fine grass and horse-hair.

Eggs, five, dull white, freckled with rusty-brown and lilac.

Wilson's Fly-catcher passes through Southern Ontario on its way to the north, in company with the Mourning Warblers and other late migrants.   Like some of the others, it has certain resting places, where it appears regularly in limited numbers every spring, but strangers, unacquainted with its haunts, might ransack the country for miles without seeing a single specimen.   The greatest number go far north to spend the summer, but it is probable that a few remain in intermediate districts, for Mr. Geo. R. White found a pair nesting in his garden in Ottawa.   This is the only record of the kind I have for Ontario.

In "New England Bird Life," Part I., page 172, is an account of a nest found by Mr. D. H. Minot on Pike's Peak, 11,000 feet up, near the timber line.   The nest and eggs were as described above.

Of its occurrence in Alaska, Mr. Nelson says: "On the Upper Yukon its nests and eggs have been taken by the 20th of May, and by the middle of that month its presence is noted on the coast of Behring Sea, where it is a summer resident, occupying the same alder thickets as the Yellow Warbler."

SYLVANIA CANADENSIS (Linn.).

291. **Canadian Warbler.** (686)

Bluish-ash; crown, speckled with lanceolate black marks, crowded and generally continuous on the forehead; the latter divided lengthwise by a slight yellow line; short superciliary line and edges of eyelids, yellow; lores black, continuous with black under the eye, and this passing as a chain of black streaks down the side of the neck, and prettily encircling the throat like a necklace; excepting these streaks and the white under tail coverts the entire under parts are clear yellow; wings and tail, unmarked; feet, flesh color; in the female and young the black is obscure or much restricted, and the back may be slightly glossed with olive. Length, about 5¼; wing, 2½; tail, 2¼.

Hab.—Eastern North America, westward to the Plains and north to Newfoundland, Southern Labrador and Lake Winnipeg, south, in winter, to Central America and Northern South America.

Nest, on the ground in a tussock of grass or weeds, composed of fibre, rootlets, leaves and pine needles.

Eggs, five, white, "beautifully marked with dots and small blotches of blended brown, purple and violet, varying in shades and tints and grouped in a wreath around the larger end."

From the 15th to the 25th of May, this species is very common in all suitable places in Southern Ontario. After the latter date the numbers are much reduced, but a few remain to spend the summer, while the bulk of the species goes farther north. When here their manners resemble those of the Green Black-cap, with whom they are often found in company, and they prefer briary thickets, through which they pass nimbly, picking up their insect fare as they go. In the fall they are less frequently seen, returning south, perhaps, by some other route. They are first seen about the middle of May, and disappear toward the end of August.

Mr. Kells found this species breeding in the low, damp woodlands near Listowel. The nests were built in the cavities of upturned roots of trees, and in the depressions in banks near pools of water.

GENUS SETOPHAGA SWAINSON.

SETOPHAGA RUTICILLA (LINN.).

292. American Redstart. (687)

Male: - Lustrous blue-black; belly and crissum, white; sides of the breast, large spot at bases of the remiges, and basal half of the tail feathers (except the middle pair), fiery orange, belly often tinged with the same. *Female:* Olivaceous, ashier on the head, entirely white below; wings and tail, blackish, with the flame color of the male represented by yellow. *Young male:*—Like the female, but browner, the yellow of an orange hue. From the circumstance that many spring males are shot in the general plumage of the female, but showing irregular isolated black patches, it is probable that the species requires at least two years to gain its perfect plumage. Length, $5\frac{1}{2}$; wing and tail, about $2\frac{1}{4}$.

HAB.—North America, north to Fort Simpson, west regularly to the Great Basin, casually to the Pacific coast, breeding from the middle portion of the United States northward. In winter, the West Indies, and from Southern Mexico through Central America to Northern South America.

Nest, in the fork of a sapling, six to twenty feet from the ground, composed of grape vine bark, grasses and weeds, and lined with fine grass, horse-hair or plant down.

Eggs, four or five, greenish or grayish-white, dotted with brown, lavender and purple.

The Redstart is one of the most active and restless little birds found in the bush, where its glowing garb of black and orange shines to great advantage among the fresh green leaves. It is generally distributed throughout Ontario, and from its manners and markings is well known to all who give any attention to the birds. In spring, it arrives from the 10th to the 15th of May, the first to appear being the adult male in full costume, after which come the females and young males in plumage nearly alike. While here, they are not high-fliers, but like to disport themselves among the middle and lower branches of deciduous trees, from which they dart off in pursuit of passing insects, making the clicking of the bill distinctly heard.

The male is so decided in his markings, that he is not likely to be mistaken for any other species. The female is plainer, but has the habit of opening and closing the tail feathers, which serves, even at a distance, to indicate the species to which she belongs.

After the end of August they are seldom seen.

Family MOTACILLIDÆ. Wagtails.

Genus ANTHUS Bechstein.

Subgenus ANTHUS.

ANTHUS PENSILVANICUS (Lath.)

293. American Pipit. (697)

Points of wings formed by the four outer primaries, the fifth being abruptly shorter. Hind claw, nearly straight, nearly or quite equal to its digit. Above, dark brown, with a slight olive shade, most of the feathers with dusky centres; eyelids, superciliary line and under parts, pale buffy or ochrey-brown, variable in shade; breast and sides of neck and body, thickly streaked with dusky; wings and tail, blackish; inner secondaries, pale-edged; one or more outer tail feathers, wholly or partly white. Length, about $6\frac{1}{4}$; wing, $3\frac{1}{4}$; tail, $2\frac{3}{4}$-3.

Hab.—North America at large, breeding in the higher parts of the Rocky Mountains and subarctic districts, and wintering in the Gulf States, Mexico and Central America. Accidental in Europe.

Nest, a cavity in the ground, lined thickly with coarse, dry grass.

Eggs, four or five, dark chocolate, with spots and streaks of grayish-brown.

In spring and fall, loose straggling flocks of Pipits are seen on the commons, either searching for food on the ground, or in short stages working their way to their breeding grounds in the far north, though how they ever get there is a wonder to any one who notices their weak and vacillating flight.

In the spring they pass along very quickly, but in the fall they are seen in flocks by the shores of muddy ponds or creeks, or in moist meadows in the open country, nervously jerking their tails after the manner of the Water-thrushes. Their only note while here is a weak, timid *cheep*, uttered while on the wing.

On the 20th of July, 1871, Mr. Allen found young birds of this species, scarcely able to fly, on Mount Lincoln, Park County, Colorado, among the snow fields above timber line.

Dr. Coues found them breeding abundantly on the coast of Labrador, and noticed their habit of resorting to the sea shore at low tide, there to ramble about in company with the Sandpipers in search of food.

Mr. Nelson says that they arrive at Fort Reliance on the upper Yukon about the 1st of May, and leave about the 5th of October. Kumlien reports that near his quarters, at the Cumberland Gulf, they nested in crevices among the rocks, but in Greenland he found them nesting in tussocks of grass like Sparrows. He tells us that

25

the Eskimo regard this bird as an enemy, and accuse it of telling
the reindeer when a man is in pursuit. It is also said to tell the
deer whether or not the man is a good shot. Kumlien adds that he
has seen an Eskimo waste his last charge in trying to kill one of
these birds, when a herd of deer was close by.

FAMILY TROGLODYTIDÆ. WRENS, THRASHERS, ETC.

SUBFAMILY MIMINÆ. THRASHERS.

GENUS MIMUS BOIE.

MIMUS POLYGLOTTOS (LINN.).

294. Mockingbird. (703)

Wings considerably shorter than tail; above, ashy-gray; below, whitish;
wings and tail, blackish, the former with two white wing-bars and large white
spot at base of primaries, latter with one to three outer feathers more or less
white. Length, 9-10; wing, about 4; tail, about 5.

HAB.—United States, south into Mexico. Rare from New Jersey, the
valley of the Ohio, Colorado and California northward.

Nest, in bushes and low trees, composed of twigs, leaves, grass, etc., put
together in a slovenly manner.

Eggs, four to six, bluish-green, heavily marked with several shades of
brown.

Among birds, as among men, individuals differ greatly in natural
ability, some being much more highly endowed than others, and
their gifts are also varied. Some, representing the architects of the
community, excel in building their homes, which have not only all
the necessary requirements for the comfort and safety of the inmates,
but exhibit a skill and taste in their construction, and in the selection and arrangement of the materials, which never fail to excite our
admiration. One of the most complete nests which has come under
my observation is that built by the Summer Yellow-bird. It is
often placed in the fork of a lilac bush near a house, and is not only
luxuriously comfortable, but so well put together that it stands the
blasts of winter. It is in good shape in the following spring, but the
birds do not use it a second season, and are seen tugging pieces out
of the old to help to build the new. Another interesting specimen
of bird architecture is the curious, pensile, purse-like nest of the
Baltimore Oriole, which is quite a familiar object, swaying at the
end of a slender twig of a drooping elm, while in the solitudes of a

cedar swamp the Winter Wren provides a wonderfully cosy home for her numerous family in the centre of a ball of green moss.

Others may be regarded as the poets, the musicians of the feathered tribe, and it would be a curious study for us to try to find out whether those who cannot sing enjoy the singing of those who can. In human ears the melody of many of the birds is as pleasing perhaps as it is to their own species, and in this respect there is none more fascinating than the Mockingbird, whose rapturous music excites admiration wherever it is heard.

One of America's most gifted poets, who evidently knew and appreciated the musical powers of the bird, thus describes it in words well worthy of the subject. The scene is on the lower Mississippi, a band of exiles is descending the river on a still evening in the early summer.

> "Softly the evening came. The sun from the western horizon
> Like a magician extended his golden wand o'er the landscape;
>
> .   .   .   .
>
> Then from a neighboring thicket, the Mockingbird, wildest of singers,
> Swinging aloft on a willow spray that hung o'er the water,
> Shook from his little throat such floods of delirious music,
> That the whole air and the woods and the waves seemed silent to listen.
> Plaintive at first were the tones and sad; then soaring to madness
> Seemed they to follow or guide the revel of frenzied Bacchantes.
> Single notes were then heard, in sorrowful, low lamentation;
> Till, having gathered them all, he flung them abroad in derision,
> As when, after a storm, a gust of wind through the tree tops
> Shakes down the rattling rain in a crystal shower on the branches."

In the Southern States the Mockingbird is a constant resident. Occasionally a pair come farther north to spend the summer, but as soon as the young are able for the journey, they again retire to the south. In the "Birds of Long Island," Mr. Giraud mentions it as an occasional summer resident there, and speaks of a pair having spent a summer near the beach at Egg Harbor. "The male," he says, "became the pet of the residents, to whom it also seemed much attached, and, as if in return for the attention they paid to his wants, he poured forth his charming melody, which on calm, bright nights, blending with the subdued voice of the ocean, rendered the time enchanting beyond the powers of description."

In Ontario, the Mockingbird is best known as a cage bird, numbers being occasionally brought from the south in captivity, and when exposed for sale are readily bought up by those who are fond of feathered pets. Even in confinement it seems to retain all its natural power and energy as a songster, and being of a sociable, familiar disposition, soon gets attached to those who are in the habit of attending to its wants.   Among American birds it has been justly styled the " Prince of Musicians," and indeed, with the exception of the British Sky-lark, whose grand, soaring flight adds greatly to the effect of its music, I know of no bird in any country possessed of such a wonderful compass of voice.   Often while exercising its powers of mimicry, it will give so correct an imitation of the notes of other birds, that the most retiring species will come from their haunts, expecting to meet their mates, when suddenly they will be driven in fear to the thicket by as correct an imitation of the harsh scream of the hawk.

The following incident gives me the privilege of claiming the species for Ontario, a pair having spent the summer of 1883 near Hamilton.

Had any one, acquainted with this neighborhood and with the habits of the bird, been asked to suggest where it was most likely to be found, he would certainly have said East Hamilton, and it was there that Mr. Eastwood first observed the male, early in the season, in one of the leafy lanes between his residence and the mountain. Mr. Eastwood was in the habit of taking exercise on horseback in the early morning, and seldom passed the place where the bird was first recognized without again seeing him on the dead branch of a low tree which he had chosen for his perch.   As the season advanced these frequent visits grew into something like personal friendship, for the bird evidently recognized his visitor, and, if absent at first, would readily respond to a call, and, mounting his usual perch, would answer in his own eloquent style.   He also caught up many of the local sounds of the neighborhood : the crowing of the rooster, the cackling of the fowls, and the notes of other birds were imitated with wonderful correctness, but sweetest of all were his own rich, full tones, which gave a new charm to that favored locality.   Only once during the season was a glimpse obtained of the female, who was evidently engaged in domestic duties, though, with the view of making the pair feel as much at home as possible, the nest was not sought for.

It was hoped that this pair or some of their family would return the following season to visit their old friends in Ontario, but, if they did so, they have not been observed.

In the " Birds of Western Ontario," mention is made of one specimen being taken by Mr. Sandys at Chatham in 1860, which is all we know of the Mockingbird in Ontario at present.

Genus GALEOSCOPTES Cabanis.

GALEOSCOPTES CAROLINENSIS (Linn.).

295. Catbird. (704)

Wings, but little shorter than tail; dark slate color, somewhat lighter below; crown of head and tail, black; under tail coverts, dark chestnut. Length, 8-9; wing, 3¾; tail, 4.

Hab.—Eastern United States and British Provinces, west to and including the Rocky Mountains, occasional on the Pacific coast. Winters in the Southern States, Cuba and Middle America to Panama. Accidental in Europe.

Nest, in a shrubbery or thicket, a few feet above the ground, composed of twigs, leaves, bark, rootlets, bits of twine or rags.

Eggs, four or five, dark bluish-green.

A very common summer resident in Southern Ontario is the Catbird; and in the North-West it is said by Prof. Macoun to be common wherever there are bushes. This is a bird well entitled to our protection ; but, unfortunately, it is the subject of an ignorant prejudice, which leads to its being persecuted, especially by boys, who would throw a stone at a Catbird with much the same feeling that they would at a cat. Perhaps one of his most familiar notes may have originated the prejudice, but outside of this, it should be remembered that he ranks high as a songster, coming next in that respect to the Mockingbird. He is one of the first to begin in the morning, and delivers his message with so much sprightliness and vivacity that we are always pleased to hear him.

In the garden he is one of our best friends, destroying an innumerable number of injurious insects, but we seldom think, when enjoying our luxuriant crop of cherries or raspberries, that we are largely indebted for such results to the much despised Catbird.

This species is common throughout Manitoba and the North-West, and was noticed by Dr. Coues as a summer resident in the Red River region and west along the boundary to Turtle Mountain.

It is a most unfortunate thing for the Catbird that he ever acquired the habit of imitating the cat-call, for it is that alone which has cast a shade over all his accomplishments, and brought upon him the derision and contempt with which he is so often regarded.

That it is an acquired habit I can well believe, for it is noticed at once to be entirely out of place in his song; but he belongs to the family of Mockers, and more than likely in ages past some mis-

chievous ancestor, while teasing a cat, picked up the cry, and trans-
mitted it till it has now become hereditary.   But for this, the bird
might have been known by some respectable name, giving him the
first place in the list of Canadian songsters, to which I believe he
is well entitled.   The only rival he has in the woods is the Brown
Thrasher, which we admit is also a grand performer.   He is a larger
bird, and his notes can be heard a long way off, but, having listened
to both attentively, I find that for variety and richness of notes,
as well as for sweetness and execution, I can cordially award the
palm to the Catbird.   In the color of his plumage there is nothing
attractive, but he is handsome in form, and of a most sociable and
kindly disposition.   He is not partial to the solitude of the woods.
His great delight seems to be to nestle near a log-house in the edge
of a clearing, where his rapturous notes are the first sounds heard
in the morning by the squatters' children, and again in the evening
they are soothed to slumber by his plaintive strains.   By all means
let us give the Catbird the encouragement and protection that may
be in our power, and we shall be well rewarded both by good music
and by good work in the garden.

Genus HARPORHYNCHUS Cabanis.

Subgenus METHRIOPTERUS Reichenbach.

HARPORHYNCHUS RUFUS (Linn.).

296. Brown Thrasher. (705)

Above, reddish-brown; below, white, with more or less tawny tinge; breast
and sides, spotted with dark brown; throat and belly, unspotted; bill, black
above, yellow below; feet, pale; iris, yellow. Length, 11; wing, 4; tail, 5-6.

Hab.—Eastern United States, west to the Rocky Mountains, north to
Southern Maine, Ontario and Manitoba, south to the Gulf States, including
Eastern Texas. Accidental in Europe.

Nest, most frequently placed in the fork of a small tree in a thicket, three
to six feet from the ground, sometimes higher, occasionally on the ground;
composed of twigs, grass, leaves and rootlets, lined with bark fibre, horse-hair
and a few feathers.

Eggs, four or five, greenish-white, thickly spotted with light reddish-brown.

The Brown Thrasher is not so abundant as the Catbird, neither
is it so confiding or familiar in its habits, seldom coming near our
dwellings. It delights in the tangled, briary thicket, in the depths
of which it disappears as soon as it is aware of being observed. Near
Hamilton it is a common summer resident, appearing regularly about
the 10th of May. At first the birds are seen stealing quietly through
the underbrush, or scratching among the withered leaves like the
Towhees, but once arrived at their breeding place, the male is heard
from the topmost twig of an isolated tree, pouring forth, morning

and evening, his unrivalled strains of music, which are heard long
ere the performer can be seen.

So far as I have observed, the Thrasher is somewhat local in its
distribution, there being certain sections of country of considerable
extent where, without apparent cause, it is entirely wanting.

It occurs throughout Ontario and crosses the boundary to Manitoba and the North-West.

During September it retires from Ontario.

GENUS THRYOTHORUS VIEILLOT.

SUBGENUS THRYOTHORUS.

THRYOTHORUS LUDOVICIANUS (LATH.).

297. Carolina Wren. (718)

Adult male:—Upper parts, brownish-red, a broad streak of yellowish-white
commencing at the nostril and passing over the eye along the side of the head,
a band of reddish behind the eye; under the eye, a spot of dusky gray; throat,
dull white: lower parts and sides of the neck, pale reddish-buff; wings and
tail, barred with blackish-brown, the outer webs of the lateral tail feathers
more distinctly barred; secondary and first row of small coverts tipped with
dull white; lower tail coverts of the same color barred with black. Length,
6 inches: extent, nearly 7 inches; tail, 2.25.

HAB.—Eastern United States (rare towards the northern border), west to
the Plains. Rare in Southern New England.

Nest, a large coarse structure, composed of grass, hay, leaves, etc., lined
with horse-hair and feathers. Found in holes in trees, wood piles, low bushes,
sometimes arched over.

Eggs, four to six, creamy-white or buff, thickly sprinkled with brownish-
pink.

I have now the pleasure of introducing a species I have long
expected to meet in Ontario. It is our near neighbor and a strong,
active bird, well able to make the journey from its usual summer
resorts to Canada, but it is evidently a shy visitor and not decidedly
migratory, many of the species remaining over the winter near the
nesting place. It is abundant in the south, common in the middle
Eastern States, and is also found, though less frequently, in New
York, Connecticut, Massachusetts and Ohio.

With this record before us, we might reasonably expect a casual
visitor, but it was not until very recently that I learned through

Mr. Saunders that a specimen of the Carolina Wren had been taken in Ontario.

This individual was shot in the town of Mount Forest in February, 1891, by Montague Smith, who had observed it singing daily near a favorite haunt for a month before it was captured. It is a true Wren, having the habit of erecting its tail and appearing and disappearing among a pile of logs, or similar cover, as nimbly as a rat or weasel. It has all the nervous irritability of the other members of the family, one evidence of this being the fact of its keeping up its song during the winter, when most of the other birds are silent. It is a special favorite with the colored people in the South, who like to hear its notes in the woods in winter, and have given it the name of "Jubilee Bird." The Carolina Wren is a very prolific species, the female turning over to the male the care of the first brood before they are able to shift for themselves, while she proceeds to deposit a second set of eggs in another nest, which the male has prepared for their reception. Family number two is turned over to the male in due course, and in this way three broods are raised during the season in a very short time. It is the largest of the Wrens, and not likely to be mistaken for any other species. Its voice is strong, sharp and clear, and can be heard at as great a distance as those of the Brown Thrasher and Catbird. It has some of the habits of the Creeper, being occasionally seen going spirally up the trunk of a perpendicular tree, examining the crevices of the bark for insects. It also resembles the House Wren in its breeding habits, the nest being often built in an outhouse or under the verandah of a dwelling, but it seems most at home in the woods, where its favorite haunts are among piles of logs or heaps of brush, on the banks of streams or ravines, where it can inspect the surroundings one instant and be entirely concealed the next. Individuals are observed to differ somewhat in color, some being rusty brown, while others are comparatively gray.

TROGLODYTES AËDON (Vieill.).

298. **House Wren.** (721)

Above, brown, brighter behind; below, rusty-brown or grayish-brown, or even grayish-white, everywhere waved with a darker shade, very plainly on wings, tail, flanks and under tail coverts; breast, apt to be darker than either throat or belly. Length, $4\frac{7}{8}$; wings and tail, about 2.

Hab.—Eastern United States and Southern Canada, west to Indiana and Louisiana.

Nest, in a hole or crevice, the neighborhood of a dwelling preferred, composed of twigs, leaves, hair, feathers, etc.

Eggs, seven to nine, white, very thickly spotted with reddish-brown.

In the thinly settled parts of the country where this Wren has been observed, it breeds in any convenient hole or crevice in a tree or fence-post by the roadside, and on account of this habit, and an imaginary superiority in point of size, those found in such places were described as a separate species, and named by Audubon the Wood Wren. The individuals procured in town and country being subsequently found to be identical, this name has for some years been allowed to drop. The birds, having taken kindly to the society of man, are nearly all furnished with houses, or, finding other suitable nesting places near our dwellings, are living almost domesticated. They are sprightly, active little birds, and do good service by the destruction of insects, which they find on the trees in the orchard, or about the outhouses. Being possessed of all the scolding propensities peculiar to the family, they resent with great spirit any intrusion in the neighborhood of their dwelling. Their greatest enemy in this respect at present is the House Sparrow, who does not hesitate to eject the Wrens, when their premises appear to suit his purpose. This habit may in time drive the Wrens back to their original mode of life in the woods.

We should be very sorry if this should ever come to pass, for the Wrens are industrious insect hunters, prying into many out-of-the-way corners which no other bird would stop to examine.

In Manitoba and the North-West, this is replaced by a closely allied species, named Parkman's Wren, which is common on the Pacific coast and east to the Mississippi Valley.

SUBGENUS ANORTHURA RENNIE.

TROGLODYTES HIEMALIS (VIEILL.).

299. Winter Wren. (722)

Deep brown above, darkest on the head, brightest on the rump and tail, obscurely waved with dusky and sometimes with whitish also; tail like rump; wings, dusky, edged with color of back, and dark barred; several outer primaries also whitish barred; a superciliary line and obscure streaks on sides of head and neck whitish; below, pale brown; belly, flanks and under tail coverts, strongly barred with dusky. Length, about 4; wing, 2 or less; tail, 1½ or less.

HAB.—Eastern North America generally, breeding from the northern parts parts of the United States northward, and wintering from about its southern breeding limit southward.

Nest, in moist places among fallen trees or upturned roots, usually a ball of green moss, warmly lined with feathers, entrance by a hole at one side.

Eggs, five to six, clear white, spotted with reddish-brown.

In Southern Ontario, the Winter Wren is most frequently seen during the periods of migration, but a few remain and raise their young in suitable places throughout the country. There is a cedar swamp in West Flamboro', made impenetrable by fallen timber, moss-grown and going to decay. In the stillness and gloom of that uninviting region, I have listened to the song of the Winter Wren in the month of June, and thought it one of the most pleasing specimens of bird music I have been privileged to hear. Tinged with melancholy it may be, but there is still a hopeful sprightliness about it that seems to rise above the gloomy surroundings and point to a brighter world outside. I have not heard of the species having been observed in winter, but it arrives from the south early in April, and lingers quite late in the fall. During the latter season, they are frequently seen in the city gardens, appearing and disappearing like mice among the roots of the bushes.

In my boyish days I was familiar with the haunts and homes of the common Wren of Britain, *Troglodytes vulgaris*, along the "Banks and Braes o' Bonnie Doon," and in song, size and color I believe it is identical with the present species.

In Manitoba this species is said to be common during the summer. Farther to the North-West it is replaced by a closely allied sub-species, named Western Winter Wren, which is the form found along the Pacific coast and in California.

GENUS CISTOTHORUS CABANIS.

SUBGENUS CISTOTHORUS.

CISTOTHORUS STELLARIS (LICHT.)

300. **Short-billed Marsh Wren.** (724)

Dark brown above, crown and middle of the back, blackish, nearly everywhere conspicuously streaked with white; below, buffy-white, shading into pale brown on the sides and behind; wings and tail, barred with blackish and light brown; flanks, barred with dusky; throat and middle of belly, whitish. Length, 4½; wing and tail, about 1¾; bill, not ½ long and very slender; tarsus, middle toe and claw, together 1¼.

HAB.—Eastern United States and Southern British Provinces, west to the Plains. Winters in the Gulf States and southward.

Nest, similar to that of the Long-billed species, but sometimes placed near the ground; no mud used in the structure, which is very compact and warmly lined with down.

Eggs, six to eight, pure white, unspotted.

Never having happened to meet with the Short-billed Marsh Wren in any of my excursions, I consider it to be either locally distributed or less abundant than the Long-billed species, which is common in all the marshes in Southern Ontario.

Throughout Northern New England, the Short-billed species is a common summer resident, and Mr. Thompson speaks of it as being "abundant all over" in Western Manitoba. It is probable, therefore, that it is a summer resident in Ontario, but so few people follow these little birds into their marshy haunts that, at present, their

history here is somewhat concealed. Mr. Saunders says it is found
in the marshes along the River St. Clair, and he has a set of eggs
which were taken in a marsh near Toronto.

It is found at certain places in Michigan, also in Minnesota, and
is abundant in the Red River Valley.   At Pembina, Dr. Coues
found it to be "erratic in distribution."   It is observed in Manitoba, but Mr. White has not seen it near Ottawa, and I have not yet
found it in Southern Ontario, from which it would appear that it is
a western species, and irregular in distribution.

Subgenus TELMATODYTES Cabanis.

CISTOTHORUS PALUSTRIS (Wils.).

301. Long-billed Marsh Wren. (725)

Above, clear brown, unbarred; back. with a black patch, containing distinct
white streaks; crown, brownish-black; superciliary line to nape, white; wings,
not noticeably barred, but webs of inner secondaries, blackish; tail, brown,
dusky barred.   Below, dull white, often quite pure, the sides alone brownish
washed, and under tail coverts somewhat varied.   Length, 4⅜-5½; wing, about
2; tail, less; tarsus, ⅞-⅞; bill, ½ or more.

Hab.—Southern British America and the United States, south in winter to
Guatemala.

Nest, a large globular mass of coarse grass and rushes loosely laced together,
sometimes plastered with mud and fastened to the reeds, warmly lined with
fine, soft grass; entrance by a hole in one side.

Eggs, six to ten, variable in shade, but usually so thickly spotted with
chocolate-brown as to appear uniformly of that color.

The Long-billed Marsh Wren is a common summer resident, found
in suitable places throughout Ontario.   Near Hamilton it breeds in
all the inlets around the Bay, and is seen from the beginning of May
till the end of August, climbing, hopping, and swaying to and fro
among the reeds in every conceivable posture.   In the spring it
appears to be constantly under great nervous excitement, which it
works off in nest-building, often constructing two or three when
only one is required.   So large a number of nests, when observed,
gives the impression that the birds breed in colonies, but I have not
noticed this to be the case.   All the nests I have seen have been so
placed that they could only be reached by wading or in a boat,
and sometimes they were among the reeds on a quaking bog where
approach was impossible.

The mode of migration of these birds is a mystery. We are accustomed to say that they retire to the south early in September, but how do they travel? Do they rise in flocks like Swallows and go off during the night, or do they make the long journey from the Saskatchewan, where they were seen by Richardson, south to Guatemala, flitting singly, or in pairs, from bush to bush? In either case it is strange that they are seldom, if ever, seen except in the marshy tracts where they spend the summer.

Mr. White has found this species breeding near Ottawa, and it is said to be common at many points in Manitoba, which seems to be its northern limit.

FAMILY CERTHIIDÆ. CREEPERS.

GENUS CERTHIA LINNÆUS.

CERTHIA FAMILIARIS AMERICANA (BONAP.).

302. Brown Creeper. (726)

Plumage above, singularly barred with dusky, whitish, tawny or fulvous-brown and bright brown, latter chiefly on the rump; below, white, either pure or soiled, and generally brownish washed behind; wings, dusky, oddly varied with tawny or whitish bars and spots; tail, plain, about 5½; wing and tail, about 2¾.

HAB.—North America in general, breeding from the northern and more elevated parts of the United States north as far as Red River settlement, migrating south in winter.

Nest, nearly always in a crevice where the bark is partially separated from the trunk of a tree. In the crevice is placed a basis of twigs, on which the nest is built, of strips of bark and moss, lined with spiders' cocoons and down.

Eggs, five to eight, dull white, spotted with reddish-brown or hazel.

This singular little bird is seen in Southern Ontario at nearly all seasons, but it is most abundant during the period of migration. About the end of April and beginning of May, it becomes quite common in the woods, and is seen flitting like a great moth from tree to tree, or winding its spiral way upward on a trunk, uttering its simple note so descriptive of the motion, *creep, creep, creep.* In summer a pair may be seen, occasionally, in more favored spots, evidently nesting, but at that season they are quite rare. Early in September they again become numerous, in company with other migrants who are travelling southward, and in the depth of winter I have occasionally seen them mixed up with a small band composed of Chickadees, Downy Woodpeckers, Nuthatches and Golden-crowned Kinglets. These birds seem to find pleasure in each other's society, when they are spending the short, sharp days of winter in some sheltered patch of evergreens.

FAMILY PARIDÆ.   NUTHATCHES AND TITS.

SUBFAMILY SITTINÆ.   NUTHATCHES.

GENUS SITTA LINNÆUS.

SITTA CAROLINENSIS (LATH.).

303. White-breasted Nuthatch. (727)

Back, rump and middle tail feathers, ashy-blue; crown and nape, glossy black, restricted or wanting in the young and many females; tail, except as above, black, spotted with white; beneath and sides of head, white; flanks and under tail coverts, rusty-brown; wings varied, black, blue and white.   Length, 6; wing, 3½; tail, 2.

HAB.—Southern British Provinces and Eastern United States to the Rocky Mountains.

Nest, a hole in a tree, sometimes a natural cavity, or again dug by the birds with great labor, lined with hair and feathers.

Eggs, six to eight, white, spotted thickly with reddish-brown.

This is one of the few birds which remain with us summer and winter.   It is quite a common species, well known to all who have occasion to be in the woods in spring, when it is seen climbing nimbly about, or hanging head downwards on the bark of a tree. In the winter time the country lads who are chopping in the bush listen with pleasure to its familiar *quank, quank*, which is often the only evidence of animal life observed.   As a climber, it has few equals, its long hind-claw enabling it to travel head downwards, a feat which even the Woodpeckers do not attempt.   Its food consists chiefly of insects, which it finds lurking in the crevices of the bark. It is also said to hide away nuts and acorns in the holes of trees, a habit which may have suggested its name.

It is rather more southern in its habitat than its Red-breasted relative.   Mr. White reports it as resident at Ottawa, and we find it so along our southern border, but it is not named among the birds of Manitoba, from which we infer that it is not found in that Province.

SITTA CANADENSIS (Linn.).

304. Red-breasted Nuthatch. (728)

Above, dark ashy-blue; tail, as in *carolinensis;* below, rusty-brown; wings, plain; crown and nape, glossy black, bordered by white superciliary line; a black line from bill through and widening beyond the eye.

Hab.—North as far as Lake Winnipeg, breeding mostly north of the United States, migrating south in winter.

Nest, in a hole in a stub, about eight inches deep, warmly lined with down and feathers.

Eggs, white, speckled and spotted with reddish-brown.

Compared with the White-bellied Nuthatch, this is the more migratory in its habits, being seen in Southern Ontario only in spring and fall, and it is not at any time numerous. I have been accustomed to think that those we get in the fall with the red breast were in full plumage, but recent observers state that, when in mature dress, the lower parts are dirty-white, slightly shaded with brown on the sides, and that only young birds have the lower parts uniform rusty-brown. While here they are very active, showing a decided partiality for the upper parts of pine trees, where they no doubt find something to suit their taste. The note resembles that of the White-bellied species, but is softer, weaker, and more frequently repeated. It arrives during the first week of May, and is soon lost sight of again till September, when it is seen passing south.

Ontario is probably its northern limit, for although it has been found in Manitoba, it is said to be exceedingly rare.

———

SUBFAMILY PARINÆ. TITMICE.

GENUS PARUS LINNÆUS.

SUBGENUS PARUS LINNÆUS.

PARUS ATRICAPILLUS (Linn.).

305. Chickadee. (735)

Above, brownish-ash; crown and nape, chin and throat, black; beneath, white, brownish on sides; wing and tail feathers, more or less whitish edged. Length, 5; wing and tail, 2¼.

Hab. Eastern North America, north of the Potomac and Ohio Valleys.

Nest, a hole appropriated or dug by the birds in a dead tree or stump, not usually very high up, lined with hair, grass, moss, wool, feathers, etc.

Eggs, six to eight, white, speckled and spotted with reddish-brown, chiefly toward the larger end.

In Southern Ontario the Chickadee is one of our most familiar resident birds. During the breeding season it retires to the woods, but at other times it is seen in little troops visiting the shade trees and orchards of the city, searching the crevices for insects, and uttering its familiar *chickadee, dee, dee*, so well known to all the boys. It has also another note, or rather two notes, one quite high which drops suddenly to one much lower, soft and prolonged, and probably both convey a meaning to the ears for which they are intended. During the severity of winter, they are most frequently seen in tamarack swamps, where they no doubt find both food and shelter.

The Chickadee breeds and is generally distributed throughout Ontario, in the northern part of which it meets with *hudsonicus* on the east and *septentrionalis* on the west, but whether or not it breeds with the latter form, I am not aware. In the "Birds of Manitoba," page 631, it is said "the Manitoba bird is not strictly *septentrionalis*, but is nearer to that form than to *atricapillus*." Thus it sometimes happens on the boundary between the region inhabited by a *species* and *sub-species* that individuals are found, of which it is difficult to say positively to what group they belong.

The Chickadee is a general favorite and well deserves to be so. Noisy, restless, familiar and cheerful, he is welcome wherever he appears. He does good work for the farmer and fruit-grower in the destruction of noxious insects, and, unlike many others in this class of workers, he keeps at it summer and winter. It has been calculated that a single pair of these little birds will kill daily five hundred insect pests.

PARUS HUDSONICUS (Forst.).

306. Hudsonian Chickadee. (740)

Crown, nape and upper parts, generally clear hair-brown or ashy-brown, with a slight shade of olive, the coloration quite the same on back and crown, and continuous, not being separated by any whitish nuchal interval; throat, quite black, in restricted area, not extending backward on sides of neck, separated from the brown crown by silky white on side of the head, this white not reaching back of the auriculars to the sides of the nape; sides, flanks and

26

under tail coverts, washed with dull chestnut or rusty-brown; other under parts, whitish; quills and tail feathers, lead color, as in other Titmice, scarcely or slightly edged with whitish; little or no concealed white on the rump; bill, black; feet, dark. Size of *P. atricapillus* or rather less.

HAB.—Northern North America, from the more elevated parts of the Northern United States (Northern New England, Northern New York, Northern Michigan, etc.) northward.

Nest, in a hole in a tree or stump, lined with fur of animals, felted firmly together.

Eggs, five or six, creamy-white, speckled with hazel.

The home of the Hudsonian Tit, as its name implies, is in the Hudson's Bay country. It is also common in Labrador, and I have seen it on the banks of the Lower St. Lawrence, travelling in little troops from tree to tree, much after the manner of our familiar Chickadee. It is truly a northern species, but as it has been found in Massachusetts, Maine and New Hampshire, I think it will yet be found in the districts of Parry Sound and Muskoka. At Ottawa, and also near Toronto, it has been found as a rare straggler. In Manitoba, Mr. Thompson says it occurs only in the coniferous forests of the north and east. It has been found throughout the wooded portions of Alaska, from its southern coast line at Fort Kenai, north throughout the Kuskoquim and Yukon River regions, to the northern tree limit, well within the Arctic circle.

It is a diligent insect hunter, but lives too far north to benefit agriculture, and for the same reason it is exempt from the persecution to which all little birds are subject in more thickly peopled districts.

FAMILY SYLVIIDÆ. WARBLERS, KINGLETS, GNATCATCHERS.

SUBFAMILY REGULINÆ. KINGLETS.

GENUS REGULUS CUVIER.

REGULUS SATRAPA (LICHT.).

307. Golden-crowned Kinglet. (748)

General color as in *calendula*; crown, bordered in front and on sides by black, inclosing a yellow and flame-colored patch (in the *male*; in the *female* the scarlet is wanting): extreme forehead and line over the eye whitish. *Young:*— If ever without traces of black on the head, may be told from the next species by smaller size and the presence of a tiny bristly feather overlying the nostril; this wanting in *calendula*. Size of *calendula*.

HAB.—North America generally, breeding in the northern and elevated parts of the United States and northward, migrating south, in winter, to Guatemala.

Nest, in appearance resembling a ball of moss; it is opened at the top, the cavity warmly lined with feathers, plant down and wool, fastened to the outer twig of a branch, six to eight feet from the ground.

Eggs, ten, ground color creamy-white, with numerous shell marks of purplish-slate and a few superficial markings of deep buff, making the whole appear of a cream color.

This is an abundant winter resident, appearing in November and remaining till April. During the severe weather in February and March, when the mercury is near zero, it is really surprising to see these tiny, feathered creatures, full of animation, flitting about among the evergreens, uttering their cheerful notes of encouragement to their companions, and digging out their insect food from the crevices of the bark. On these occasions they are usually accompanied by Chickadees, Downy Woodpeckers and White-bellied Nuthatches, making a merry company, nowise discouraged by the severity of the weather.

The Gold-crest is known to breed in Northern New England, a nest containing young having been found by Mr. H. D. Minot in a forest of evergreens and birches on the White Mountains of New Hampshire, on the 16th of July, 1876. I once met with a pair, evidently mated, who were located in a swamp in West Flamboro' about the end of June. I did not persevere in seeking the nest, though I felt sure it was close at hand. That is the only time I have seen the species here in summer.

It is generally distributed throughout Ontario, but is rare in Manitoba, and in Alaska is replaced by the western form (*Regulus satrapa olivaceus*). Mr. Brewster found it breeding in Worcester County, Mass., where he secured three nests with eggs, a detailed account of which is given in the *Auk*, Vol. V., pages 337-344.

REGULUS CALENDULA (LINN.).

308. **Ruby-crowned Kinglet.** (749)

Above, greenish-olive; below, whitish; wings and tail, dusky, edged with greenish or yellowish; wing coverts, whitish tipped; crown, with a rich scarlet patch in both sexes (but wanting in both the first year), no black about head; bill and feet, black. Length, 4-4½; wing, 2¼-2½; tail, 1¼-1¾.

Hab.--North America, south to Guatemala, north to the Arctic coast, breeding mostly north of the United States.

Nest, large for the size of the bird, a mass of matted hair, grass, moss and feathers, placed on the bough of a tree.

Eggs, five to nine, whitish or pale buff faintly speckled with light brown at the larger end.

In Southern Ontario, the Ruby-crown is a regular migrant in spring and fall, but in summer or winter it has not been observed.

During the latter part of August and beginning of September, these little birds are exceedingly abundant, although from their small size and the weak, lisping note they utter at this season, their numbers can be estimated only by close observation. I was once caught in the rain in the woods in the month of April, and took shelter in a clump of evergreens, which I found was in possession of a flock of Ruby-crowns. When the clouds passed away and a light breeze shook the sparkling drops from the foliage, I was delighted to hear some of the Kinglets indulge in a song of considerable compass and duration. It was more full, soft and musical than anything I ever heard from so small a bird. At that season their stay is short; sometimes they are seen only during two or three days, but in the fall they travel more leisurely. Their breeding ground is far north, and for many years their nests were sought for in vain. Now their discovery is reported from many and far distant points. Colorado, Montana and Oregon are named as supplying the little domicile, and Mr. Chamberlain tells us of one which was taken at Lennoxville, Quebec, May 15th, 1882. This was pensile, and was attached to the branch of a small tree. It contained nine eggs, one of which was a Cowbird's.

SUBFAMILY POLIOPTILINÆ. GNATCATCHERS.

GENUS POLIOPTILA SCLATER.

POLIOPTILA CÆRULEA (LINN.).

309. Blue-gray Gnatcatcher. (751)

Above, ashy-blue, bluer on the head, lighter on the rump; forehead and line over eye, black, wanting in the female; ring around the eye and under parts, whitish; outer tail feather, except at base, two-thirds the second and tip of third, white, rest of tail, black. Length, 4½; wing, 2; tail, 2¼.

Hab.--Middle and southern portions of the United States, from the Atlantic to the Pacific, south in winter to Guatemala, Cuba and the Bahamas; rare north toward the Great Lakes, Southern New York and Southern New England, straggling north to Massachusetts and Maine.

Nest, a model of bird architecture, compact-walled and contracted at the brim, elegantly stuccoed with lichens fixed to slender twigs at a height varying from ten to fifty or sixty feet from the ground.

Eggs, four or five, greenish-white, speckled with reddish, umber-brown and lilac.

The Gnatcatcher is, I believe, a regular summer resident in Southern Ontario, though apparently locally distributed and not very abundant. There is one particular patch of bush where I usually see this species every spring, but elsewhere I have not observed it. Mr. Dickson finds it regularly at Waterdown, and Mr. Saunders reports it as not very rare near London. It keeps mostly to the tops of tall trees, and might readily be overlooked by anyone not acquainted with its habits.

In the breeding season it is said to have a pleasing song, and it shows considerable spirit in driving off intruders from the neighborhood of its nest.

In Ontario this species is apparently limited to the south-west border, north of which I have not heard of its having been observed. Mr. White has not met with it at Ottawa, neither is it mentioned in any of the local lists north of those named.

FAMILY TURDIDÆ. THRUSHES, SOLITAIRES, STONECHATS, BLUEBIRDS, ETC.

SUBFAMILY TURDINÆ. THRUSHES.

GENUS TURDUS LINNÆUS.

SUBGENUS HYLOCICHLA BAIRD.

TURDUS MUSTELINUS (GMEL.)

310. Wood Thrush. (755)

Above, bright tawny, shading into olive on rump and tail; beneath, white, everywhere except throat and belly, with large distinct spots of dusky; bill, dusky above, yellowish below; legs, flesh-colored. Length, 7½ inches; wing, 4; tail, 3.

HAB.—Eastern United States to the Plains, north to Southern Michigan, Ontario and Massachusetts, south, in winter, to Guatemala and Cuba.

Nest, in a sapling or low tree, seldom more than twenty feet from the ground, composed of twigs, leaves, grass, rootlets and moss, cemented together with clay.

Eggs, three or four, deep greenish-blue.

The Wood Thrush is a shy, retiring songster, little known except to those who are fond of rambling in the woods in spring-time. The favorite resort of the species is in moist beech woods, where the clear, flute-like notes of the male may be heard in the early morning, and also toward sunset, during the months of May and June. Were the song of the Wood Thrush continuous, the bird would take the highest rank among the songsters of the grove. Its tones are loud and full of liquid tenderness, but they suddenly break off short, which to us is a matter of regret.

Early in May they arrive from the south, and are soon generally distributed over Southern Ontario, but they are somewhat fastidious in their choice of a summer residence, and are absent from many clumps of bush in which we expect to find them. They avoid the dwellings of man, and seem most at home in the retirement of the woods, where they raise their young. During September they all move off to the south.

In the east Mr. White has found this species at Ottawa, but in the west I have no record of it at any distance to the north of our southern boundary. Even in its usual habitat, it is so retiring that its full, liquid notes are seldom heard save by the few who are out in search of some of the beautiful souvenirs which Nature scatters so profusely throughout the woods in our lovely Canadian spring-time. When will some divinely gifted Canadian appear to sing the praises of our native birds, as men of other lands have done for theirs?

Hogg and Shelley have eulogized the Sky Lark in strains so musical that they rival those of the birds they have sought to honor.

TURDUS FUSCESCENS (Steph.).

311. Wilson's Thrush.   (756)

Above, uniform tawny; below, white, olive-shaded on sides and strong fulvous tint on breast; breast and sides of neck, with small dusky spots. Length, about 7; wing, 4; tail, 3.

HAB.—Eastern United States to the Plains, north to Manitoba, Ontario, Anticosti and Newfoundland.

Nest, on or near the ground, composed of grass, leaves and rootlets, rather loosely put together.

Eggs, four or five, greenish-blue, unspotted.

With the exception of the Robin, the Veery is the most numerous of the Thrushes which visit Southern Ontario. It arrives here during the first week in May, and for a few days is quite common in the woods everywhere. Many soon pass on farther north to breed, but some remain and locate themselves among the undergrowth in moist, uncleared places, where they spend the summer. On their first arrival, they remain for a few days quietly in the woods, but, as soon as nesting begins, the clear, loud *veery* is heard at all hours of the day. The song has a sharp metallic ring, and at first is pleasant to listen to, but when heard in some favored locality, where several males are answering each other, it becomes monotonous through frequent repetition. It is rather a tender bird, and is one of the first to move off in the fall. The young are able to shift for themselves in August, and by the end of September all are gone.

Dr. Coues found this species breeding abundantly in the vicinity of Pembina in June, and Mr. Thompson also reports it as common throughout Manitoba, but farther north I have not heard of its having been observed.

TURDUS ALICIÆ (BAIRD.).

312. **Gray-cheeked Thrush.** (757)

Similar to the Olive-backed Thrush, but without any buffy tint about head, or yellowish ring around eye; averaging a trifle larger, with longer, slenderer bill.

HAB.—Eastern North America, west to the Plains and Alaska, north to the Arctic coast, south, in winter, to Costa Rica. Breeds chiefly north of the United States.

Nest, in a low tree or bush, compactly built of fine sedges, leaves, stems and dry grass, interwoven and lined with fine grass. Sometimes mud is used as in the Robin's nest.

Eggs, three or four, greenish-blue, marked with spots of reddish-brown.

It is still a question with many ornithologists whether or not this should be separated from the Olive-backed, or regarded as only a variety of that species. The Committee of the American Ornitho-

logical Union decided to separate it as above, and I quite agree with
the decision, for the few I have found could be identified at once by
the description.   When seen in the woods, however, it resembles the
Olive-backed so closely that, till well acquainted with its appearance,
it is difficult to tell the one from the other.   On this account, we
cannot with certainty say which is the more numerous, but so far
as I can judge, the proportion of the Gray-cheeked species which
passes this way is not more than one to two of the other.   Dr. Coues
regards it as the northern form of the Olive-backed, and suggests
that this difference in the breeding range produces the change in
size and color, which are regarded as specific distinctions.   Like all
the other Thrushes, it is most likely musical at home, but here it
comes and goes in silence.

The home of this species is in the far north, and while in Ontario
it is only a migrant in spring and fall.   In the "Natural History of
Alaska," Mr. Nelson says : "This species is common throughout all
the northern portion of Alaska, wherever willow and alder thickets
afford it shelter.   Its western range extends to Behring Straits and
beyond, and it has been recorded from Kamschatka.   Along the
entire Yukon and other streams bordered by trees or bushes in this
region, it is present in great abundance during the breeding season."
They are usually very shy, but "as soon as the breeding season is
over, they become less retiring and frequent the vicinity of villages
and more open spots, where many are killed by the native boys,
armed with their bows and arrows.   Their skins are removed and
are hung in rows or bunches to dry in the smoky huts, and are preserved as trophies of the young hunters' prowess.   In the winter
festivals, when the older hunters bring out the trophies of their skill,
the boys proudly display the skins of these Thrushes and hang them
alongside."

TURDUS USTULATUS SWAINSONII (Cab.).

313. Olive-backed Thrush. (758a)

Above, uniform greenish-olive; below, white, olive-shaded on sides; sides
of head, throat, neck and breast, strongly tinged with buff; breast and
throat, thickly marked with large dusky-olive spots.   Length, about 7; wing,
3¾; tail, 3.

HAB.—Eastern North America and westward to the Upper Columbia

River and East Humboldt Mountains, straggling to the Pacific coast. Breeds mostly north of the United States.

Nest, in a tree or bush, six or eight feet from the ground, composed of rootlets, leaves and moss.

Eggs, three or four, greenish-blue, freckled with brown.

In Southern Ontario, the Olive-backed Thrush is a regular visitor during the season of migration, appearing in small companies about the 10th of May, and remaining till about the 25th of the same month, after which none are seen till they return in the fall. While here they frequent low, moist woods, and spend much of their time on the ground, where their food is evidently obtained. When at home, near their nest, the male is said to have a very pleasing song, which he takes delight in repeating, but while here they have only a low, soft call-note, easily recognized in the woods, but difficult to describe.

The return trip begins toward the end of September, and continues for about three weeks. At this time the birds move leisurely, and as they fare sumptuously on different sorts of wild berries, they get to be in excellent condition, both as regards flesh and plumage. We occasionally fall in with individuals of this species much below the average size, and with the lower parts more deeply suffused with buff. Dr. Wheaton has also observed these little fellows, and suggests that they may be a local southern-bred race. In Southern Ontario, none have been observed except in spring and fall.

They are said to breed in Manitoba, and Prof. Macoun reports them as summer residents in the North-West Territory. Of Alaska, Mr. Nelson says : " From the observations and collections made on the upper Yukon, the Olive-backed Thrush appears to be a common summer resident there, and thus extends its breeding range within the Arctic circle. It appears to be influenced to a great extent in its range by the presence or absence of woods, and its northern limit may be marked as coinciding with the tree limit. The Gray-cheeked Thrush, on the contrary, extends beyond this, wherever a bunch of dwarf willows will give it shelter, to the very shores of the Arctic and Behring Seas."

TURDUS AONALASCHKÆ PALLASII (Cab.).

314. Hermit Thrush. (759*b*)

Above, olive, shading into rufous on rump and tail; below, white, olive-shaded on sides; sides of head, eyelids, neck and breast, strongly tinged with buff; throat and breast, marked with large dusky-olive spots. Length, about 7¼; wing, 3¼; tail, 3.

Hab.—Eastern North America, breeding from the Northern United States northward, and wintering from the Northern States southward.

Nest, on the ground, sometimes slightly above it, composed of weeds, leaves, rootlets and grass.

Eggs, three or four, greenish-blue, without spots.

The Hermit Thrush is a regular visitor in spring and fall, arriving a few days before the Olive-backed, and making but a short stay, although it probably does not go so far north to breed as the latter species. Referring to the Hermit, the following occurs in the "List of Birds of Western Ontario": "Found common in full song in a large swamp, June 22nd, 1882. No nest found, although it was undoubtedly breeding. None observed in summer in any other locality."

At home, the habits of the Hermit are in keeping with its name. Among the dense shrubbery in some retired spot, it builds its nest and raises its young. There, too, it pours forth its sweet song on the "desert air," where very few have been privileged to hear it. During the seasons of migration the birds come more into the open country, but they are at all times shy and fond of concealment. On these occasions they have only a simple call-note, apparently used to tell their companions where they are.

This species is found on the Island of Anticosti during summer. Mr. Thompson reports it as a summer resident in Manitoba, and Prof. Macoun found it breeding at Lake Mistassini.

In the Rocky Mountain region it is replaced by Audubon's Hermit, and on the Pacific coast by the Dwarf Hermit, both closely allied but still differing slightly from the present species.

Genus MERULA Leach.

MERULA MIGRATORIA (Linn.).

315. American Robin. (761)

Above, dark olive-gray, blackish on head and tail; below, reddish-brown; throat, vent and under tail coverts, white, throat with black streaks; outer pair of tail feathers, white-tipped; bill, dusky above, yellow below; feet, dark;

very young birds spotted above and below. Length, 9½ inches; wing, 5¼; tail, 4¼.

HAB. -Eastern North America to the Rocky Mountains, including Eastern Mexico and Alaska. Breeds from near the southern border of the United States northward to the Arctic coast; winters from Southern Canada and the Northern States (irregularly) southward.

Nest, in a tree, frequently an apple tree in an orchard; large and rough looking, composed of twigs, grass and weeds cemented together with mud, lined with fine grass.

Eggs, four or five, plain greenish-blue, without spots.

The Robin is well known and widely distributed throughout Ontario. In the south it is most abundant during the period of migration, but great numbers breed all over the Province, and along the southern border it is no uncommon thing to meet with individuals spending the winter in sheltered hollows, from which they are ready to start out and hail the first indications of returning spring. As the season advances, northern-bound individuals of this species arrive from the south and pass on with little delay, but those which are satisfied to remain at once become engaged in the great business of the season, that of raising their young. The males are the first to arrive, and are occasionally heard rehearsing their summer song, evidently somewhat out of practice. In a few days the females make their appearance and receive every attention.

The site for the nest is soon selected, and both birds work diligently till the structure is completed. The first set of eggs is laid in April, and during the tedious days of incubation the male often mounts his perch to cheer his faithful mate with what to her may seem delightful strains of music. To human ears the song does not rank as a brilliant performance, but it is given with great earnestness and liberality, and is welcomed as the prelude to the grand concert of bird music which is soon to be heard in the woods and fields all over the country. At this season the food of the Robin consists chiefly of worms and various insects. It is a fine exhibition of bird-life to see him, early in the dewy morning, hop daintily over the newly cut grass to where an earth worm is exposing himself near the surface. With his head on one side, the bird watches every wriggle of the worm with intense interest. If it is well out of the ground, it is seized, and with a jerk thrown clear of its hole, but if only a part of the worm is exposed, the course is different. It is then caught quickly and held firmly while it struggles hard to get into its hole. Robin knows that now a sudden jerk will part the animal and give him only a portion, but he knows how much strain the material will bear, and he holds on till the exhausted worm relaxes its hold, is tossed out and pounded till fit for use.

As the season advances a second and even a third brood of young may be raised. The birds acquire a fondness for fruit, and now come the charges against them of robbing the cherry-tree. No doubt they do take a few for themselves and families, but after all they are entitled to some consideration on account of the numbers of noxious insects which they destroy in the garden, and for my own part I would sacrifice a good many cherries rather than have the Robins banished from around the house.

Those which travel to the far north have a different experience. Dr. Richardson tells us that " the male is one of the loudest and most assiduous songsters which frequent the Fur Countries, beginning his chant immediately on his arrival. Within the Arctic circle the woods are silent during the bright light of noonday, but towards midnight when the sun travels near the horizon, and the shades of the forest are lengthened, the concert commences, and continues till 6 or 7 in the morning. Nests have been found as high as the 54th parallel of latitude about the beginning of June. The snow even then partially covers the ground, but there are in these high latitudes abundance of berries of *vaccinium ngliginosum* and *vites idea, arbutus alpina, empetsum nigrum,* and of some other plants, which, after having been frozen up all winter, are exposed by the first melting of the snow, full of juice and in high flavor. Thus is formed a natural *cache* for the supply of the birds on their arrival, and soon afterwards their insect food becomes abundant."

In Southern Ontario large numbers are seen congregating together, feeding on the berries of the mountain ash, poke weed, red cedar, etc. If the weather is mild, they remain till November, but usually we have a cold blast from the north in October, which hurries the bulk of them off to their winter-quarters in the south.

Genus SIALIA Swainson.

SIALIA SIALIS (Linn.).

316. Bluebird. (766)

Male:—Uniform sky-blue above, reddish-brown below ; belly, white. *Female:*—Duller. *Young:*—Spotted.

Hab.—Eastern United States to the eastern base of the Rocky Mountains, north to Manitoba, Ontario and Nova Scotia, south, in winter, from the Middle States to the Gulf States and Cuba. Bermudas, resident.

Nest, in natural or artificial holes in trees, stubs or posts, or in bird-boxes, composed of miscellaneous material, loosely put together.

Eggs, four to six, pale blue, unmarked.

In former years the Bluebirds were among our most abundant and familiar birds, raising their young near our dwellings, and returning year after year to occupy the boxes put up for their accommodation. Since the advent of the English Sparrow, they have been gradually decreasing in numbers, and are now seldom seen near their old haunts, from which they have been driven by that pugnacious tramp, *Passer domesticus*. They are still common throughout the country, where they are everywhere welcomed as early harbingers of spring, and in the fall they linger till late in October, as if loath to depart. This species was a special favorite with Wilson, on account of which it is often spoken of as Wilson's Bluebird, to distinguish it from the Indigo bird, and one or two other species to which the name is sometimes applied.

That enthusiastic lover of birds has made it the subject of one of his pleasing poetical effusions, in which he faithfully describes many of its habits, amongst others its early arrival in spring and reluctant departure in the fall. With a short extract from this production, I shall say good-bye, for the present, to the "Birds of Ontario":

"When winter's cold tempests and snows are no more,
　　Green meadows and brown furrow'd fields reappearing,
The fishermen hauling their shad to the shore,
　　And cloud-cleaving Geese to the lakes are a-steering;
When first the lone butterfly flits on the wing,
　　When red glow the maples, so fresh and so pleasing,
O then comes the Bluebird, the herald of spring,
　　And hails with his warblings the charms of the season.

"When all the gay scenes of the summer are o'er,
　　And autumn slow enters, so silent and sallow,
And millions of warblers, that charmed us before,
　　Have fled in the train of the sun-seeking Swallow,
The Bluebird forsaken, yet true to his home,
　　Still lingers and looks for a milder to-morrow,
Till, forced by the horrors of winter to roam,
　　He sings his adieu in a lone note of sorrow."

ADDENDA.

ŒSTRELATA HASITATA (Kuhl).

317. Black-capped Petrel. (98)

Top of head, and upper parts generally, except upper tail coverts, uniform dusky, the back and scapulars paler, with perceptibly still paler terminal margins to the feathers; upper tail coverts, basal half (approximately) of tail, head and neck, except top of the former (and sometimes the hind neck also), together with lower parts, including axillars and under wing coverts, pure white; the sides of the chest sometimes with a brownish-gray wash. Length, 14-16 inches; wing, 11.40-11.75; tail, 4.80-5.30.

HAB.—Middle Atlantic, straggling to coasts of North America and Europe.

On the 30th of October, 1893, the dead body of a Black-capped Petrel was picked up on the shore of the Island, at Toronto, and brought to Mr. Spanner's shop, where it was seen by Mr. K. C. McIlwraith and others, and correctly identified.

This is a bird which rears its young on the lonely islands of the sea, and, except in such places, is rarely seen on land anywhere. The individual now referred to was, from some unknown cause, entirely out of its reckoning, and most likely died for lack of suitable nourishment.

GLOSSARY

OF TECHNICAL TERMS USED IN THE PRECEDING DESCRIPTIONS.

MEASUREMENTS.

LENGTH.   Distance between the tip of the bill and the end of the longest tail feather.

EXTENT.   Distance between the tips of the outspread wings.

LENGTH OF WING.   Distance from the angle formed at the (carpus) bend of the wing to the end of the longest primary.

LENGTH OF TAIL.   Distance from the roots of the tail feathers to the end of the longest one.

LENGTH OF BILL.   From the tip of the upper mandible to the point where it meets the feathers of the forehead.

LENGTH OF TARSUS.   Distance from the point where the tarsus joins the leg above, to the point where it joins the middle toe below.

LENGTH OF TOES.   Distance from the point where the leg joins the foot along the top to the root of the claw.

LENGTH OF CLAWS.   Distance in a straight line from the root to the tip of the claw.

A

ABERRANT.   Deviating from ordinary character.

ACUMINATE.   Tapering gradually to a point.

ALBINISM.   State of whiteness, complete or partial, arising from deficiency or entire lack of pigment in the skin and its appendages.

ALULA.   Literally, little wing.   The bastard wing, composed of the feathers which are set on the so-called thumb.

ATTENUATE.   Slender and tapering toward a sharp point.

AXILLARS.   Elongated feathers on the sides of the body under the wings.

B

BAND OR BAR.   Any color mark transverse to the long axis of the body.

BEND OF WING.   Angle formed at carpus in the folded wing.

C

CALCAREOUS.   Chalky.

CANTHUS.   Corner of eye where the lids meet.

CAROTID.   The principal blood-vessel of the neck.

CARPAL ANGLE.   Prominence at the wrist joint when wing is closed.   From this point to the end of the longest quill constitutes the "length of wing."

CERE.   Fleshy covering of the base of the bill.

CERVICAL. Pertaining to the hind neck.
CHIN. Space between the forks of the lower jaws.
CLAVICLE. Collar bone.
COMMISSURE. Line where the two mandibles meet.
CRISSUM. Under tail coverts.
CULMEN. Ridge of upper mandible.
CUNEATE. Wedge-shaped. A cuneate tail has the middle feathers longest.

D

DECIDUOUS. Temporary; falling early.
DECOMPOSED. Separate; standing apart.
DENTIROSTRAL. Having the bill toothed or notched.
DIAGNOSTIC. Distinctively characteristic.
DORSAL. Pertaining to the back.

E

EMARGINATE. Notched at the end; slightly forked.
ERYTHRISM. A peculiar reddish state of plumage.

F

FALCATE. Sickle-shaped.
FEMORAL. Pertaining to the thigh.
FERRUGINOUS. Rusty red.
FISSIROSTRAL. Having the bill cleft far beyond the base of its horny part.
FORFICATE. Deeply forked.
FULIGINOUS. Sooty-brown.
FULVOUS. Of a brownish-yellow color.
FURCATE. Forked.
FUSCOUS. Of a dark grayish-brown color.

G

GIBBOUS. Swollen; protuberant.
GONYS. Keel or outline of the bill so far as united.
GRADUATED. Changing length at regular intervals.
GULAR. Pertaining to the upper fore neck.
GUTTATE. Having drop-shaped spots.

H

HALLUX. The hind toe.

I

IMBRICATED. Fixed shingle-wise; overlapping.
INTERSCAPULAR. Between the shoulders.

J

JUGULUM. Lower throat.

L

LAMELLA. A thin plate or scale, such as are seen inside a duck's bill.
LANCEOLATE. Shaped like the head of a lance.
LARYNX. Adam's apple; a hollow cartilaginous organ; a modification of the windpipe.

LOBE. Membraneous flap, chiefly on the toes.

LORE. Space between the eye and the bill.

M

MAXILLAR. Pertaining to the upper jaw.

MELANISM. State of coloration arising from excess of dark pigment; a frequent condition of Hawks.

MEMBRANE. Soft skinny covering of the bill of some birds.

N

NUCHA. The upper part of the hind neck next the hind head.

O

OSCINES. A group of singing birds.

OCCIPUT. The hind head.

P

PALMATE. Web-footed.

PARASITIC. Habitually making use of other birds' nests.

PECTINATE. Having tooth-like projections, like those of a comb.

PECTORAL. Pertaining to the breast.

PLUMBEOUS. Lead color.

PRIMARIES. The large stiff quills growing on the first bone of the wing; usually nine or ten, sometimes eleven, in number.

R

REMIGES. Quills of the wing.

RETRICES. Quills of the tail.

RICTUS. Gape of the mouth.

S

SAGGITATE. Shaped like an arrow head.

SCAPULARS. Long feathers rising from the shoulders and covering the sides of the back.

SECONDARIES. Quills which grow on the second bone of the wing.

SECONDARY COVERTS. The wing feathers which cover the bases of the secondary quills.

SEMIPALMATE. Having the feet half webbed.

SERRATE. Toothed like a saw.

SPECULUM. A brightly colored spot of the secondaries, especially of ducks.

SPURIOUS QUILL. The first primary when very short.

SUPERCILIARY. Pertaining to the eye-brows.

T

TAIL COVERTS. The small feathers underlying or overlying the base of the tail.

TARSI. The shanks of the legs.

TERTIALS. Feathers which grow from the second bone of the wing at the elbow joint.

TIBIA. Principal and inner bone of leg between knee and heel.

.

INDEX.

28

.

www.ingramcontent.com/pod-product-compliance
Lightning Source LLC
Chambersburg PA
CBHW021347210326
41599CB00011B/780